Climate Change and Food Insecurity

Climate Change and Food Insecurity

Editors

Christopher Robin Bryant
Andrea Vitali
Azzeddine Madani

MDPI • Basel • Beijing • Wuhan • Barcelona • Belgrade • Manchester • Tokyo • Cluj • Tianjin

Editors

Christopher Robin Bryant
School of Environmental Design and Rural Development,
University of Guelph
Guelph
Canada

Andrea Vitali
Department of Agriculture and Forest Sciences DAFNE
University of Tuscia
Viterbo
Italy

Azzeddine Madani
Department of Human Sciences
Khemis-Miliana University
Khemis-Miliana
Algeria

Editorial Office
MDPI
St. Alban-Anlage 66
4052 Basel, Switzerland

This is a reprint of articles from the Special Issue published online in the open access journal *Climate* (ISSN 2225-1154) (available at: www.mdpi.com/journal/climate/special_issues/climate_food).

For citation purposes, cite each article independently as indicated on the article page online and as indicated below:

LastName, A.A.; LastName, B.B.; LastName, C.C. Article Title. *Journal Name* **Year**, *Volume Number*, Page Range.

ISBN 978-3-0365-7441-7 (Hbk)
ISBN 978-3-0365-7440-0 (PDF)

Contents

Tassadit Kourat, Dalila Smadhi and Azzeddine Madani
Modeling the Impact of Future Climate Change Impacts on Rainfed Durum Wheat Production in Algeria
Reprinted from: *Climate* **2022**, *10*, 50, doi:10.3390/cli10040050 1

Barbara Sawicka, Piotr Barbaś, Piotr Pszczółkowski, Dominika Skiba, Farhood Yeganehpoor and Barbara Krochmal-Marczak
Climate Changes in Southeastern Poland and Food Security
Reprinted from: *Climate* **2022**, *10*, 57, doi:10.3390/cli10040057 19

Adham Badawy, Amgad Elmahdi, Sayed Abd El-Hafez and Ali Ibrahim
Water Profitability Analysis to Improve Food Security and Climate Resilience: A Case Study in the Egyptian Nile Delta
Reprinted from: *Climate* **2022**, *10*, 17, doi:10.3390/cli10020017 45

Kallol Barai, Rafa Tasnim, Bruce Hall, Parinaz Rahimzadeh-Bajgiran and Yong-Jiang Zhang
Is Drought Increasing in Maine and Hurting Wild Blueberry Production?
Reprinted from: *Climate* **2021**, *9*, 178, doi:10.3390/cli9120178 59

Andrew Hanley, Galina Brychkova, Wilson John Barbon, Su Myat Noe, Chan Myae and Phyu Sin Thant et al.
Community-Level Impacts of Climate-Smart Agriculture Interventions on Food Security and Dietary Diversity in Climate-Smart Villages in Myanmar
Reprinted from: *Climate* **2021**, *9*, 166, doi:10.3390/cli9110166 79

Nisha Koppa and Giriraj Amarnath
Geospatial Assessment of Flood-Tolerant Rice Varieties to Guide Climate Adaptation Strategies in India
Reprinted from: *Climate* **2021**, *9*, 151, doi:10.3390/cli9100151 95

Noor Ahmad Akhundzadah, Salim Soltani and Valentin Aich
Impacts of Climate Change on the Water Resources of the Kunduz River Basin, Afghanistan
Reprinted from: *Climate* **2020**, *8*, 102, doi:10.3390/cli8100102 111

Mélanie Doyon and Juan-Luis Klein
Non-Conventional Agricultural Spaces and Climate Change: The Cases of Le Grenier boréal and Lufa Farms in Quebec, Canada
Reprinted from: *Climate* **2021**, *9*, 148, doi:10.3390/cli9100148 131

Vuyiseka A. Myeki and Yonas T. Bahta
Determinants of Smallholder Livestock Farmers' Household Resilience to Food Insecurity in South Africa
Reprinted from: *Climate* **2021**, *9*, 117, doi:10.3390/cli9070117 147

Mary Funke Olabanji, Thando Ndarana and Nerhene Davis
Impact of Climate Change on Crop Production and Potential Adaptive Measures in the Olifants Catchment, South Africa
Reprinted from: *Climate* **2020**, *9*, 6, doi:10.3390/cli9010006 163

Abubakar Hadisu Bello, Mary Scholes and Solomon W. Newete
Impacts of Agroclimatic Variability on Maize Production in the Setsoto Municipality in the Free State Province, South Africa
Reprinted from: *Climate* **2020**, *8*, 147, doi:10.3390/cli8120147 183

Dula Etana, Denyse J. R. M. Snelder, Cornelia F. A. van Wesenbeeck and Tjard de Cock Buning
Trends of Climate Change and Variability in Three Agro-Ecological Settings in Central Ethiopia: Contrasts of Meteorological Data and Farmers' Perceptions
Reprinted from: *Climate* **2020**, *8*, 121, doi:10.3390/cli8110121 . . . . **201**

climate MDPI

Article

Modeling the Impact of Future Climate Change Impacts on Rainfed Durum Wheat Production in Algeria

Tassadit Kourat [1,*], Dalila Smadhi [2] and Azzeddine Madani [3]

1 Hall de Technologie-Kharrouba, Ecole Supérieure d'Agronomie de Mostaganem, Mostaganem 27000, Algeria
2 Division Bioclimatologie et de l'Hydraulique Agricole, Station Mehdi Boualem, Baraki, Institut National de Recherche Agronomique, Algiers 16000, Algeria; dalila.smadhi@inraa.dz
3 Faculty of Social and Human Sciences, Khemis Miliana University, Miliana 44000, Algeria; azzeddine.madani@univ-dbkm.dz
* Correspondence: t.kourat@edu.ensa.dz

Abstract: The predicted climate change threatens food security in the coming years in Algeria. So, this study aims to assess the impact of future climate change on a key crop in Algeria which is rainfed durum wheat. We investigate the impact of climate change on rainfed durum wheat cultivar called Mexicali using AquaCrop crop model and the EURO-CORDEX climate projections downscaled with the ICHEC_KNMI model under RCP 4.5 and RCP 8.5. A delta method was applied to correct the incertitudes present in the raw climate projections of two experimental sites located in Sétif and Bordj Bou Arreridj (BBA)'s Eastern High plains of Algeria (EHPs). AquaCrop was validated with a good precision (RMSE = 0.41 tha^{-1}) to simulate Mexicali cultivar yields. In 2035–2064, it is expected at both sites: an average wheat grain yield enhances of +49% and +105% under RCP 4.5 and RCP 8.5, respectively, compared to the average yield of the baseline period (1981–2010), estimated at 29 qha^{-1}. In both sites, in 2035–2064, under RCP 4.5 and RCP 8.5, the CO_2 concentrations elevation has a fertilizing effect on rainfed wheat yield. This effect compensates for the negative impacts induced by the temperatures increase and decline in precipitation and net solar radiation. An increase in wheat water productivity is predicted under both RCPs scenarios. That is due to the water loss drop induced by the shortening of the wheat-growing cycle length by the effect of temperatures increase. In 2035–2064, early sowing in mid-September and October will lead to wheat yields improvement, as it will allow the wheat plant to benefit from the precipitations increase through the fall season. Thus, this early sowing will ensure a well vegetative development and will allow the wheat's flowering and grain filling before the spring warming period.

Keywords: climate change; rainfed durum wheat; AquaCrop; delta method; CO_2 fertilizing effect

Citation: Kourat, T.; Smadhi, D.; Madani, A. Modeling the Impact of Future Climate Change Impacts on Rainfed Durum Wheat Production in Algeria. *Climate* **2022**, *10*, 50. https://doi.org/10.3390/cli10040050

Academic Editor: Steven McNulty

Received: 1 February 2022
Accepted: 18 March 2022
Published: 23 March 2022

1. Introduction

Non-climatic stressors (e.g., demographic and income growth, demand for animal products) and climate change (CC) influence the food system. These climatic and non-climatic stressors have an effect on the four pillars of food security (availability, access, use, and stability) [1]. From the beginning of the 1990s, the Intergovernmental Panel on Climate Change (IPCC), showed that over the period from 1850 to 2012. the global average temperature had experienced warming of 0.78 °C This global warmingcould be induced by the change in the carbon dioxide (CO_2) concentration in the atmosphere from 278 ppm to 379 ppm. The prediction for the end of the 21st century is a global warming that will range between 1.5 °C and 2 °C [2]. In Africa, in recent decades, temperatures have increased at a rate somewhat faster than the global average temperature. Thus, the 2019 year was identified as one of the three warmest years on this continent [3]. CC is a consequence of global warming, which has adverse effects on fluctuations in annual total precipitations, average temperature, global increase in atmospheric CO_2, and sea-level rise. These are some of the major manifestations of CC, which have direct and indirect socio-economic negative impacts on plant development and crop yield (reduction in crop yields by up to 70%) ([4,5]). In general, crop yields will increase in cold areas where low temperature currently limits crop growth. However, heat stress on crops and water scarcity will lead to a decline in yields in warm environments. Warm temperatures and precipitation variability associated with a high frequency of extreme climate events (e.g., droughts, floods, heat waves, etc.) have worsened food insecurity in several regions of the world, especially in Africa ([3,6]). Information related to climate impacts on crops is important for understanding their macroeconomic implications for food security. This climate information allows us to choose e the appropriate adaptation strategies supported by knowledge of the processes that lead to changes in yield, under average and extreme weather conditions [7]. Identifying the drivers of changes and variability in yields can enable the development of targeted adaptation measures such as: (i) insurance solutions against specific weather [8], (ii) support the planning of long-term investments in irrigation infrastructure [9], or (iii) improve reproductive efficiency as the suitability of adaptive traits changes with CC and elevated CO_2 concentrations in the atmosphere [10].

In Algeria, 50% of non-irrigated agricultural lands are cultivated with cereals, especially durum wheat, with a low national average grain yield estimated at 17 qha^{-1} (2000–2020 period). However, these lands are mainly located in the High Plains region, known for its semi-arid climate [11]. Moreover, the decrease in the national production of meat and milk had affected their prices and caused an increase in demand for cereal products. The last are characterized by their subvention prices by the government, especially wheat, which is considered as the main source of protein in the diet of Algerian people [12]. Thus, Algeria meets its national needs for durum wheat with massive imports, with an average annual bill of $1 billion. These food bills are paid, thanks to oil rent [11].

It is projected in Algeria by the future horizon 2030: (i) an increase in temperatures of +0.9 to +1.3 °C and their variability, (ii) an intensification of the frequency of heat waves, and (iii) an accentuation of the variability of precipitations, which will result in an increase in dry and wet episodes by +10% and will be accompanied by a decrease in precipitation of −9 up to −14% [13]. In the future decades, the harmful impacts of the above projected CC will manifest themselves above all else by the increase in the frequency and severity of droughts. This projected drought will threaten crop production, mainly rainfed crops yields, such as durum wheat. Thus, by the future, under the projected CC, the demographic surge will lead to an increase in national wheat needs. Thus, with the fall in oil prices, the satisfaction of national demand for wheat could become a real concern to economic balance and food security in Algeria [14].

Given the importance of wheat in human nutrition and global trade, many studies (e.g., [15–17]) are carried out across the world to assess the impact of CC on wheat yield. These studies used crop models and the Representative Concentrations Pathway scenarios (RCPs) [18]. The results of the above studies could not apply directly in Algeria. Because the CC's impacts on wheat production are specific to each region in the world according to its local climate and to its financial and technical capacity to face the CC impacts. According to [19], in Algeria, the negative impacts of CC on water resources were assessed with the UKHI model (United Kingdom Meteorological Office High Resolution). This assessment study carried out by seasonal climate forecasting showed a decrease in the rainiest area and an increase in the driest ones in Algeria. So, the rainfed crop is very vulnerable to future climate change. As a consequence of the above-cited projected CC in Algeria, a decline in crop yields by −10 to −30% was predicted by 2030 [20]. Despite the strategic role of rainfed wheat in national food security and its high level of vulnerability to the projected CC, the studies of CC assessment impacts on wheat production, using crop models and associated with RCPs scenarios, are very rare in Algeria. With the exception of the study carried out by Rouabhi et al. [21], who used a statistical model to predict durum wheat yields under RCP 2.6 and RCP 8.5, in 2070, at the Setif region. So, the main aim of this research is to improve the available knowledge related to the future CC negative impacts on rainfed wheat in Algeria. This study could help the farmers better understand the CC issue and its impacts on wheat. This study could help the farmers to choose more resilient and CC-adapted agricultural practices in the future. Furthermore, these kinds of studies will be important to the agricultural stakeholders in preparing adapted policies which will accompany the farmers in their quest to prevent wheat yields losses induced by CC. Thus, these types of studies are very important in preparing a national strategy to adapt rainfed wheat against the projected CC negative impacts. This could help them to protect the national economic balance from the potential negative impacts of food insecurity induced by low national wheat production.

2. Materials and Methods

2.1. Study Area

The High Eastern plains of Algeria (HEPs) of Algeria are located in the Northeast of the country between the latitudes 35–36.5° N and longitudes 4.5–8.3° E (Figure 1). The HEPs are characterized by altitudes that vary between 900 and 1200 m. The HEPs are limited to the North and West by the eastern part of Tellian Atlas mountain ranges, to the South by the salt lake called Chott Hodna and the eastern part of the Saharan Atlas mountain ranges, and to the East by Tunisia [22]. In this research, the study of the CC impacts on wheat yield at the level of the semi-arid HEPs is established at two experimental sites. The last are located in two wilayas (or departments) called Sétif and Bordj Bou Arreridj (BBA), regions known for the practice of rainfed durum wheat production. According to the precipitation map established by the National Agency of Water Resources in 1993, the annual average precipitation in the HEPs can reach 500 mm in the North and decrease up to 300 mm in the South and can even reach less than 200 mm in the salt lakes areas. During the period 1995–2009, in the HEPs, the Tmax ranges between 32 and 37 °C, while the Tmin varies between 0 and 5 °C [23]. The HEPs are more exposed to sunshine radiation thanks to the continental climate and high altitude. This topography made the HEPs well suited to rainfed cereal crops. However, the high variability of annual total precipitation results in extremely variable yields from year to year. The choice of the location of both sites of Sétif and BBA for this study is justified by the availability of soil and climate data, and the availability of phenological data for a cultivar of durum wheat called "Mexicali" at the Setif site.

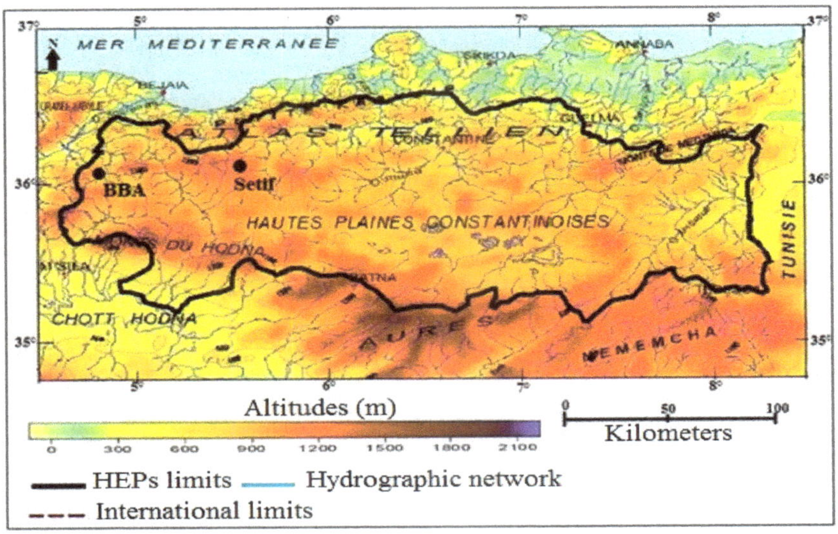

Figure 1. Geographical location of the study area (source [23]).

2.2. Observed Baseline and Projected Future Climate Data

The daily climate data: Tmax and Tmin (in °C), precipitations (P in mm), wind speed (V in m/s), relative humidity (Hr in %) and sunshine duration (S in hours), observed during the baseline (or reference) period (BP) 1981–2010 are collected from the professional Meteorological stations The last belong to the National Meteorological Office (ONM) of Setif and BBA. The Setif station is located in the Soummam watershed, and the BBA station is located in the Hodna watershed; the geographic coordinates and elevation of these two stations are shown in Table 1.

Table 1. Geographical coordinates and altitudes of the of the Setif and BBA meteorological stations.

Station	Latitude (°C)	Longitude (°C)	Altitude (m)
BBA	36.06° N	4.66° E	957
Setif	36.16° N	5.31° E	1015

The global simulated future climate data of P, Tmax, Tmin, net sunshine radiation (Nr), Hr, and V used in this study, comes from the Coordinated Regional Climate Downscaling (CORDEX) experiment, Europe domain. It must be mentioned that these data were downloadable from the website https://euro-cordex.net/060378/index.php.en (accessed on 5 January 2022). They are simulated under the RCPs scenarios: RCP 4.5 and RCP 8.5 during the future period 2035–2064. The RCPs are four greenhouse gas concentration trajectories adopted by the IPCC on its Fifth Assessment Report (AR5). The numerical values of the RCPs (2.6, 4.5, 6.0, and 8.5 W m^{-2}, respectively) refer to radiative forcing in 2100 [24,25]. These projected radiative forcings are estimated based on the forcing of greenhouse gases, mainly CO_2 and other forcing agents. The above four selected RCPs were considered to be representative of the literature and included one mitigation scenario leading to a very low forcing level (RCP2.6), two medium stabilization scenarios (RCP4.5 and RCP6), and one very high greenhouse emission scenario(RCP8.5), induced by a massive use of fossil energy and a high change in land use [26]. Generally, in CC impact studies, the RCP 4.5 and RCP 8.5 are used.

Then, to obtain the local future climate data projected for Setif and BBA meteorological stations, the above projected global climate data were downscaled by applying a dynamic downscaling method on a grid with a very fine resolution of 0.11° (~11 km). This downscaling is performed by the use of a Regional Circulation Model (RCM) called KNMI forced by a Global Circulation Model (GCM) called ICHEC. The choice of this combination GCM/RCM: ICHEC_KNMI is justified by its best simulation of climate data observed during the BP in Algeria [27]. Then, the climate data simulation errors (or bias) present in the raw downscaled future climate data of Setif and BBA stations are corrected using the delta method [28]. According to these last authors, the basic principle of this method is the addition and/or the multiplication of the anomalies of the simulated future climate data to the daily observed climate data during the BP at Setif and BBA, as indicated with Equations (1) and (2).

$$T^{*}_{Fcor}(d) = T_{obs}\ (d) + \mu_m(T_{Fraw}(d)) - \mu_m(T_{eval}(d)) \quad (1)$$

$$\boldsymbol{P^{*}_{Fcor}(d) = P_{obs}\ (d)\ \frac{\mu_m(P_{Fraw}(d))}{\mu_m(P_{Eval}(d))}} \quad (2)$$

where $T^{*}_{Fcor}(d)$ and $\boldsymbol{P^{*}_{Fcor}(d)}$ are the daily bias-corrected future temperature and precipitation, $T_{obs}\ (d)$ and $\boldsymbol{P_{obs}\ (d)}$ are the daily observed temperature and precipitation during the BP, $\mu_m(T_{Fraw}(d))$ and $\boldsymbol{\mu_m(P_{Fraw}(d))}$ are the monthly averages of daily raw future temperature and precipitation, and $\mu_m(T_{eval}(d))$ and $\boldsymbol{\mu_m(P_{Eval}(d))}$ are the monthly averages of daily temperature and precipitation simulated for the BP, respectively.

The bias-correction methods such as the quantile mapping method widely used in hydrological impact studies can be difficult to validate in semi-arid climates. This is due to the limited number of rainy days, especially during summer. Moreover, the high variability of precipitations year to year, which is an atypical characteristic of the Mediterranean climate, made the quantile mapping method not easy to use in the HEPs of Algeria [29]. In contrast, the delta method did not rely on the stationary assumption of model bias and did not modify the results of the climate model [30]. Therefore, it can be considered more robust and should be preferred in cases where other approaches cannot be satisfactorily validated. Thus, these precedents results justify the choice of the deltas method for the correction of the uncertainties of the future climatic data simulated by the ICHEC_KNMI climate model in this study. Thus, the two future CC scenarios, Sc_1 and Sc_2 which refer to the scenarios: RCP 4.5 in 2035–2064 and RCP 8.5 in 2035–2064, respectively, are evaluated in this study.

2.3. *The Aquacrop Model*

2.3.1. AquaCrop Model Description

AquaCrop is a crop water productivity model developed by FAO's Land and Water Division in Rome, Italy. This crop model was created to address food security issues and assess the effect of environment and field water management on agricultural production (www.fao.org/aquacrop/overview/en/, accessed on 15 January 2022). This model simulates the response of herbaceous crop yields to water. It is particularly well suited to conditions in which water is a limiting factor in agricultural production [31]. AquaCrop combines precision, simplicity, and robustness; it is widely used around the world given the limited number of inputs required for its simulation process [32]. AquaCrop can also simulate crop growth under CC scenarios by taking into consideration different CO_2 concentrations scenarios. However, AquaCrop does not take into account the negative impacts of pests, diseases, and weeds on yields [33].

2.3.2. AquaCrop Model Input Data

Observed Historical Climatic Data and Projected Future

To simulate the durum wheat's grain yield for the BP (1981–2010) with the AquaCrop crop model for the experimental sites of BBA and Setif, it is necessary to introduce into this model the daily climate data of Tmin, Tmax, and P. These data were observed during the BP at the professional meteorological stations of Setif and BBA as it is necessary to introduce into the AquaCrop model, the daily data of the reference evapotranspiration (ET_0). This last was calculated in advance by the ET_0 calculator software according to the Penman–Monteith equation [34]. The CO_2 concentration is also required by AquaCrop to simulate durum wheat grain yield. Thus, the global annual averages of CO_2 concentrations from the Mauna Loa observatory in Hawaii are attached to the AquaCrop package, so they are used to simulate durum wheat yields. The same method was applied to simulate durum wheat's future yields in 2035–2064, under RCP 4.5 and RCP 8.5, using the corrected daily climate data simulated by the ICHEC-KNMI climate model for the future horizon 2035–2064, under both RCPs scenarios.

Soil Data

The values of Setif and BBA experimental site's soil organic matter, clay, and sand content, obtained by laboratory analysis, were introduced into the SPAW model [35].Thus, this last, in turn, simulates the permanent wilting point (PWP), the field capacity (FC), the total quantity of available water contained in the soil (TAW), and saturated hydraulic conductivity (SAT). The values of PWP, FC, and SAT are simulated by the SPAW model by applying a pedotransfer function. These four soil parameters are essential for the AquaCrop model run [36].

2.3.3. AquaCrop Model Calibration and Validation

Before using the AquaCrop crop model to simulate grain yields of a local durum wheat cultivar called "Mexicali" in the future horizon 2035–2064 (under RCP 4.5 and RCP 8.5). This crop model was first calibrated using Mexicali cultivar phenological data, observed by [37] at a field test during the 2010/2011 agricultural campaign at the experimental farm of Sétif (belonging to the National Institute of Agronomic Research). This farm is located at 36.15° N latitude and 5.37° E longitude, and at an altitude of 970 m. In order to calibrate AquaCrop, according to the climatological conditions of this above Setif experimental site, the daily climate data (Tmax, Tmin and P) observed during the growing season 2010/2011, the daily ET_0 data throughout this growing season, calculated using ET_0 calculator, and also the soil data (PWP, FC, SAT and soil horizon deeps) are injected into this model.

The durum wheat's non-conservative parameters in the AquaCrop model are given in Table 2, such as: the sowing density, the length (in days) of the stages of emergence, maximum leaf expansion, maximum roots depth, flowering, seed formation, and the maturity observed during the (2010/2011) growing season, are also introduced into the AquaCrop model. Then, the last was validated with the values of the Mexicali cultivar's final grain yields and above-ground biomass, observed throughout the experimentations, carried out by [37], during the period of the three growing seasons: (2010/2011), (2011/2012) and (2012/2013), at the Setif experimental farm.

Table 2. Calibrated AquaCrop model specific parameters for Mexicali cultivar of wheat.

Non-Conservative Crop Parameters	Value
Length to emergence (day)	10
Reference harvest index (HI) (%)	57
Length to building up HI (day)	37
Duration of flowering (day)	29
Length to max cc (day)	70
Length max root depth (day)	49
Length to flowering (day)	61
Length to start canopy senescence (day)	82
Length to maturity (day)	106
Initial canopy cover (%)	4.5
Maximum canopy cover (%)	90
Plant density (plant/m^2)	300
Canopy decline coefficient at senescence	0.405% GDD
Canopy growth coefficient	0.669% GDD
Max effective root depth (m)	1
Crop transpiration coefficient	0.98
Water productivity (kg/m^3)	1.35
SWDT for canopy expansion, upper limit	0.2 TASW
SWDT for canopy expansion, lower limit	0.6 TASW
SWDT for stomatal closure, upper limit	0.6 TASW
SWDT for canopy senescence, upper limit	0.7 TASW
Shape factor of canopy expansion	5
Shape factor of stomatal closure	2.5
Shape factor early senescence	2.5
Base temperature (°C)	0
Max temperature (°C)	26

GDD, growing degree-day; SWDT, soil water depletion threshold; TASW, total available soil water.

Finally, the values of the Mexicali cultivar's non-conservative parameters in AquaCrop are calibrated several times in order to obtain values of simulated Mexicali cultivargrain yield and final above-ground biomass, close to those observed by [37] during these above three experimentation growing seasons. Thus, the final values of the Mexicali cultivar's non-conservative (indicated in Table 2) and conservative parameters are applied in all simulation scenarios on AquaCrop.

2.4. Statistical Correlation between Durum Wheat Grain Yields and Growing Season Length with Temperature, Rainfall and Net Solar Radiation Changes

In order to detect any possible sensitivity of durum wheat grain yield to the projected future CC, the Pearson correlation test [38] was applied between the time series of 30 years of the Mexicali cultivar grain yields, simulated by AquaCrop for the Sc_1, Sc_2, and BP scenarios, with the time series of 30 years of the averages seasonal: mean temperatures (T_S), cumulative precipitation (P_S) and net incident solar radiation (Nr_S) projected under both the future CC scenarios and those recorded during BP. Furthermore, this test was applied between the time series of 30 years of the growing season length (GSL) of Mexicali cultivar simulated by AquaCrop for the Sc_1, Sc_2, and BP scenarios, with the time series of 30 years of T_S, P_S, and Nr_S, simulated by the ICHEC-KNMI climate model under the last three scenarios. Thus, this test allowed r to detect the impact of T_S, P_S, and Nr_S changes on the GSL. The season considered in this study is the period coinciding with the Mexicali cultivargrowing cycle.

3. Results

3.1. Assessment of the Quality of the Simulated Climate Data for the Baseline Period

The monthly averages of P (mm), Tmax and Tmin (°C), and Nr (MJm^{-2}) were recorded on the two meteorological stations of BBA and Setif departments, during the BP (1981–2010) are indicated in Figure 2. These recorded monthly averages are compared with those simulated by the climate model ICHEC_KNMI for this same BP. Additionally, in Figure 2, the RMSE values for each of the above four climate parameters are indicated. Thus, for the BBA station, Tmax, Tmin, and Nr are simulated with great precision as shown by the low values of their respective RMSE 1.998 °C, 2.029 °C, and 2.193 MJm^{-2}, (Figure 2a,c,g). However, the P were simulated with less precision (RMSE = 6.674 mm), as indicated in Figure 2e. Thus, for the months of the period from October to March, the simulated P are overestimated. However, the simulated P for the months April to September are underestimated compared to the P recorded on the BBA station during the BP Figure 2b,d,h show that Tmax, Tmin, and Nr, respectively, at the Sétif station, are simulated with good precision as indicated by the low values of their respective RMSEs (0.439 °C, 1.381 °C, and 2.785 MJm^{-2}). However, the P simulated with relatively lower accuracy as shown in Figure 2f and the relatively higher RMSE value (9.875 mm).

3.2. Projected Climate during the Mexicali Cultivar Growing Season

In Table 3, in order to detect the impact of future CC on Mexicali cultivar's LGS, the thirty years average of the last simulated by AquaCrop for the BP are compared to those simulated under Sc_1 and Sc_2 by this crop model. Furthermore, in Table 3, the thirty years average of T_S (°C), P_S (mm), Nr_S ($M\,J\,m^{-2}$) recorded during the BP are compared to thosesimulated by the ICHEC_KNMI model under Sc_1 and Sc_2 at the Setif and BBA stations. This comparison aims to detect any relationship between LGS and CC.

Table 3. Comparison of the thirty years averages of T_S, P_S, Nr_S between the BP and the futurescenarios Sc_1 and Sc_2.

Future Simulation Scenario	T_S	T_S Change (°C)	P_S	P_S Change (mm)	(%)	Nr_S ($M\,J\,m^{-2}$)	Nr_S Change ($M\,J\,m^{-2}$)	(%)
Setif								
BP	10.3		263			1059.1		
Sc_1	13.8	3.5	244	−18.8	−7.1	712.1	−347	−32.8
Sc_2	10	−0.3	329	65.6	24.9	922.3	−136.8	−12.9
BBA								
BP	10.7		220			1058.9		
Sc_1	15.2	4.5	153	−67.1	−31	468.7	−590.2	−55.7
Sc_2	10.5	−0.1	285	64.3	29.2	1016.5	−42.4	−4

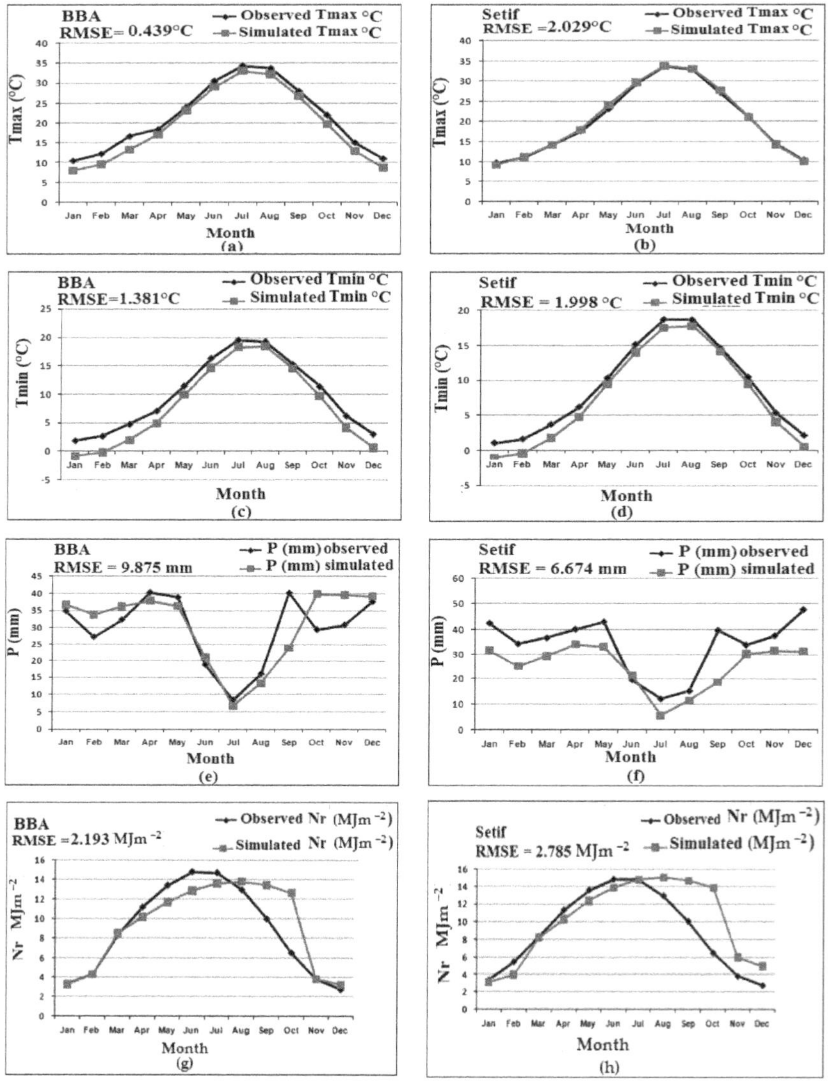

Figure 2. Comparison of the degree of agreement between the observed and simulated Tmin, Tmax, P, and Nr monthly means at Setif and BBA stations during the BP.

By taking into account that the Mexicali cultivar sowing date is fixed on 28 November in AquaCrop. For the BP, Sc_1, and Sc_2 scenarios. At Setif, the average Ts during the BP is 10.3 °C. Thus, the ICHEC_KNMI model predicted an increase in T_S by +3.5 and its decrease by −0.3 °C under Sc_1 and Sc_2 scenarios, respectively. At BBA, this climate model predicted the same T_S trend predicted at Setif, so an increase in T_S by +4.5 and its decline by −0.1 °C are projected under Sc_1 and Sc_2, respectively. According to these results, the T_S elevation is more pronounced under Sc_1 than under Sc_2 at both stations Setif and BBA. This last result could be attributed to the fact that under RCP 4.5, the T_S increase is accentuated during the months coinciding with the Mexicali cultivar growing season (especially during March and April). Meanwhile, under RCP 8.5, the T_S increase will be more accentuated during the summer and autumn (period from June to October), so it does not coincide with the

Mexicali cultivar growing season. During the BP, the P_S average at Setifstation is 263.1 mm, so a decline of −18.8 mm (−7.1%), and an increase of 65.6 mm (+24.9%) are projected under Sc_1 and Sc_2, respectively. At BBA, the ICHEC-KNMI model predicted the same trend of P_S projected at Setifbut with a more accentuated degree. Thus, a decrease in P_S of −67.1 mm (−31%) and its increase of +64.3 mm (+29.2%) are predicted by this climate model under Sc_1 and Sc_2, respectively.

The averages Nr_S observed during the BP are 1059 and 1058 MJm^{-2} at the Setif and BBA stations, respectively. They are projected to drop by −347 MJm^{-2} (−32.8%) and −590 MJm^{-2} (−55.7%) under Sc_1, and by −136.8 MJm^{-2} (−12.9%) and −42.4 MJm^{-2} (−4%) under Sc_2 scenario at Setif and BBA, respectively.

3.3. Evaluation of AquaCrop Model Performance in Simulation Wheat Grain Yield and Above-Ground Biomass

Figure 3 shows the comparison between the Mexicali cultivar's final yields and above-ground biomasses observed at Setif experimental site during the three tests growing seasons: 2010/2011, 2011/2012, and 2012/2013, with respect to those simulated by AquaCropin these same growing seasons. Thus, according to Figure 3a, the Mexicali cultivar's yields were simulated with good precision for the three years. However, as shown in Figure 3b, AquaCrop overestimates the simulation of the above-ground biomass for the 2012/2013's growing season. This could be due to an error in the biomass measurement at the field test. Generally, the averages of statistical indicators of the model's performance, for the three growing seasons were better in predicting yield (RMSE = 0.41 tha^{-1}, NRMSE = 8.81% and d = 0.80), than in prediction above-ground biomass (RMSE = 2.25 tha^{-1}, NRMSE = 21.65% and d = 0.54) (Table 4). In Brazil, Rosa et al. [39] validated the AquaCrop model to predict wheat grain yields with an estimated RMSE = 0.6 tha^{-1} and a Willmott agreement index of (d) $\geq$ 0.80.

Table 4. Comparison of Mexicali cultivar's yields and above-ground biomasses observed and simulated by AquaCrop in the 2010/2011, 2011/2012 and 2012/2013 growing seasons.

Statistical Indices	RMSE		NRMSE		Willmott Agreement Index (d)	
	Yield (tha^{-1})	Biomass (tha^{-1})	Yield (%)	Biomass (%)	Yield	Biomass
Three years average	0.41	2.25	8.81	21.65	0.80	0.54

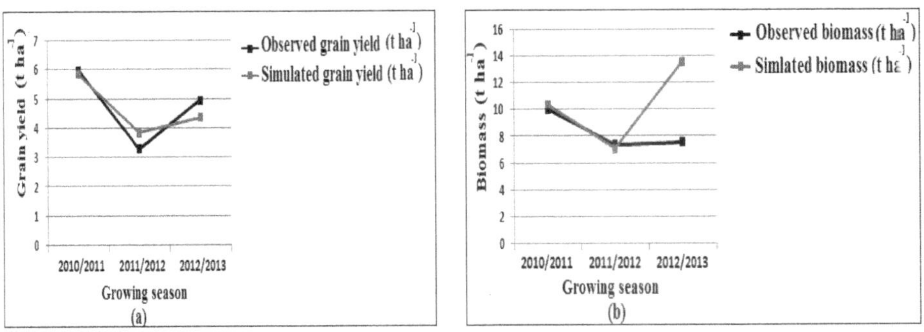

Figure 3. (**a**,**b**) Comparison between the observed and simulated of Mexicali cultivar's grain yields and above ground-biomass for the three growing seasons: 2010/2011, 2011/2012 and 2012/2013 at Setif experimental site.

3.3.1. Impact of Future Climate Change on Durum Wheat Grain Yield

With the aim of showing the projected impacts of changes in T_S, P_S, and the atmospheric CO_2 concentration, on rainfed durum wheat grain yields, the charts in Figure 4 were carried out. So, for the BP, as shown in Figure 4a,b, the average grain yield of Mexicali cultivar simulated by AquaCrop crop model are estimated to 34.7 and 23.3 qha^{-1} at Setif and BBA experimental sites, respectively. So, a Mexicali cultivar grain yields enhancements estimated at (+82 and +76.6) and (+16 and +133%) are projected under the Sc_1 and Sc_2 scenarios, in Setif and BBA field tests, respectively.

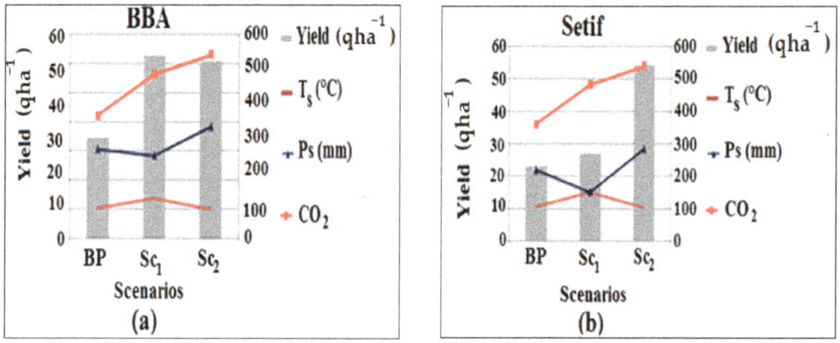

Figure 4. (**a**,**b**) Impact of future change in seasonal temperatures, precipitations, and CO_2 concentrations on Mexicali cultivar grain yield at BBA and Setif sites.

3.3.2. Wheat Growing Season Length, Reference Evapotranspiration and Water Productivity Prediction under Future Climate Change Scenarios

Figure 5 summarizes the results of AquaCrop simulations of the thirty years averages of GSL, WP, and ET_0, under the BP, Sc_1, and Sc_2 scenarios at BBA and Setif experimental sites. Figure 5 also shows the relationship between GSL, WP, and ET_0 variations with the changes in seasonal temperatures and CO_2 concentrations projected under Sc_1 and Sc_2 with respect to their averages simulated for the BP.

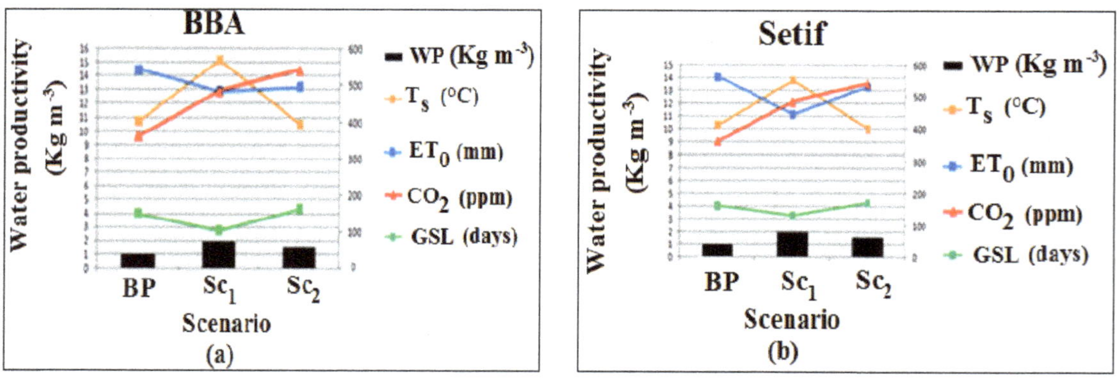

Figure 5. (**a**,**b**) Impact of temperature and CO_2 concentrations futures changes on reference evapotranspiration and wheat water productivity and the growing cycle length.

As indicated in Figure 5a,b, the Mexicali cultivar's GSL follows an inverse evolution of T_S one under Sc_1 and Sc_2 scenarios at BBA and Setif sites, respectively. For the BP, AquaCrop simulated a Mexicali cultivar's GSL of 161 and 151 days at Setif and BBA sites, respectively. It is predicted that a shortening of the Mexicali cultivar's GSL by 30 and 47 days occurs in Sc_1; meanwhile, the Mexicali cultivar's GSL lengthening of +10 and +13 days are predicted under the Sc_2 conditions at Setif and BBA sites, respectively. In comparison with the BP, the shortening of the Mexicali cultivar's GSL under Sc_1 by 30 and 47 days under Sc_1 is due to the increase in T_S of +3.5 and +4.5 °C at Setif and BBA sites, respectively. However, the Mexicali cultivar's GSL lengthening by 10 and 13 days in Sc_2 is due to the drop in T_S of −0.3 and −0.1 °C at Setif and BBA sites, respectively.

In addition to these above results, the Pearson correlation test revealed the existence of a negative and statistically significant correlation between this Mexicali cultivar's GSL and T_S during the BP, under both Sc_1 and Sc_2. However, this test proved that there is a positive and statistically significant correlation between the Mexicali cultivar's GSL and P_S during the BP and under both Sc_1 and Sc_2 scenarios at theSetif and EBA sites. Thus, the lengthening of the Mexicali cultivar's GSL by 10 days under Sc_2 at the Setif site could alsobe explained by the P_S increase by +65.6 mm. So, despite the shortening of Mexicali cultivar's GSL by the effect of the expected T_S increase, Mexicali cultivar's grain yield is projected to be enhanced under Sc_1.

Moreover, according to the results of this study reported in Figure 5a,b, the AquaCrop model simulated thirty years average water productivity (WP) of 1 and 0.7 kgm^{-3} of the Mexicali cultivar for the BP at the experimental sites of Setif and BBA, respectively. Thus, this crop model predicted WP enhance, estimated at (+1 and +0.3) and (+0.6 and +0.9) kgm^{-3}, corresponding to WP enhancement rate of (+100 and +43) and (+60 and +129)% under Sc_1 and Sc_2, at Setif and BBA experimental sites, respectively.

Furthermore, according to Figure 5a,b, the ET_0 simulated by the AquaCrop crop model for the BP is estimated to be 562.5 and 542.8 mm, for the Setif and BBA sites, respectively. Thus the drops of (−118 and −58.6) and (−32 and −47) mm, corresponding to decline rates of (−21 and −11) and (−6 and −8)%, are projected under the Sc_1 and Sc_2, at the Setif and BBA sites, respectively. So, these above ET_0 declines projected under Sc_1 could be induced by the shortening of MC's GSL. However, under Sc_2, the lengthening of the Mexicali cultivar's GSL did not prevent the ET_0 decline.

3.4. Adaptation of Durum Wheat Cultivation to Future Climate Change by Adjusting a Sowing Date

To adapt the rainfed durum wheat crop to the projected CC throughout its growing season, a CC adaptation strategy based on the sowing dates adjustment was tested in the AquaCrop model. Thus, five sowing dates on: 15 September, 15 October, 15 November, 30 November, and 15 December are tested in the AquaCrop crop model to simulate the Mexicali cultivar's grain yields under the BP, Sc_1 and Sc_2 scenarios. So, the average Mexicali cultivar's grain yields simulated by applying the above sowing dates in BP, Sc_1, and Sc_2 scenarios are reported in Figure 6. So, at the BBA site, as shown in Figure 6a, for the BP, the best simulated yield is estimated at 47.9 qha^{-1} by applying a sowing date of 15 October. However, Mexicali cultivar's grain yield losses estimated at −37.5, −35, and −7% are projected in Sc_1 with early sowing on 15 September, 15 October, and 15 November, respectively at the BBA site. However, late sowings on 30 November and 15 December allow grain yields a gain of +13% and +27%, respectively, under Sc_1 at the BBA site. As shown in Figure 6. a, at this last experimental site, the Mexicali cultivar's grain yield gains are projected to decline under Sc_2 with the delay of the sowing date. Thus, the best estimated grain yield is 70.5 qha^{-1} simulated by applying an early sowing date of 15 September. In the case of the Setif site (Figure 6b), with early sowing on 15 October, the best Mexicali cultivar's grain yields simulated with AquaCrop are 54,69, and 73 qha^{-1} for the BP, Sc_1, and Sc_2, respectively. So, as indicated in Figure 6b, the future Mexicali cultivar's grain yields in the Setif site tend to improve with early sowing (in September and October) and decrease with late sowing (in November and December).

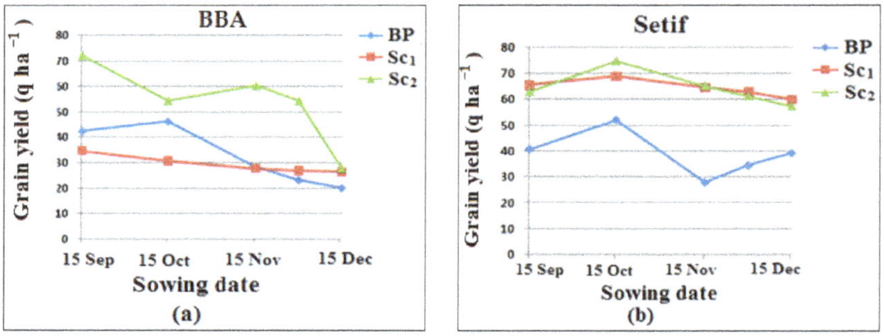

Figure 6. (**a**,**b**) Wheat grain yield simulated by AquaCrop by adjusting the sowing dates to adapt wheat crop to the future climate change projected under Sc_1 and Sc_2 at BBA and Setif sites.

4. Discussion

The precipitations of the BP simulated by the climate model ICHEC_KNMI are underestimated compared to those observed at theSetif and BBA stations during the BP. Thus, the simulated precipitations are underestimated for all months of the year except June, where P were overestimated. Romeraet al. [40] suggested that the high variability of rainfall and the weak network of observation stations in the Maghreb region made the rainfall simulations in the EURO-CORDEX database full of uncertainties (or bias). The underestimation of Nr in spring and summer at both stations was possibly caused by the climate model overestimation of cloud cover as suggested by [41].

The decline in Ps under Sc_1 compared to the BP at both stations is attributed to the shortening of the Mexicali cultivar's GSL. This is induced by the increase in T_S, as P_S is expected to increase throughout the Mexicali cultivar's GSL. In addition, the increase in spring's precipitations and the lengthening of the Mexicali cultivar's GSL under Sc_2 explain why P_S is expected to increase in this scenario compared to the BP. The Nr_S decreases in Sc_1 could be due to the shortening of the Mexicali cultivar's GSL and the aerosol pollution projected under these two RCPs scenarios, as suggested by [42]. Meanwhile, the decrease in Nr_S in Sc_2 could be attributed to the atmospheric pollution caused by the presence of aerosols in the atmosphere as the Mexicali cultivar's GSL was prolonged in this last scenario.

The wheat grain yield increase projected under both RCPs scenarios is due to the fertilizing effect of the air enrichment with CO_2. This result is more consistent at the BBA site, where the evolution of the grain yield curve perfectly follows that of CO_2 concentrations. Thus, the Mexicali cultivar's grain yields projected in Sc_2 are higher than those projected in Sc_1 at the BBA site because the CO_2 concentrations expected under RCP 8.5 are higher than those projected under RCP4.5. However, at the Setif site, the average grain yield simulated in Sc_1 is higher than that simulated under Sc_2. This could be explained by the decline in T_S by −0.3 °C in Sc_2 because it is possible that the lower temperatures reduce the fertilizing effect of CO_2 on durum wheat grain yields. However, under $Sc_{1,}$ the projected grain yields at Setif are better by comparison with those projected at the BBA site. This result can be explained by the combined negative effect on wheat grain yield of very severe water stress and the thermal stress projected under Sc_1 at the BBA site (Table 3). According to the Pearson correlation test results, in Sc_1 and $Sc_{2,}$ there is a negative and statistically significant correlation between T_S and Mexicali cultivar's grain yields at the Setif site. Meanwhile, at the BBA site, the correlation is negative and statistically insignificant. On the contrary, the P_S and Nr_S's correlation is positive and statistically significant with Mexicali cultivar's grain yield simulated by AquaCrop by both future scenarios.

Therefore, the fertilizing effect of CO_2 offsets the negative effects of rising T_S and decreasing P_S and Nr_S on durum wheat yields projected, under Sc_1 and $Sc_{2,}$ at the Setif and BBA experimental sites. These results are compatible with the conclusions of recent studies carried out by ([43–45] respectively, in China, Germany, and Morocco. Likewise,

Pugh et al. [46] reported that rainfed wheat in arid and semi-arid regions located in low latitudes would benefit from the fertilizing effect of CO_2, but less in temperate regions located in high latitudes. Moreover, Long et al. [47] reported that 550 ppm high CO_2 concentration in experiments in a Free Air CO_2 Enrichment (FACE) and in closed chamber experiments, resulted in a wheat yield enhancement of +31 and +13%, respectively, in both these experimental devices. Tubiello et al. [48] suggested that the fertilizing effects of a high CO_2 concentration might be overestimated in crop models because their simulation of yield enhancement induced by a high CO_2 concentration was much greater than that observed in FACE studies. However, they suggested that the magnitude of these effects is still under debate. In China, the study [49], proved that the fertilizing effect of the CO_2 enrichment in the atmosphere slows the negative effects of warm air temperatures and precipitation decline on wheat yield under RCP 4.5 and RCP 8.5, at the beginning of the 21st century in China. Likewise, Xiong et al. [50] found that the CO_2 enrichment in the atmosphere enhanced wheat yield by +0.9% by offsetting the negative effect of the drop in solar radiation, but without this fertilizing effect of CO_2, the wheat yield would decrease by −9.7%. Under RCP 8.5, in Egypt, the increase in atmospheric CO_2 concentrations will act as a growth stimulant. The simulated irrigated wheat yield across Egypt was projected to increase slightly (2.4%) in the 2030s and will decline slowly toward the end of the century (−1.7% by 2050s and−4.0% by 2080s). This result was attributed to the increase in negative impacts of the projected warm temperature [51]. In Jordan, the ESCWA [52] reported that under RCP 4.5 and RCP 8.4, with a fixed CO_2 concentration, the rainfed wheat yields simulated with the AquaCrop model will increase by an average of about +33.8 and +48.3% in 2025 and 2045 future periods. Meanwhile, with elevated CO_2 concentration, the simulated wheat yield will be enhanced by +53.5 and +81.6%inboth above future periods, respectively, with respect to the baseline yield.

The Pearson correlation test revealed the existence of a negative and statistically significant correlation between Mexicali cultivar's GSL and T_S during the BP and under both Sc_1 and Sc_2.This result is compatible with that of [45] in Morocco and [53] in the entire Mediterranean region. Furthermore, in Palestine and Jordan, the rainfed wheat growth cycle period simulated under RCP 8.5 is projected to shorten by 2030 and 2050 [52].

In addition to these above results, the Pearson correlation test proved that there is a positive and statistically significant correlation between Mexicali cultivar's GSL and P_S during the BP and under both Sc_1 and Sc_2 at the Setif and BBA sites. Thus, the lengthening of the Mexicali cultivar's GSL by 10 days under Sc_2 at the Setif site could also be explained by the increase in P_S by +65.6 mm. Despite the shortening of Mexicali cultivar's GSL by the effect of the expected increase in T_S. The Mexicali cultivar's grain yield is projected to enhance under Sc_1.Tao et al. [54] and Liu et al. [55], explained that the shortening of wheat's GSL is due to the vegetative development stage's length reduction, meanwhile the duration of the reproductive stage remained intact. So, this negated the yield losses reported by Zheng et al. [16], who recommended wheat cultivars flower early in order to prevent wheat crops from the risk of yield loss which could be induced by very warm temperatures in late spring throughout the period of grain formation.

The ET_0 drop predicted under Sc_1 is due to the shortening of the Mexicali cultivar's GSL. This result is consistent with that of [56] on rice in Bangladesh and [45] on wheat in Morocco. However, under Sc_2, the lengthening of the Mexicali cultivar's GSL did not avoid the ET_0 decrease, which could be explained by the stomatal regulatory effect ofdurum wheat, which made it possible to reduce water losses by evapotranspiration, as it was suggested by [45], under the fertilizing effect of the elevated CO_2 concentrations in the atmosphere, projected under Sc_2.

WP is the ratio of the amount of durum wheat biomass produced (in kg) to the amount of water lost by evapotranspiration during durum wheat's growing cycle. Thus, according to the AquaCrop model simulations, the WP enhancement under both Sc_1 and Sc_2 scenarios is due to the increase in Mexicali cultivar's above-ground biomass induced by the fertilizing effect of the enrichment of the atmosphere with CO_2. This induced the acceleration of photosynthetic activity and the decrease in water loss by evapotranspiration, thanks to stomatal regulation under Sc_2, and the shortening of the Mexicali cultivar's GSL induced by the T_S increase under Sc_1. According to [52], the AquaCropcrop model, with a fixed CO_2 concentration, predicted an enhancement of rainfed wheat's WP by an average of +17.8 and +30% for the 2025 and 2045 future periods, whereas in the case of elevated CO_2, an increase of +3 and +56% are projected by both future horizons, respectively, under RCP 4.5 and RCP 8.5.

According to [46], the fertilizing effect of a high CO_2 concentration can improve the WP of C_3 plants (such as wheat) by stimulating their photosynthetic activity. Ainsworth et al. [57] reported that across a range of FACE experiments, with a variety of plant species, the growth of plants at elevated CO_2 concentrations of 475–600 ppm leads to increasing leaf photosynthetic rates by an average of +40%. These last authors explained that CO_2 concentrations are essential in regulating the openness of stomata, the pores that allow plants to exchange gasses with the exterior environment. Thus, open stomata allow CO_2 to diffuse into leaves for photosynthesis, but also provide a way for water to circulate out of leaves. Plants, therefore, regulate the degree of stomatal opening, a measure called stomatal conductance, which is used as a compromise between the aims of maintaining high rates of photosynthesis and low rates of water loss. So, as CO_2 concentrations increase, plants can maintain high photosynthetic rates with relatively low stomatal conductance. Added to that, they also reported that across a multitude of FACE experiments, growth under elevated CO_2 concentrations decreases stomatal conductance of water by an average of −22%. However, Taub et al. [58] suggested that generally, the magnitude of the effect of CO_2 on crop water use would depend on how it affects other determinants of plant water use, such as plant size, morphology, and leaf temperature.

At both experimental sites, under the climate conditions projected by 2035–2064, under RCP 4.5 and RCP 8.5, the earlier sowings of mid-September and mid-October lead to the best yields because the earlier sowing dates allow the wheat crop to take advantage of the increase in precipitations predicted in the late summer and early fall. That will allow the achievement of the vegetative development stage of the Mexicali cultivar's plant from November until the beginning of February. In addition, this early sowing allows the flowering stage to take place between the period from the end of February until the beginning of April, which allows the Mexicali cultivar's plant to avoid the high temperaturesin May and June. These results are compatible with those of [59], who also predicted the adaptation of wheat to early sowings in 2031–2060 climate conditions, under both RCP 4.5 and RCP 8.5 scenarios in the Mediterranean area. However, late sowing in mid-November and mid-December resulted in poor yields, as they led to the achievement of the durum wheat's flowering and grain-filling stages through the high-temperature period of the mid-April and early June period, which induce durum wheat's grain yield losses by scalding. This result is in concordance with those of [60], who reported that the early maturing cultivar did not show a yield reduction on any sowing dates, thanks to the earliness of the anthesis stage, on which risk of crop exposure to heat stress during the sensitive grain-filling stage is decreased or avoided. In Ethiopia, [61], reported that by the middle and the end of the 21st century, one wheat cultivar is adapted to late sowing date, under low CO_2 emissions of RCP 4.5, meanwhile another cultivar is well adapted to early sowing date, under elevated CO_2 emissions of RCP 8.5. So, it is important to assess the adaptation of wheat sowing dates under future CC scenarios by simulating different wheat cultivars yields with the crop model in order to select the best-adapted wheat cultivar to the projected CC.

5. Conclusions

This study has shown the strategic interest of referring to climate and crop modeling for the prediction of the impacts of future climate change on the rainfed durum wheat yield in the High eastern plain of Algeria. For the baseline period 1981–2010, the climate model ICHEC_KNMI used in this study has proved its reliability in simulating temperatures and net solar radiation with good precision. However, precipitations are simulated with less certainty, given the high variability of precipitation in Algeria, which made the simulation very complicated. The AquaCrop crop model was used to assess the impacts of future climate change on grain yields, length of the growth cycle, and the water productivity of the durum wheat cultivar, Mexicali, in 2035–2064, under RCP 4.5 and RCP 8.5. This study showed that the effect of the increase in CO_2 concentrationsin the atmosphere made it possible to avoid the drop yields of the rainfed Mexicali cultivar. This decline in grain yield could be induced by the negative effects: of the drop in precipitation and net solar radiation and by the increased air temperatures projected over the growing season of this cultivar in 2035–2064, under RCP 4.5 or RCP 8.5. Moreover, this study proved that the increase in temperature expected in 2035–2064 under the above scenarios, causes a shortening of the duration of the growth cycle of the Mexicali cultivar. However, an increase in yields and water productivity of this cultivar are projected by this future horizon, thanks to the fertilizing effect of the enrichment of the atmosphere with CO_2. This study has made it possible to plan a strategy for adapting rainfed durum wheat to rising temperatures by applying early sowing in October, which avoids the loss of yield during the wheat growth cycle very sensitive stage, of grain filling in spring. However, further research is needed, using climate projections from an ensemble of climate models instead of a single model, to reduce the observed uncertainties in the precipitation simulation. It is necessary to evaluate, under RCPs scenarios, with other crop models, the effect of supplemental irrigation and fertilization on the adaptation of this Mexicali cultivar and other durum wheat cultivars to future climate change. Thus, the association of climate and crop modeling proves to be a relevant tool that meets the needs of farmers in terms of choice of farming practices and cultivars, in order to reduce the negative impacts of climate change on crops yields, as it enables decision-makers in the agricultural sector to plan sustainable and effective policies to help farmers to face the projected climate change and avoid crop yield losses, thereby maintaining food security in Algeria.

Author Contributions: T.K. collected, corrected, analyzed, simulated crop data with AquaCrop, interpreted the crop and climate data and wrote the initial draft of the paper. D.S. provided some data and supervised and corrected the paper, she also, reanalyzed and reinterpreted some data and contributed in the improvement of the manuscript contain and quality. A.M. read and contributed in improvement the English editing of the paper. All authors have read and agreed to the published version of the manuscript.

Funding: This research received no external funding.

Data Availability Statement: The data presented in this study are available on request from the corresponding author. The data are not publicly available due to restriction imposed by the National office of meteorology in Algeria.

Conflicts of Interest: The authors declare no conflict of interest.

References

1. Mbow, C.; Rosenzweig, C.; Barioni, L.G.; Benton, T.G.; Herrero, M.; Krishnapillai, M.; Liwenga, E.; Pradhan, P.; Rivera Ferre, M.G.; Sapkota, T.; et al. Food Security. In *Climate Change and Land 2019, an IPCC Special Report on Climate Change, Desertification, Land Degradation, Sustainable Land Management, Food Security, and Greenhouse Gas Fluxes in Terrestrial Ecosystems*; Shukla, P.R., Skea, J., CalvoBuendia, E., Masson-Delmotte, V., Pörtner, H.-O., Roberts, D.C., Zhai, P., Slade, R., Connors, S., van Diemen, R., et al., Eds.; *in press*.
2. IPCC. Intergovernmental Panel on Climate Change. *Global Warming of 1.5 °C. An IPCC Special Report on the Impacts of Global Warming of 1.5 °C above Pre-industrial Levels and Related Global Greenhouse Gas Emission Pathways, in the Context of Strengthening the Global Response to the Threat of Climate Change, Sustainable Development, and Efforts to Eradicate Poverty*; Masson-Delmotte, V., Zhai, P., Pörtner, H.-O., Roberts, D., Skea, J., Shukla, P.R., Pirani, A., Moufouma-Okia, W., Péan, C., Pidcock, R., et al., Eds.; *in press*.
3. WMO World Meteorological Organization. *The Global Climate in 2015–2019*; World Meteorological Organization Publication: Geneva, Switzerland, 2019; p. 20.
4. Raza, A.; Razzaq, A.; Mehmood, S.S.; Zou, X.; Zhang, X.; Lv, Y.; Xu, J. Impact of climate change on crop adaptation and strategies to tackle its outcome: A review. *Plants* **2019**, *8*, 34. [CrossRef] [PubMed]

5. Tandzi Ngoune, L.; Mutengwa, S.C. *Factors Affecting Yield of Crops, Agronomy, Climate Change & Food Security*; Khan, A., Ed.; IntechOpen, 2020; Available online: https://www.intechopen.com/chapters/70658 (accessed on 30 January 2022). [CrossRef]
6. IPCC. Intergovernmental Panel on Climate Change. Africa. In *Climate Change 2014: Impacts, Adaptation, and Vulnerability. Part B: Regional Aspects. Contribution of Working Group II to the Fifth Assessment Report of the Intergovernmental Panel on Climate Change*; Barros, V.R., Field, C.B., Dokken, D.J., Mastrandrea, M.D., Mach, K.J., Bilir, T.E., Chatterjee, M., Ebi, K.L., Estrada, Y.O., Genova, R.C., et al., Eds.; Cambridge University Press: Cambridge, UK; New York, NY, USA, 2014; pp. 1199–1265.
7. Webber, H.; Ewert, F.; Olesen, J.E.; Müller, C.; Fronzek, S.; Ruane, A.C.; Bourgault, M.; Martre, P.; Ababaei, B.; Bindi, M.; et al. Diverging importance of drought stress for maize and winter wheat in Europe. *Nat. Commun.* **2018**, *9*, 4249. [CrossRef] [PubMed]
8. Panda, A. *Climate Change and Agricultural Insurance in the Asia and Pacific Region*; The Asian Development Bank: Mandaluyong, Philippines, 2021; 44p.
9. Iglesias, A.; Garrote, L. Adaptation strategies for agricultural water management under climate change in Europe. *Agric. Water Manag.* **2015**, *155*, 113–124. [CrossRef]
10. Snowdon, R.J.; Wittkop, B.; Chen, T.-W.; Stahl, A. Crop adaptation to climate change as a consequence of long-term breeding. *Theor. Appl. Genet.* **2021**, *134*, 1613–1623. [CrossRef]
11. Kourat, T. Évaluation du Changement Climatiqueet Impact sur la Production du blé dur Pluvial dans les Hautes Plaines Orientales de L'algérie. Ph.D. Thesis, ENSA El Harrach, Algiers, Algeria, 2021. Available online: http://localhost:8080/xmlui/handle/123456789/2489 (accessed on 3 January 2022).
12. Bessaoud, O.; Pellissier, J.P.; Rolland, J.P.; Khechimi, W. *Rapport de Synthèse sur L'agriculture en Algérie*; Research Report; CIHEAM-IAMM: Montpellier, France, 2019; p. 82.
13. MATE. Ex Ministère de l'Aménagement de Territoire et de l'Environnement. In *Etude Diagnostique sur la Biodiversité et les Changements Climatiques en Algérie*; Cooperatioproject between the Algerian Ministry of Land Planning and the Environment (MATE), United Nations Development Program (UNDP) and Global Environment Facility (GEF); Final Report; MATE: Algiers, Algeria, 2015; p. 111.
14. Harrag, M.; Boulfred, Y. La sécurité alimentaire en Algérie Une étude analytique sur les céréales Food security in Algeria An analyticalstudy on cereals. *Rev. L'économie Financ. Aff.* **2019**, *2*, 162–188. Available online: Chrome-extension://efaidnbmnnnibpcajpcglclefindmkaj/viewer.html?pdfurl=https%3A%2F%2Fwww.asjp.cerist.dz%2Fen%2FdownArticle%2F109%2F3%2F2%2F92285&clen=170135 (accessed on 11 January 2022).
15. Rapacza, M.; Macko-Podgórnib, A.; Jurczyka, B.; Kucharc, L. Modeling wheat and triticale winter hardiness under current and predicted winter scenarios for Central Europe: A focus on deacclimation. *Agric. For. Meteorol.* **2021**, *313*, 108739. [CrossRef]
16. Zheng, Z.; Cai, H.; Wang, Z.; Wang, X. Simulation of Climate Change Impacts on Phenology and Production of Winter Wheat in Northwestern China Using CERES-Wheat Model. *Atmosphere* **2020**, *11*, 681. [CrossRef]
17. Qian, B.; Jing, Q.; Cannon, A.J.; Smith, W.; Grant, B.; Semenov, M.A.; Xu, Y.-P.; Ma, D. Effectiveness of using representative subsets of global climate models in future crop yield projections. *Sci. Rep.* **2021**, *11*, 20565. [CrossRef]
18. Agriconsulting. Support to Climate Change Mitigation and Adaptation in the ENPI South Region. Algeria Country Report, ENPI/2012/308-787. 2013. Available online: https://www.climamed.eu/wp-content/uploads/files/Algeria-Country-Report.pdf (accessed on 5 December 2021).
19. Climat-Med. Policy Fiche: Managing the Impact of Climate Change on Agriculture. 2018. Climat-Med. 13 Pages. Available online: https://www.climamed.eu/wp-content/uploads/files/Agriculture-CC-Algeria.pdf (accessed on 20 December 2021).
20. Wayne, G. The Beginner's Guide to Representative Concentration Pathways. 2013. Available online: https://skepticalscience.com/docs/RCP_Guide.pdf (accessed on 13 December 2021).
21. Rouabhi, A.; Adouane, C.; Felloussia, A.L. Assessment of climate change and its effects on cereal production by 2070 in north Algeria. *Rev. Agric.* **2019**, *10*, 27–39. Available online: file:///C:/Users/USER/Downloads/Rouabhi-al.pdf (accessed on 2 January 2022).
22. Kourat, T.; Medjerrab, A. Analyse et cartographie des pluies et l'incidence de leursvariabilitéspatiotemporelle sur la délimitation des zones céréalières dans les hautesplainesorientales de l'Algérie. *Révue. Agric.* **2016**, *1*, 220–229. Available online: //efaidnbmnnnibpcajpcglclefindmkaj/viewer.html?pdfurl=https%3A%2F%2Frevue-agro.univ-setif.dz%2Fdocuments%2Fnumerospecial%2FSession6%2FKourat%2520et%2520Medjerb.pdf&clen=1619913&chunk=true (accessed on 5 December 2021).
23. Kourat, T. Analyse et Cartographie des Pluies et L'incidence de leurs Variabilite Spatio-Temporelle sur la Délimitation Des zones Céréalières dans les Hautes Plaines Orientales de L'algérie. Master's Thesis, Faculté des Sciences de la Terre, De Geographie et de L'amenagement de Territoire, Université des Sciences et de la TechnologieHouari Boumediene, Algiers, Algeria, 2015.
24. Moss, R.H.; Babiker, M.; Brinkman, S.; Calvo, E.; Carter, T.; Edmonds, J.A.; Elgizouli, I.; Emori, S.; Lin, E.; Hibbard, K.; et al. *Towards New Scenarios for Analysis of Emissions, Climate Change, Impacts, and Response Strategies*; Intergovernmental Panel on Climate Change: Geneva, Switzerland, 2008.
25. Van Vuuren, D.P.; Edmonds, J.; Kainuma, M.; Riahi, K.; Thomson, A.; Hibbard, K.; Hurtt, G.C.; Kram, T.; Krey, V.; Lamarque, J.-F.; et al. The representative concentration pathways: An overview. *Clim. Chang.* **2011**, *109*, 5. [CrossRef]
26. Saeed Al Blooshi, F.; Alyan, S.; Joshua, N.J.; Ksiksi, T.K. Modeling Current and Future Climate Change in the UAE using Various GCMs in MarksimGCM^{R}. *Open Atmos. Sci. J.* **2019**, *13*, 56–64. [CrossRef]
27. Arbadi, R.; Bacciu, V.; Benkheira, A.; Bouazzaoui, A.; Bouzid, B.W.; Brachemi, O.; Ziani Cherif, S.M.; Ghouari, N.; Salis, M.; Tefiani, W. *Réduction D'échelle et ModélisationClimatique Avec Une Application a la Gestion des ForetsenAlgérie*; Technical Paper; ClimaSouth: Algiers, Algeria, 2018; p. 56.

28. Navarro-Racines, C.E.; Tarapues, J.; Thornton, P.; Jarvis, A.; Ramirez-Villegas, J. High-resolution and bias-corrected CMIP5 projections for climate change impact assessments. *Sci. Data* **2020**, *7*, 7. [CrossRef] [PubMed]
29. Foughali, A.; Tramblay, Y.; Bargaoui, Z.; Carreau, J.; Ruellan, D. Hydrological Modeling in Northern Tunisia with Regional Climate Model Outputs: Performance Evaluation and Bias-Correction in Present Climate Conditions. *Climate* **2015**, *3*, 459–473. [CrossRef]
30. Tramblay, Y.; El Adlouni, S.; Servat, E. Trends and variability in extreme precipitation indices over Maghreb countries. *Nat. Hazards Earth Syst. Sci.* **2013**, *13*, 3235–3248. [CrossRef]
31. Latief, A.; Syed, S.M. Satelite Farming. In *An Information and Technology Based Agriculture*; Springer: Berlin/Heidelberg, Germany, 2018; 181p.
32. Steduto, P.; Hsia, T.C.; Raes, D.; Fereres, E. AquaCrop—The FAO Crop Model to Simulate Yield Response to Water: I. Concepts and Underlying Principles. *J. Agron.* **2009**, *101*, 426–437. [CrossRef]
33. Raes, D.; Steduto, P.; Hsiao, T.C.; Fereres, E. *AquaCrop*; Version 4.0; Manuel d'utilisation; FAO, Land and Water Division: Rome, Italy, 2014.
34. Allen, R.G.; Pereira, L.S.; Raes, D.; Smith, M. *Crop Evapotranspiration. Guidelines for Computing Crop Requirements. FAO Irrigation and Drainage Paper*; FAO: Rome, Italy, 2015; Volume 56, 15p.
35. Saxton, K.E.; Willey, P.H. *The SPAW Model for Agricultural Field and Pond Hydrologic Simulation*; CRC Press: Boca Raton, FL, USA, 2006.
36. Kaly, E.; Malou, R.; Akpo, L.E. Évaluation du modèleAquaCrop de la FAO enriziculturei rriguée dans le Delta du FleuveSénégal. *J. Anim. Plant Sci.* **2016**, *30*, 4781–4796.
37. Khebbat, Z. Etat Hydrique du sol et Performance D'une Collection de Variétés de blé dur. Ph.D. Thesis, Faculté des Sciences de la Nature et de la Vie, Université Ferhat Abbas, Setif, Algeria, 2015; 83p.
38. Gray, C.D.; Kinnear, P.R. *IBM SPSS Statistics 19 Made Simple*; Psychology Press: New York, NY, USA, 2012; p. 688.
39. Rosa, S.L.K.; Souza, J.L.M.D.; Tsukahara, R.Y. Performance of the AquaCrop model for the wheat crop in the subtropical zone in Southern Brazil. *Pesqui. Agropecuária Bras.* **2020**, *55*, e01238. [CrossRef]
40. Romera, R.; Sánchez, E.; Domínguez, M.; Gaertner, M.; Gallardo, C. Evaluation of present-climate precipitation in 25 km resolution regional climate model simulations over Northwest Africa. *Clim. Res.* **2015**, *66*, 125–139. [CrossRef]
41. Freidenreich, S.M.; Ramaswamy, V. Analysis of the biases in the downward shortwave surface flux in the GFDL CM2.1 general circulation model. *J. Geophys. Res.* **2011**, *116*, D08208. [CrossRef]
42. Liu, M.; Matsui, H. Aerosol radiative forcings induced by substantial changes in anthropogenic emissions in China from 2008 to 2016. *Atmos. Chem. Phys.* **2021**, *21*, 5965–5982. [CrossRef]
43. Erda, L.; Wei, X.; Hui, J.; Yinlong, X.; Yue, L.; Liping, B.; Liyong, X. Climate change impacts on crop yield and quality with CO_2 fertilization in China. *Philos. Trans. R. Soc. B Biol. Sci.* **2015**, *360*, 2149–2154. [CrossRef]
44. Degener, J.F. Atmospheric CO_2 fertilization effects on biomass yields of 10 crops in northern Germany. *Front. Environ. Sci.* **2015**, *3*, 48. [CrossRef]
45. Bouras, E.; Jarlan, L.; Khabba, S.; Er-Raki, S.; Dezetter, A.; Sghir, F.; Tramblay, Y. Assessing the impact of global climate changes on irrigated wheat yields and water requirements in a semi-arid environment of Morocco. *Sci. Rep.* **2019**, *9*, 19142. [CrossRef] [PubMed]
46. Pugh, T.; Müller, C.; Elliott, J.; Deryng, D.; Folberth, C.; Olin, S.; Schmid, E.; Arneth, A. Climate analogues suggest limited potential for intensification of production on current croplands under climate change. *Nat Commun.* **2016**, *7*, 12608. [CrossRef] [PubMed]
47. Long, S.P. Food for Thought: Lower-than-expected crop yield stimulation with rising CO_2 concentrations. *Science* **2006**, *312*, 918–1921. [CrossRef]
48. Tubiello, F.N.; Amthor, J.S.; Boote, K.; Donatelli, M.; Easterling, W.; Fischer, G.; Gifford, R.M.; Howden, S.; Reilly, J.; Rosenzweig, C. Crop response to elevated CO_2 and world food supply: A comment on "Food for Thought" by Long et al. Science 312:1918–1921. *Eur. J. Agron.* **2007**, *25*, 215–223. [CrossRef]
49. Xiao, D.; Bai, H.; Liu, D.L. Impact of future climate change on wheat production: A simulated case for china's wheat system. *Sustainability* **2018**, *10*, 1277. [CrossRef]
50. Xiong, W.; Holman, I.; Lin, E.; Conway, D.; Li, Y.; Wu, W. Untangling relative contributions of recent climate and CO_2 trends to national cereal production in China. *Environ. Res. Lett.* **2012**, *7*, 044014. [CrossRef]
51. Asseng, S.; Kheir, A.M.S.; Kassie, B.T.; Hoogenboom, G.; Abdelaal, A.I.N.; Haman, D.Z.; Ruane, A.C. Can Egypt become self-sufficient in wheat? *Environ. Res. Lett.* **2018**, *13*, 094012. [CrossRef]
52. ESCWA. Economic and Social Commission of Western Asia. *Modelling Climate Change Impacts on Agricultural Productivity Using AquaCroptools and RICCAR Datasets. Climate Change & Natural Resources Sustainability Cluster*; Virtual Workshop for the Arab Region; ESCWA: Beirut, Lebanon, 2020.
53. Saadi, S.; Todorovic, M.; Tanasijevic, L.; Pereira, L.S.; Pizzigalli, C.; Lionello, P. Climate change and Mediterranean agriculture: Impacts on winter wheat and tomato crop evapotranspiration, irrigation requirements and yield. *Agric. Water Manag.* **2015**, *147*, 103–115. [CrossRef]
54. Tao, F.; Zhang, Z. Climate change, wheat productivity and water use in the North China Plain: A new super-ensemble-based probabilistic projection. *Agric. For. Meteorol.* **2013**, *170*, 146–165. [CrossRef]

55. Liu, H.; Li, Y.; Josef, T.; Zhang, R.; Huang, G. Quantitative estimation of climate change effects on potential evapotranspiration in Beijing during 1951–2010. *J. Geogr. Sci.* **2013**, *24*, 93–112. [CrossRef]
56. Acharjee, T.K.; Ludwig, F.; van Halsema, G.; Hellegers, P.; Iwan, S. Future changes in water requirements of Boro rice in the face of climate change in North–West Bangladesh. *Agric. Water Manag.* **2017**, *194*, 172–183. [CrossRef]
57. Ainsworth, E.A.; Rogers, A. The response of photosynthesis and stomatal conductance to rising (CO_2): Mechanisms and environmental interactions. *Plant Cell Environ.* **2007**, *30*, 258–270. [CrossRef] [PubMed]
58. Taub, D. Effects of Rising Atmospheric Concentrations of Carbon Dioxide on Plants. *Nat. Educ. Knowl.* **2010**, *3*, 21.
59. Moriondo, M.; Bindi, M.; Brilli, L.; Costafreda-Aumedes, S.; Dibari, C.; Leolini, L.; Padovan, G.; Trombi, G.; Karali, A.; Varotsos, K.V.; et al. Assessing climate change impacts on crops by adopting a set of crop performance indicators. *Euro-Mediterr. J. Environ. Integr.* **2021**, *6*, 45. [CrossRef]
60. Yang, C.; Fraga, H.; van Ieperen, W.; Trindade, H.; Santos, J.A. Effects of climate change and adaptation options on winter wheat yield under rainfed Mediterranean conditions in southern Portugal. *Clim. Chang.* **2019**, *154*, 159–178. [CrossRef]
61. Demelash, T.; Amou, M.; Gyilbag, A.; Tesfay, G.; Xu, Y. Adaptation Potential of Current Wheat Cultivars and Planting Dates under the Changing Climate in Ethiopia. *Agronomy* **2022**, *12*, 37. [CrossRef]

climate

MDPI

Article

Climate Changes in Southeastern Poland and Food Security

Barbara Sawicka [1,*], Piotr Barbaś [2], Piotr Pszczółkowski [3], Dominika Skiba [1], Farhood Yeganehpoor [4] and Barbara Krochmal-Marczak [5]

1 Department of Plant Production Technology and Commodity Science, University of Life Sciences in Lublin, 20-950 Lublin, Poland; dominika.skiba@up.lublin.pl
2 Department of Potato Agronomy, Institute of Plant Breeding and Acclimatization, National Research Institute, 05-870 Błonie, Poland; p.barbas@ihar.edu.pl
3 Experimental Station for Cultivar Assessment of Central Crop Research Centre, Uhnin, 20-950 Lublin, Poland; p.pszczolkowski.inspektor@coboru.gov.pl
4 Department of Plant Eco-Physiology, University of Tabriz, Tabriz 51368, Iran; farhoodyeganeh@yahoo.com
5 Department of Food Production and Safety, Krosno State College, Rynek 1, 38-400 Krosno, Poland; barbara.marczak@kpu.krosno.pl
* Correspondence: barbara.sawicka@up.lublin.pl

Citation: Sawicka, B.; Barbaś, P.; Pszczółkowski, P.; Skiba, D.; Yeganehpoor, F.; Krochmal-Marczak, B. Climate Changes in Southeastern Poland and Food Security. *Climate* **2022**, *10*, 57. https://doi.org/10.3390/cli10040057

Academic Editors: Christopher Robin Bryant, Andrea Vitali and Azzeddine Madani

Received: 31 January 2022
Accepted: 4 April 2022
Published: 6 April 2022

Abstract: The conducted research is of particular importance for the country's food security in the context of climate change in Southeastern Poland. The aim of the research was to determine the influence of climate on the variability of the appearance and the rate of spread of potato blights as the main factor limiting the potato yield in the conditions of Central and Eastern Europe. Combined statistical and simulation modeling methods were used. A mixed effect model was used to detect the effects of temperature, humidity, rainfall and wind speed on potato yield, and partial regression analysis models were used. The natural, agricultural and economic conditions in terms of suitability for potato cultivation were assessed, and factors influencing the fluctuation of the cultivated acreage, yield and harvesting of potatoes were identified. The forecast was based on empirical data from 2000 to 2019. It has been proven that potato cultivation in Southeastern Poland is more vulnerable to climate change than in the rest of the country. The results obtained from analyzing multi-annual results can help policymakers to develop strategies to increase the stability of future potato production and the safety of the crop. This will enable the better use of generated data and methodological approaches to analyze the role of climate, both on a regional and global scale.

Keywords: drought; explorations; forecasting models; food security; food access; late blight; potato production; rainfall; stress; cultivars temperature yield

1. Introduction

Climate change is one of the most important issues in science today. It is also an extremely complex and multidisciplinary aspect [1,2]. Climate change is a general threat to food production. The two regions of the world with the highest growth due to malnutrition (PoU) are Asia (418 million) and Africa (282 million people affected by hunger) in 2020 [2,3]. Food insecurity is growing steadily around the world, especially in the African region. Severe global food insecurity is witnessing a huge increase from 8.3% (605 million) in 2014 to 11.9% (928 million) in 2020 [4]. Yields of wheat, maize and other crops are slowing in many countries due to extreme heat, severe weather conditions and droughts [1]. According to some estimates, in the absence of effective adaptation, global yields could fall by as much as 30% by 2050. Global warming, according to the FAO [2], will lead to rising costs of bread, market shocks and political unrest. In recent years, climate change in many countries of the world has been associated with food insecurity, especially in developing countries; for example, this is observed in many African countries such as Nigeria, Senegal, Guinea and others [4]. Serious food insecurity problems also exist in many developed countries in Europe and North America, particularly in some rural areas away from major urban centers.

In these areas, human populations may be unsecured from the food side, as this population is usually poor [3,4]. Reducing precipitation will strengthen the desertification process, mainly in the outskirts of the Sahara and Southern Africa. In the case of Asia, climate change will affect the diversification of the production potential of agriculture in individual countries. The most unfavorable changes will occur in poor countries, especially in the coastal areas of the monsoon zone. Rising sea levels will flood some farmlands and worsen the availability of fresh water [3,4]. Kulig et al. [5] and Ziernicka-Wojtaszek [6,7] point out that if the warming trend continues, significant changes in the structure of crops may also occur in Europe, especially in Southeastern Europe, including Poland. For example, on the one hand, it will be possible to introduce plants with increased thermal requirements (maize, sugar sorghum, sweet potato and grapevine) on a larger scale, and on the other hand, the acreage and yields of certain agricultural cultures will be significantly reduced. These climate changes can have a serious impact on food production and have the potential to limit the production of staple food raw materials, such as potatoes [5,8,9]. The temperate climate zone in which Europe is located shows the greatest variation in climatic conditions among all zones. It distinguishes a group of cool climates and a group of warm climates, including maritime, transitional and continental climates. Due to the favorable conditions for agriculture, this zone is the food basin of the world. Strong warming has been observed in the last three decades. Central Europe, including Poland, is also experiencing this. The forecasts for Poland also predict further warming, as well as changes in the spatial and seasonal distributions as well as the amount of precipitation. However, climate models do not agree on the direction of these changes. Rainfall in Poland is expected to decrease in summer (it depends on the model) and increase in winter. Consequently, there is still considerable uncertainty about the likely impact of climate change on Poland's water resources. In general, changes in thermal characteristics, precipitation and air humidity will have an impact on changes in water balance and, thus, on the productivity of crops and the country's food security. Due to the insufficient consistency between climate models, the scope of changes may differ depending on the model and biotic and abiotic factors. Climate change trends in Poland, in terms of temperature, precipitation and relative humidity, in the southeastern part of Poland based on global the scenarios for climate and emissions (according to IPCC AR5) are presented in the studies by Kulig et al. [5], Ziernicka-Wojtaszek [6,7] and Mezghani et al. [10]. In general, all climate models show a systematic upward trend average air temperature, both in the short term and as well as the future. Significant regional differences between simulations and seasons were found. However, some simulations were not very good at recreating the temperature gradient from the northwest to the southeast of Poland. This applies, inter alia, to the topographic influence of the mountains in the south of the country, which can be seen, for example, in orographic and convective rainfall [10]. Data from a common subset of global climate models show greater changes in precipitation and less warming than the average based on the full set of GCM models. This means that the predictions based on a subset of global climate models used in Poland cover a limited range of possible climate change, compared to the entire set of GCMs. This is especially true of air temperature, while in the case of precipitation there is almost the same range of variation. Total rainfall is likely to increase by 2035, then changes will stabilize and will be approximately + 6% by the end of the 21st century. In the study of climate change, apart from the course of precipitation and air temperatures, it is also analyzed [7,10]. The average daily value of the Humidex Thermal Discomfort Index, which combines the influence of temperature and air humidity. High values of this index may affect the feeling of thermal discomfort even in healthy people, worsening the health of people suffering from heart and circulatory system diseases, and may accelerate the outbreak of fungal diseases, including potato blight epidemics, and affect the rate of its spread [5,10,11]. However, there is a well-founded fear that with the increase in sensitive meteorological data and the effects of these changes, the risk of unfavorable climate changes in the region of Central and Eastern Europe may begin to increase again, which may be particularly conducive to the spread of dangerous potato diseases [5,11–13].

The impact of climate change on Polish agriculture was studied by focusing on the potato, assessing the unfavorable growing conditions in the time horizon of 2000–2019. Among the many pests that attack potato crops in almost all regions of the world, potato blight (*Phytophthora infestans* Mont de Barry) dominates, a fungus-like organism that is responsible for the most dangerous disease in potato plantations and was responsible for the famine in Ireland in 1843–45 and continues to cause worldwide devastation with respect to potatoes. Moreover, this disease reappears in the form of different genotypes and causes huge losses in potato yields [11–13]. Its harmfulness consists in destroying the aerial parts of plants, which in turn results in a reduction in the assimilation area and, thus, the quantity and quality of the yield of progeny tubers. The development of the disease is closely related to the meteorological conditions in potato plantations. The disease develops most rapidly in conditions of high humidity (prolonged rainfall or long-lasting fog) and air temperatures between 12 and 18 °C. In such favorable conditions, if chemical protection is not applied, up to 10% of the assimilation area of potato plants can be destroyed daily—while the destruction of more than 50% of leaves and stems stops the accumulation of tuber yield under the bush. Crop losses on unprotected plantations in Europe reach 70–80%. In some years, the disease appears on plantations very early, even in May. If the meteorological conditions accompanying primary infections are favorable for the development of this disease, then we are dealing with very early epidemics of potato blight and the premature destruction of tops across the field, which significantly affects the size and quality of the crop. In the case of very early infections, potato yield losses may reach even 100%, which directly threatens the country's food security [13–16]. The occurrences and intensities of *P. infestans* on the aerial parts of plants and then on tubers are strictly dependent on the meteorological conditions and the source of the pathogen in the field. The periods of increased air humidity, caused by long rains or long morning mists or dews (RH > 90%) and low temperatures (approx. 15 °C), favor the development of the pathogen causing potato blight. If this type of humid weather persists for several days in June or early July, massive plant contamination can be expected. Under such conditions, *P. infestans* spores turn into zoosporangia, containing 6–16 tidal spores, which are easily released into the environment, causing massive plant infections and resulting in a high reduction in tuber yield [16]. Lower humidity and air temperatures above 18 °C contribute to *P. infestans* spores directly by germinating and infecting neighboring plants.

The further development of this pathogen is not only the most intense at temperatures above 20 °C but also at increased ambient humidity. Under such conditions, potato cultivation may be destroyed by *P. infestans* within a week, and in extreme cases, even within 2–3 days. The pathogen's spores can spread with wind or rain for a distance of even several dozen kilometers, which facilitates the immediate spread and spread of this disease [17,18]. The harmfulness of potato blight is related both to the decrease in the yield obtained and to the direct infection of tubers. The reduction in the yield is the result of the disease destroying the above-ground part (assimilating surface), leading to the inhibition of tuber growth, which directly threatens the food security of the country's inhabitants [10,15,16].

It is possible to estimate the risk of a decrease in potato yields as a result of potato blight (*Phytopthora infestans* Mont de Bary), which is the reason for a significant drop in the yield on the basis of historical data [8]. The rarity of unfavorable meteorological phenomena and the short period of their recording, however, may not capture the real risk [5]. One way to solve this problem is to simulate the values of air temperature and humidity, rainfall and wind speed to obtain a broader picture of the risk of this disease, which can significantly reduce potato yields.

The 1996 World Food Summit established that food security exists when all people, at all times, have physical and economic access to sufficient, safe and nutritious food that meets the nutritional needs and preferences of an active person and a healthy lifestyle [1,2]. This widely accepted FAO definition identifies the following pillars of food security:

- **Food availability** refers to the availability of a sufficient quantity of food of appropriate quality, either domestically produced or imported (including food aid). What is particularly important is the access to food as a result of sudden shocks (e.g., economic crisis, refugee or climate crisis) or cyclical events (e.g., seasonal food insecurity),
- **Access to food** means the availability of natural persons to adequate food resources; entitlements to acquire a sufficient amount of food for the preparation of a nutritious diet. Powers are defined as the collection of all commodity packages over which an individual can establish command, taking into account the legal, political, economic and social patterns of the community in which he lives (including traditional rights such as access to shared resources),
- **Utilization** means using food through an appropriate diet, clean water, proper sanitation and health care in order to achieve a state of good nutrition in which all human physiological needs are met. This also underlines the high importance of non-food inputs for food security,
- **Stability** means that people must have access to appropriate food at all times in order to ensure food security for the entire public, for households or for an individual person. People should not risk losing their health. The concept of stability can, therefore, refer to both the dimensions of accessibility and access to food security [2,3].

Food uncertainty is likely to deteriorate drastically in many situations and in many countries, especially in Central and Eastern Europe, in the near future [1,3,4]. According to FAO [1], the main factors driving the food crisis include armed conflicts usually followed by extreme weather conditions and climate variability.

Potato production is related to food safety and food security. Several different aspects of food security have been developed in recent years. Food security can not only be considered in various aspects—international and national—but also from the point of view of the household. In the international dimension of food security, the need to combat hunger is indicated [3,4], where food is perceived in terms of a public good. On the other hand, in the national dimension, the emphasis is on the appropriate institutional policy. The aim is for each country to improve its own food law, making the idea of food security a reality.

Hence, the aim of the study was to find the relationship between climate change in Southeastern Poland and the threat to the safety of the potato crop as the basic food raw material [11,12,18]. On this basis, an alternative research hypothesis has been formulated, which assumes that the earlier forecasts of the potential impact of adverse climate changes conducive to the outbreak of *P. infestans* and the development of this pathogen may allow emergency managers who plan ahead for the occurrence of high risk to prioritize conservation measures plants and save the yield, in view of the null hypothesis that these activities do not affect the outbreak of the epidemic and cannot prevent a decrease in potato yields due to the development of *P. infestans*.

2. Material and Methods

For the purpose of the work, meteorological data from the experimental stations of the Central Research Center for Cultivated Plants in Southeastern Poland and some selected data from the meteorological stations of the Institute of Meteorology and Water Management (IMGW) in the province of Podkarpackie from 2000 to 2019 were used. Statistical data were also used, such as the following: cultivation area, yields and crops of potatoes (WUS). In addition, the research was based on our own observations and monitoring of potato blight in the years 2000–2019.

2.1. Climatic Conditions

The climate of Podkarpacie is transitional between oceanic and continental climates. It has 3 climatic zones: lowland, submontane and mountain. Average annual temperatures range from 8.3 to 9.5 °C. Annual rainfall ranges from 600 to 1000 mm [19].

2.2. Soil Conditions

The area of Southeastern Poland is characterized by high soil variability, which is related to the very diverse topography. The northern part is flat, and the southern part is mountainous. In lowland agricultural areas, there are mainly fallow and brown soils made of sands, loams, loams and silt deposits. Acidified soils (66%) prevail in this area, including soils that are very acidic 35%, acid 31%, slightly acid 20%, neutral and alkaline 14%. Regardless of the parent rock type and grain size composition, the acidification level is similar (high) both in the northern and southern parts of the voivodeship. Most soils with a pH below 5.5 (very acidic and acidic) are found in the following districts: Bieszczadzki, Brzozowski, Dębicki, Kolbuszowski, Leski, Niski and Sanocki (72–91%). The state of soil abundance in available macroelements and microelements is related to the geochemical composition of the soil, but at the same time, it is an indicator of the level of plant production. As much as 54% of agricultural soils in the voivodship show a deficit of phosphorus. The greatest deficiency of assimilable phosphorus is found in the soils of mountain areas (Bieszczady, Leski, Jasielski, Krośnieński and Sanocki counties) at 78–90% and soils from the Niski and Stalowa Wola counties. Very low and low K_2O contents are shown in 45% of the studied soils (in the following districts: Bieszczadzki, Dębicki, Kolbuszowski and Niżański). Seventy-five percent of agricultural land has a satisfactory magnesium content, and only twenty-five percent has a very low and low content. In the northern part of this area of Poland, fawn and brown whitewashed soils formed from water–glacial deposits are dominant. In the central part, there is a predominance of less than brown and leached soils made of sands and tills and brown soils formed from loess and loess-like formations, and the southern part is dominated by acidic brown soils and leached from flysch rocks [20].

2.3. Meteorological Conditions

Meteorological data for the years 2000–2019 were obtained from the meteorological stations at the Variety Assessment Experimental Plants belonging to the Central Research Center for Cultivated Plants (Dukla, Nowy Lubliniec, Przecław and Skołoszów), in the region of Southeastern Poland and from the Agrometeorological Bulletins of the Institute of Meteorology and Water Management [21]. The CLIMGEN model [22] was used to analyze the data.

The station in Dukla (49°33′ N, 21°41′ E) is located in the southwestern part of the Podkarpackie Province at an altitude of 324 m above sea level, with an average annual temperature of 7.3 °C and annual rainfall of 887 mm. Unlike other stations, it was characterized by high rainfall throughout the growing season. The highest amount of rainfall was recorded there in 2001 (830 mm) and the lowest in 2008 (476 mm) (Table 1).

The warmest year was 2008, with an average air temperature of 15.0 °C, while the coldest one was 2005 with an average air temperature of 13.2 °C. All the years were wet, and the hydrothermal coefficient oscillated between 3.1 and 1.7. Optimal hydrological and meteorological conditions occurred only in 2007 [24,25] (K = 1.3).

Table 1. Descriptive statistics of average air temperature, precipitation totals and the hydrothermal coefficient of Sielianinov in 2000–2019 according to COBORU meteorological stations in Dukla, Nowy Lubliniec Przecław and Skołoszów for the potato vegetation period (April–September).

Specification	*2000*	*2001*	*2002*	*2003*	*2004*	*2005*	*2006*	*2007*	*2008*	*2009*	*2010*	*2011*	*2012*	*2013*	*2014*	*2015*	*2016*	*2017*	*2018*	*2019*
Average air temperature																				
Mean	15.2	15.0	14.7	15.6	15.2	14.0	14.6	15.3	15.7	14.6	15.3	15.1	16.1	15.3	15.4	15.1	15.9	15.7	15.7	17.0
Median	16.4	16.3	14.7	16.9	17.2	15.2	15.0	15.6	18.1	16.9	16.9	17.4	18.0	17.0	17.5	16.8	17.5	17.2	17.3	16.8
Stand. deviation	3.8	2.9	4.1	4.5	4.5	3.8	3.6	3.9	4.3	3.8	4.1	4.5	4.2	3.9	3.9	3.8	4.2	3.4	3.9	2.7
Kurtosis	−1.1	−1.4	−0.9	−0.6	−0.2	−1.3	−0.5	−0.8	−1.0	−1.1	−0.8	−1.4	−1.5	−0.6	−1.3	−0.9	−0.9	−1.0	−1.0	−1.3
Skewness	−0.4	−0.4	−0.3	−0.8	−1.1	−0.4	−0.6	−0.3	−0.7	−0.6	−0.6	−0.5	−0.3	−0.8	−0.6	−0.6	−0.6	−0.6	−0.6	−0.1
Range	11.5	8.4	13.0	13.9	13.5	10.8	11.9	13.1	12.2	11.6	12.3	12.2	12.1	11.3	10.6	11.3	12.9	10.3	11.6	8.4
Minimum	8.8	10.7	7.4	7.1	6.1	7.6	7.9	8.4	8.0	7.9	8.1	7.9	9.3	8.0	8.8	8.1	8.4	9.7	8.8	12.4
Maximum	20.3	19.1	20.4	21.0	19.6	18.4	19.8	21.5	20.2	19.5	20.4	20.1	21.4	19.3	19.4	19.4	21.3	20.0	20.4	20.8
V (%)	24.8	**19.1**	27.9	**28.5**	**29.7**	26.8	24.4	25.3	27.6	25.7	26.8	29.6	26.0	25.3	25.3	25.3	26.7	21.6	25.0	**16.1**
Sum of rainfall																				
Mean	78	82	102	78	63	82	77	72	71	79	73	89	71	66	68	77	66	66	80	71
Median	73	72	90	67	61	65	74	75	54	73	72	93	54	73	66	75	75	56	83	83
Stand. deviation	30	54	55	45	35	54	36	50	43	31	29	40	51	24	42	21	32	37	31	35
Kurtosis	0	1	6	7	2	3	−1	3	0	−1	−1	0	1	0	−1	0	−1	2	−1	−1
Skewness	1	1	2	2	1	1	0	1	1	0	0	−1	1	1	1	1	−1	2	−1	0
Range	111	188	252	214	154	230	138	223	156	108	105	145	178	96	132	79	107	143	108	115
Minimum	37	19	40	25	10	19	11	7	19	28	20	6	19	33	18	43	8	24	21	16
Maximum	148	207	292	239	164	249	149	230	175	136	125	151	197	129	150	122	115	167	129	131
V (%)	39	66	54	57	56	66	47	**69**	61	40	40	45	**72**	37	63	**27**	48	55	38	48
HCH *																				
Mean	1.9	1.8	2.4	1.7	1.5	1.9	1.8	1.6	1.5	1.9	1.5	1.9	1.6	1.4	1.5	1.7	1.5	1.4	1.7	1.3
Median	1.6	1.7	2.2	1.6	1.4	1.7	1.7	1.4	1.2	1.8	1.6	1.8	1.4	1.4	1.5	1.6	1.5	1.4	1.7	1.3
Stand. deviation	1.0	1.0	1.1	0.7	0.8	0.9	0.7	0.9	1.1	0.7	0.4	1.0	1.2	0.5	0.8	0.4	0.6	0.6	0.7	0.5
Kurtosis	3.7	0.1	0.1	1.4	−0.1	3.1	−0.2	4.9	3.5	0.4	1.0	0.1	3.3	0.6	−1.1	−0.1	0.5	−0.0	−0.1	−0.5
Skewness	1.7	0.7	0.8	0.8	0.5	1.3	0.2	1.4	1.9	0.5	−0.6	0.6	1.8	0.6	−0.0	0.8	−0.8	0.7	0.0	0.3
Range	4.2	3.7	4.1	3.1	3.1	4.2	3.2	4.8	4.6	3.0	1.9	3.9	5.1	1.8	2.4	1.4	2.5	2.5	2.8	2.2
Minimum	0.7	0.3	0.9	0.6	0.2	0.5	0.3	0.2	0.5	0.5	0.5	0.2	0.3	0.7	0.3	1.2	0.1	0.4	0.3	0.4
Maximum	4.9	4.1	5.0	3.7	3.2	4.8	3.4	4.9	5.1	3.5	2.4	4.1	5.4	2.5	2.7	2.6	2.6	2.9	3.1	2.6
V (%)	51.1	52.3	45.0	40.9	51.7	47.9	42.0	**60.0**	**72.1**	38.9	**28.2**	50.3	76.9	31.4	51.8	**21.7**	43.6	45.0	38.8	41.8

* HTC—Hydro-thermal Coefficient of Sielianinov expressed as the quotient of the monthly sum of atmospheric precipitation and the sum of average daily air temperatures in a given decade or month for the period in which the average daily temperature exceeds 10 °C [23].

The station in Nowy Lubliniec (50°17′ N 23°05′ E) is located at an altitude of 217 m above sea level and is characterized by an average annual air temperature of 7.3 °C and a total rainfall of 665 mm (Table 1). The average long-term air temperature in Nowy Lubliniec ranged from 13.8 °C in 2005 to 15.5 °C—in 2003. The highest amount of rainfall occurred in July 2003 and amounted to 239 mm, while the lowest value of this feature was recorded in 2006—11 mm (Table 2). The sum of atmospheric precipitation in the analyzed years ranged from 563 mm in 2000 to 352 mm in 2006. The hydrothermal index ranged from 1.2 to 2.2, with the optimal years being 2007–2008, while they were wet for the remaining years [24,25] (Table 1).

Table 2. Statistical characteristics dependent and independent variables.

Traits	Y	X_1	X_2	X_3	X_4	X_5
Minimum	16.9	1.42	16.01	14.31	23.52	0.61
Maximum	29.0	17.5	19.62	19.63	133.1	3.52
Median	18.15	11.32	18.03	15.6	67.02	1.94
Mean	22.96	9.46	17.96	16.97	78.31	2.07
Standard deviation	3.66	2.17	0.98	0.89	22.12	0.65
Skewness	0.06	−1.81	−0.15	2.34	0.42	0.25
Kurtosis	−1.33	12.52	−0.99	10.56	1.12	0.11
Coefficient of variation V [%]	15.95	22.94	5.46	5.24	28.25	31.41

Y—yield; X_1—temperature of the April–May; X_2—temperature of June–July; X_3—temperature of August–September; X_4—rainfalls of April–September; X_5—indicators of hydrothermal of April–September.

The meteorological station in Przecław, located in the northern part of the region (53°22′ N; 14°28′ E), at an altitude of 185 m above sea level, was characterized by different meteorological conditions and characterized by an average temperature of 8.1 °C and annual rainfall in the amount of 644 mm. The warmest year in this region was 2008, with an average temperature of 8.7 °C. In 2004, the temperature and precipitation conditions during the potato growing season were optimal (K = 1.2). Most of the years analyzed were wet, with the index ranging from 1.3 to 1.9. The highest amount of rainfall in the April-May period was recorded in 2001 (523 mm), and the lowest in 2004 [24,25] (326 mm) (Table 1).

According to the weather station in Skołoszów (49°55′ N 22°48′ E), located at an altitude of 204 m above sea level, the average annual air temperature was 8.3 °C, and the annual rainfall was 667 mm. The warmest year was 2018, and the average air temperature in the period from April to September was 17.0 °C, while the coldest was 2005, with an average temperature of the growing season of 14.0 °C. The lowest amount of precipitation during the vegetation period was recorded in 2004 (297 mm), and the highest was recorded in 2002 (591 mm). The least favorable distribution of precipitation took place in 2008. The years 2000, 2003 and 2008 can be described as quite dry, 2005–2007 can be descripted as optimal, 2008 can be described as quite humid and 2002 is described as wet [24,25] (Table 1).

2.4. Monitoring of Potato Blight

The experimental plots were set up in a randomized block design in 4 replications. Chemical protection against the plague was not carried out during the growing season of potato. Observations on each cultivar were carried out from the moment of potato emergence, every 10 days until the disease appeared on all cultivars. The time when 25% of the plants are already in the field was assumed as the date of potato emergence. In the period of greater risks of plague (prolonged rainfall or fog), observations were carried out more often (1–2 times a week).

The severity of the plague attack was rated on a scale of 9°, where 9° means single necrotic spots and the destruction of the assimilation surface up to 0.5%, and 1° means the complete destruction of the plant. The above data, together with the degree of plant development at the time of observation (according to the BBCH scale), were uploaded to the server thanks to the Pi-monitoring program. Additional information on the location and soil type of the field, cultivated variety and date of emergence of the plants was also entered into the database. After the end of the growing season, the rate of spread of potato blight was determined and calculated as the increase in the damage to tops per unit of time according to van der Plank [26].

2.5. Statistical Analyses

The work uses meteorological data on a year-round scale to generate a real drop in potato yields under the influence of climate change. An earlier prediction of the outbreak of the disease epidemic and an appropriately earlier response to the epidemic of potato blight, which is the cause of a significant drop in the potato yield, is possible thanks to the use of a simulator of records of adverse events in several localities by comparing modeled and observed events over a period of strong intensification of unfavorable meteorological conditions [27].

Based on the analysis of the size of yield characteristics and the course of the weather, the variability of the potato cultivation area, yielding and harvest was determined. Moreover, the partial models of the usefulness of the simulation of atmospheric conditions in the period of 20 years for this region were presented, comparing the observed cases. For this purpose, a combination of statistical methods and simulation modeling was used by using modeling techniques. The forecasts were developed for the same data model using classical statistical methods. Based on the analysis of the size of the cultivation, yield, harvest of potatoes and the course of the weather, a model for forecasting potato yielding in this part of Poland was searched. It was based on empirical data from 2000 to 2019 [28–31].

The variability of the analyzed yield characteristics and meteorological data was analyzed mainly by means of descriptive statistics (SPSS). On the basis of the diversified course of meteorological conditions during the growing season, an attempt was made to estimate the area of potato cultivation, its yield and harvest by means of a multivariate regression analysis, assuming weighted averages of selected meteorological elements for the examined localities (Table 1). The models used agronomic, phenological and meteorological data, and the correctness of their impact was verified on the basis of data sets not involved in the construction of the models.

Since the magnitude of a given phenomenon is influenced by many simultaneous factors, an attempt was made to build a model forecasting the area and harvest of potato. Such a prediction methodology is useful for optimizing the responses of the independent variables. In this case, the increase in variable y is a response and a function of the yield and the cultivated area. This function can be expressed with the following general formula.

$$y = f(x,x) + e, \tag{1}$$

It was assumed that the variables xj are the predictors on which the answer y depends. The dependent variable y is a function of xj and xj2, and the experimental error was denoted by e, which means any measurement error [19]. In order to determine the dependence of potato yield on meteorological indicators, a multivariate regression analysis was used, the parameters of which were determined by the least squares method. As a measure of adjusting the regression function to empirical data, the coefficient of determination R^2 was used [27]. Regression analysis models were computed according to the following general formula:

$$y = a + bjx\,j \tag{2}$$

where y denotes the dependent variable, a denotes intercept, b denotes the value of the regression coefficient and x denotes independent variable. Partial regressions were used to examine quantitative relationships between the potato yield and individual independent variables. Partial regression coefficients (bj) indicate how much the yield changes as a given factor increases by a unit. The described relationships were considered in terms of the standard deviation from the arithmetic mean [27]. All statistical analyses were performed using SAS 9.1.

The results of the blight tests were statistically analyzed using the analysis of variance. The significance of the sources of variation was determined by the "F" Fisher–Snedecor test, and the significance of the differences was determined by Tukey's test [19]. The rate of spread of potato blight as a function of the date of observation was calculated by regression calculus. The observation dates were coded for the calculations, taking the first date as "0", the second as "10" and the third as "20", etc. Leaf infection was expressed in logarithmic values corresponding to 9° degrees of the scale and using the van der Plank formula [26]:

$$y = \log \frac{x}{1 - x}, \tag{3}$$

where x denotes the values expressed in one hundred parts. They make it possible to express the percentage of damage to the leaf surface in the form of a straight line. The rate of spread of potato blight was considered to be a unit increase in infection over time.

3. Results

3.1. Cultivation Area

The potato cultivation area in 2000–2019 indicated a systematic decline in this value starting from 2002. In 2019, the potato cultivation acreage was only 26.2 thousand hectares, whereas 20 years earlier, there were three times more. The results of the regression analysis showed a downward trend according to the second-degree parabolic regression (Figure 1). The relationship between the analyzed features is usually characterized by the correlation coefficient R, assuming values in the range [−1.1]. It determines the strength of the relationship between the variables. However, the measure of the fit of the regression line to empirical data is the coefficient of determination R^2, taking values in the range [0.1] or [0%, 100%]. Moreover, the coefficient of determination can be corrected by the number of degrees of freedom, which increases its value.

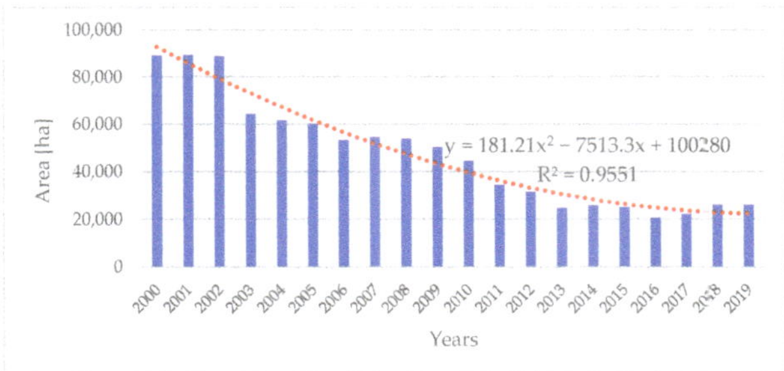

Figure 1. The actual general area of potato cultivation and the tendency of changes in this area in the southeastern part of Poland, 2000–2019.

3.2. *Shaping the Efficiency of Potatoes*

The total potato yield varied depending on the course of weather conditions in the years of the study. The highest value of this feature was recorded in 2015 (29.0 tha^{-1}), and the lowest was in 2000 (16.9 tha^{-1}). These fluctuations resulted mainly from the changing course of weather conditions during the growing season and differences in the level of soil factors. The regression analysis of the mean potato yield values for this region showed a curvilinear dependence of the fourth degree tuber yield on the years of the research (Figure 2). The coefficient of determination of this equation was 61.9%, which, according to Kranz [32], ensures its high credibility.

Figure 2. Variability of the total yield of potato tubers in Southeastern Poland in 2000–2019.

Potato yields in cultivar experiments were characterized by even greater yield variability in this part of Poland. In terms of the total and commercial yield, the yield of very early cultivars was the most diverse, ranging from 30 to 65 t·ha^{-1} in the case of the total yield and from 37 to 62 t·ha^{-1} in the case of the commercial yield. The most stable in yieldings were medium-late and late cultivars, and their total yield ranged from 40 to 58 t·ha^{-1}, and the marketable yield ranged from 38 to 52 t·ha^{-1} (Figure 3).

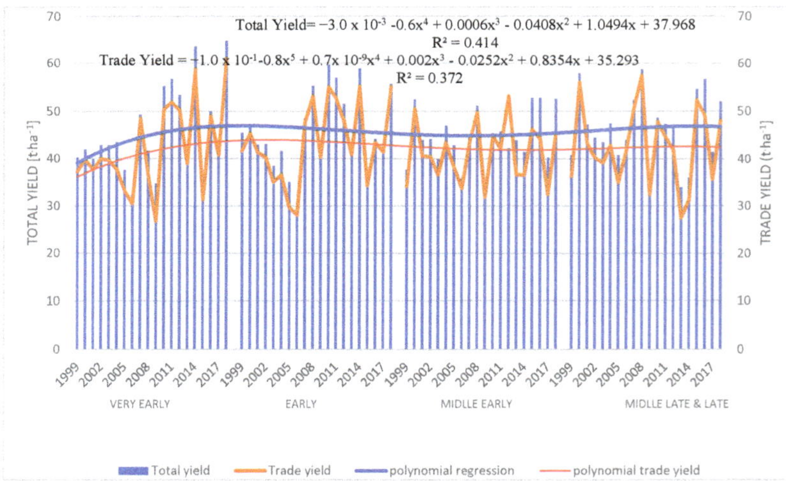

Figure 3. The total and trade yield of potato in Southeastern Poland in the Experimentals of Stations for Cultivar Assessment of Central Crop Research Centre, by groups of earliness of varieties.

Determining the trend of changes indicates periodic averages for the time series. Normal moving averages are used only for one-year data or for other time series with no seasonal variation. Two types of averages were used to smooth time series with moving averages: regular or centered. So, by using moving averages, you can smooth series containing only trend and random fluctuations. These fluctuations can be eliminated by replacing the original values of the series with a series of means calculated from several adjacent components of the time series. On the basis of the series y1, y2, . . . , yn, two-, three- and five-period means can be calculated. These formulas were written as follows.

$$f(z) = q + 1, q + 2, \ldots n - q \tag{4}$$

The function (f (z)) belongs to holomorphic functions, where any function, f (z) with complex values, can be written as follows:

$$f(z) = P(x\,,y) + iQ(x,y) \tag{5}$$

where x, y ∈ R, P (x, y) R and Q (x, y) ∈ R. It has been found that both real and unreal parts of holomorph functions are satisfied by CR equations (i.e., Cauchy Riemann) and are described above (derivation of CR formulas assuming a holomorphism of the function). In this way, random fluctuations have been eliminated to a greater extent from the time series. The new, smoothened series is four words shorter due to improved smoothing. The moving average values were recorded at the level of the middle period (for k = 3 at the level of the second period, and for k = 5 at the level of the third period, etc.). Hence, a secondary series of moving averages was obtained, which is shorter than the empirical series, a primary series of 2 for the 3-year mean, 4 for the 5-year mean, 6 values for the 7-year mean, etc., because only an odd number of words can be assigned a score to a specific period. This allows for further analysis to compare the original time series terms with the smoothened series terms. The longer the moving average used to smoothen the series, the better the smoothing is, but at the same time, the more the periods for which no trend values are obtained are lost. Hence, the selection of the length of the moving averages requires some moderation [5,20].

3.3. Potato Harvest Variability

The evaluation of the yield, made on the basis of the actual results obtained from WUSP [2000–2019], indicates a systematic decrease in potato yields in the analyzed period. Large differences in the value of this feature between consecutive seasons resulted from changes in the cultivation area and the size of the potato yield. The regression analysis of the value of this feature showed a third-degree relationship with the years of research (Figure 4). The highest value characterizing the potato harvest was recorded in 2001, and the lowest was recorded in 2016. This indicates a clear downward trend in potato harvest over the last 20 years. The coefficient of determination of this equation (R^2 = 91.56%) indicates its very high reliability [32].

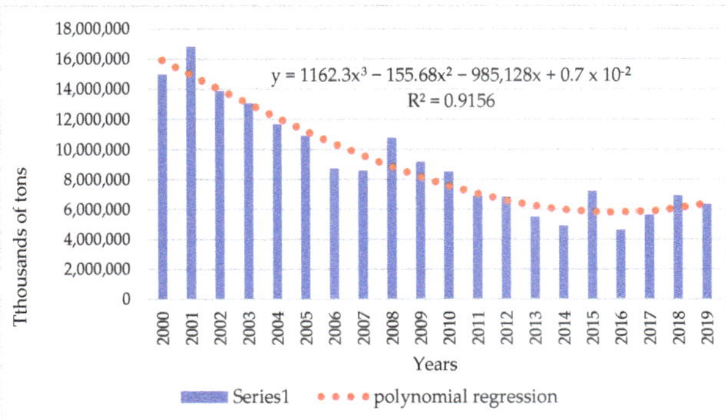

Figure 4. Changes in the total potato harvest in the southeastern part of Poland in 2000–2019.

3.4. Potato Harvest Simulation

The simulation of the potato harvest shows that the increase in this value continued until 2018, after which it decreased. There was observed a significant influence of the cultivation area on the harvesting potatoes. The following equation was adopted in the potato area and harvest forecast model.

$$y = 9493x^6 + 556.45x^5 - 11.925x^4 + 115.889x^3 - 51.300x^2 + 993.431x + 0.7 \times 10^{-8} \tag{6}$$

The coefficient of determination of this equation was R^2 = 84.07%, which makes it highly significant and reliable [32].

Forecasting crops yields can be used to plan the structure of their sowing, both on a microscale, i.e., a farm, and on a macroscale, e.g., a country. On this basis, it is also possible to estimate the profitability of growing a given plant. For potato cultivation, forecasting starch yields would be even more important as starch production is determined by law in the EU. Exceeding it reduces the profit of the grower and starch plants. Therefore, the use of modern prognostic techniques can bring measurable financial benefits and improve the profitability of growing a given species.

Table 2 shows certain regularities of the analyzed yield characteristics and meteorological data. The yield is characterized by a high maximum and minimum value. A high maximum and a relatively low minimum were also observed for rainfall in the April–September period. All meteorological data, with the exception of the April–September air temperature, showed a low skewness coefficient lower than one, which means that it takes negative values for distributions with left-hand asymmetry.

Kurtosis for most of the variables was positive, in a wide range from 0.11 to 10.56, but with a distribution close to normal, which means the more frequent occurrence of extreme values but at the same time a greater probability of the expected values. For the variables of tuber yield and air temperature in the April–shadow–May period, the kurtosis value ranged from −0.99 to −1.33, which means a greater share of values close to the median than in the normal distribution. The results in this case are less focused around the midpoint (Table 2). The standard deviation of the examined variables showed relatively little differentiation throughout the year. The highest values of the standard deviation were recorded for the tuber yield and the total rainfall in the period April–September, and the lowest in the case of the April–September hydrothermal coefficient and air temperature in August–September. The dispersion of the obtained results was characterized by the coefficient of variation, which, being the quotient of the absolute measure of trait variability, made it possible to compare the differentiation of several communities in terms of the same feature and the same data set in terms of several functions. The smaller the value of the coefficient of variation, the more stable the function is. The highest variability of the features described by the coefficient of variation was characteristic for the sum of precipitation and the Sielianinov hydrothermal coefficient for the April–September period, while the air temperature in June–July and August–September turned out to be the most stable (Table 2).

The regression analysis was based on the analysis of Pearson's simple correlation coefficients for the investigated dependent and independent variables (Table 3).

Table 3. Correlation coefficients of dependent (y) and independent (x) variable.

Variables	Y	X_1	X_2	X_3	X_4	X_5
Y	1.000					
X_1	0.316 *	1.000				
X_2	0.270 *	0.090	1.000			
X_3	0.380 **	0.096	0.399 **	1.000		
X_4	0.488 **	−0.183	−0.301 *	−0.222	1.000	
X_5	0.496 **	−0.211	−0.447 **	−0.296 *	0.838 **	1.000

Y—total yield; X_1—temperature of April–May; X_2—temperature of June–July; X_3—temperature of August–September; X_4—rainfalls of April–September; X_5—indicators of hydrothermal of April–September; * significant at $p \leq 0.05$ ** significant at $p \leq 0.01$.

The data that most strongly correlated with each other were analyzed using the multiple, polynomial, linear and partially nonlinear regression methods, which allowed the determination of the influence of many independent features on one selected dependent feature and to build an appropriate regression model. Multiple regression was preceded by the analysis of the determination coefficient R^2 for the examined features and the determination of the probability coefficient for the absolute statistic t, verified at two significance levels $p_{0.05}$ (statistically significant difference) and $p_{0.01}$ (statistically significant difference) [19].

3.5. Influence of Air Temperature on Tuber Yield

The temperature conditions in the April–May period and the precipitation and temperature-precipitation conditions described by the Sielianinov hydrothermal index in April–September had a decisive influence on the potato yield in the southeastern part of Poland. The influence of air temperature in this period was described by the power function equation in the following form.

$$y = 13.52x^{0.1349} \tag{7}$$

This indicates that a positive influence of air temperature in April–May is decisive for plant emergence, rooting and the formation of stolons and tubers. The value characterizing the yield increased with the temperature increase in this period. The coefficient of determination of this equation was 88.63%, which makes it significant and reliable (Figure 5).

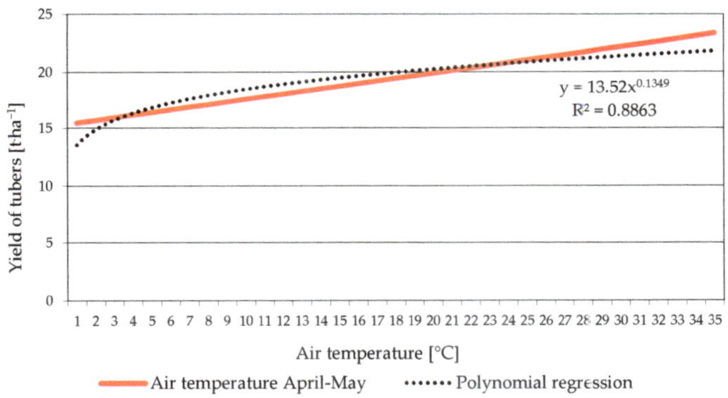

Figure 5. Partial dependence of tuber yield on air temperature in the April–May period.

Similar relationships were proven for the June–July period. During this period, there was also a positive impact of ever higher air temperature on the shaping of the potato yield described by the equation on Figure 6.

The temperature in the first part of the growing season had a positive effect on potato yields. This fact is confirmed by the coefficient of determination (R^2 = 87.4%).

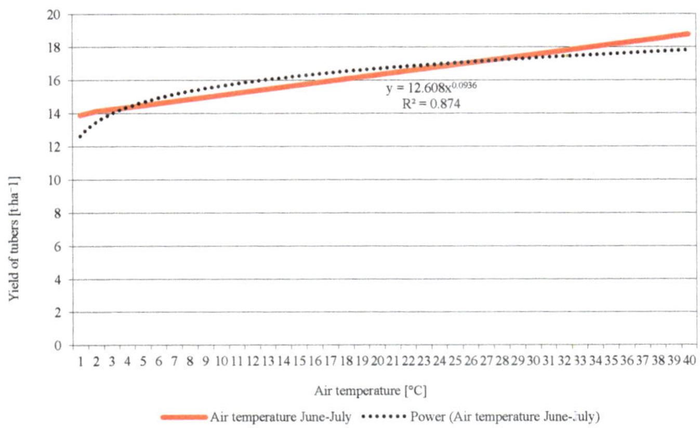

Figure 6. Partial dependence of tuber yield on air temperature in the June–July period.

The results of the regression analysis of the yield with air temperature in the second part of the growing season (August–September) are described by the formula in Figure 7. As a result of the air temperature increase in this period of time, there was a systematic decrease in yield value. The value of the coefficient of determination, amounting to 84.67%, makes the equation highly reliable [19].

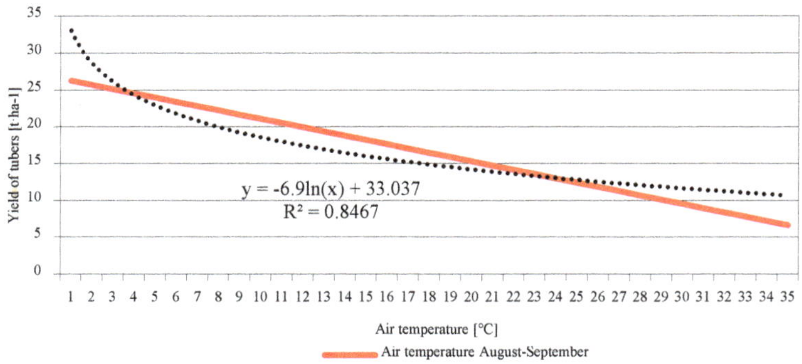

Figure 7. Partial dependence and potato tuber yield on air temperature in August–September.

3.6. Influence of Precipitation on the Formation of Tuber Yield

The factor used to assess the influence of meteorological conditions on potato yielding was also the sum of atmospheric precipitation. Among the periods in which their impact on the variability of potato yielding was the most frequent, the rainfall in the April–May period turned out to be the most impactful. The regression model described by the following equation.

$$y = 21{,}936x^{-0.051}, \quad (8)$$

This indicates a negative impact of increasing sum of rainfall in the analyzed time interval on tuber yield (Figure 8). The coefficient of determination of this equation amounting to 77.02% indicates its high credibility. It was shown that the potato yield assessed on the basis of data from the Central Statistical Office, limited by excess water in April–May, turned out to be 39% lower than the actual yield and as much as 65% lower than the potential yield in the experiments obtained in the Central Research Center for Testing in Poland.

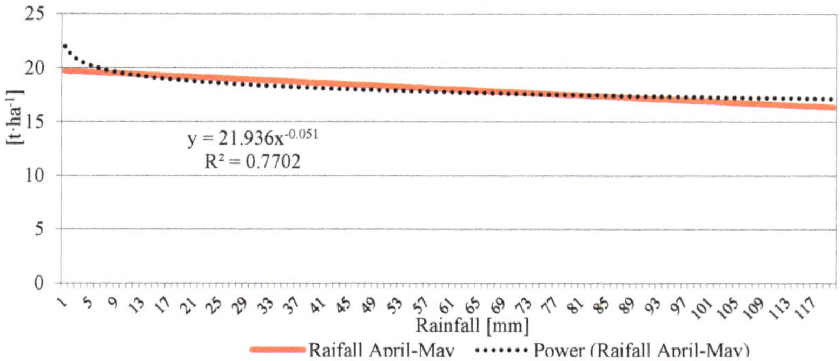

Figure 8. Partial dependence of tuber yield on total rainfall in April–May.

The results of polynomial regression analysis of tuber yield versus rainfall total, performed with the stepwise method, in the case of other time intervals turned out to be insignificant. The coefficients of determination of these equations were also low, which means that other environmental factors not considered in the analysis could have contributed to tuber yields.

3.7. *Influence of Hydrothermal Conditions on the Formation of Tuber Yield*

The impact of these conditions was similar to that of the rainfall in April–May. From the regression equation described by the formula in Figure 9 results, it can be observed that thermal and precipitation conditions negatively influenced the amount of potato yield. With the increase in the value of the hydrothermal index, the value of the yield decreased systematically. Of all the meteorological elements described by the coefficient of determination, the hydrothermal index turned out to be the most reliable (R^2 = 87.72%).

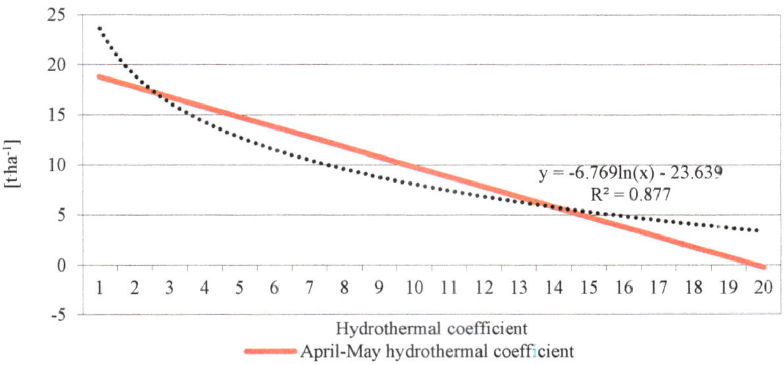

Figure 9. Partial dependence of the potato tuber yield on the value of the hydrothermal coefficient in April–May.

In Figure 10, the values of changes in precipitation are correlated with the average air temperature. Regression analysis showed that these data accounted for 38–49% of the variability in rainfall levels. The diagram also shows the correlation between the yield and meteorological elements in individual localities. The diagram presents selected meteorological parameters, and Student's *t*-test showed that rainfall during the growing season was the most statistically significantly ($p \leq 0.01$) correlated with the air temperature in June–July.

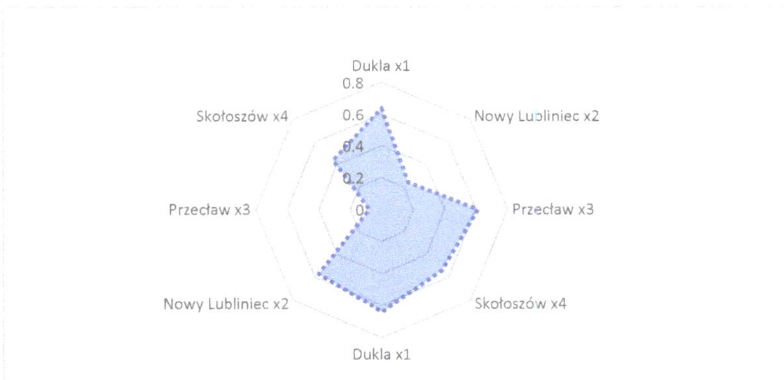

Figure 10. The relationship between fluctuations in rainfall level and various meteorological parameters. Y—yield; X_1—temperature of the April–may; X_2—temperature of June–July; X_3—temperature of August–September; X_4—rainfalls of April–September; X_5—indicators of hydrothermal of April–September.

3.8. *The Rate of Spread of Potato Blight*

The first symptoms of potato blight were usually observed 42–73 days after planting, depending on the study year and 44–81 days depending on the variety. The resistance traits of the cultivars studied determined the date of appearance of *P. infestans* on plants and the infection of 50% of the area of potato leaf blades and the rate of disease spreading on the aerial parts of plants.

Table 4 shows the rate of potato late blight spreading by groups of earliness. Very early varieties showed the fastest spread of this pathogen, while the slowest spread was observed in medium-late and late varieties. The average time of destruction of 50% of the assimilation area for varieties with increased resistance (5–6°) was 22 days—ranging from 16 (very early varieties) to 31 days (late varieties). The theoretical term of stopping the yield for late cultivars resistant to *P. infestans* (resistance 6–7° on a scale of 9°) differed significantly from the other assessed cultivars by about 7 days.

The meteorological conditions in the years of the study were the most decisive factor in the time pace of *P. infestans* spread. The fastest spread of this pathogen, in all groups of early varieties, was in 2007, the flood year, and the slowest was in the dry year, 2011 (Table 4).

Table 4. Infection coefficients of *Phytophthora infestans* in the time.

Years	Earliness Group				Mean
	Very Early	Early	Medium Early	Medium Late and Late	
2000	0.275	0.204	0.196	0.075	0.188
2001	0.218	0.201	0.132	0.089	0.160
2002	0.278	0.189	0.111	0.078	0.164
2003	0.290	0.226	0.128	0.069	0.178
2004	0.301	0.187	0.151	0.070	0.177
2005	0.412	0.325	0.273	0.189	0.300
2006	0.523	0.356	0.298	0.176	0.338
2007	0.568	0.486	0.378	0.311	0.436
2008	0.218	0.178	0.143	0.073	0.153
2009	0.290	0.226	0.156	0.069	0.185
2010	0.356	0.334	0.279	0.168	0.284
2011	0.189	0.079	0.075	0.073	0.104
2012	0.209	0.184	0.156	0.071	0.155
2013	0.298	0.191	0.151	0.089	0.182
2014	0.314	0.243	0.181	0.083	0.205
2015	0.191	0.206	0.122	0.069	0.147
2016	0.521	0.378	0.295	0.179	0.343
2017	0.351	0.181	0.120	0.073	0.181
2018	0.246	0.148	0.114	0.073	0.145
2019	0.231	0.134	0.110	0.056	0.133
Mean	0.314	0.233	0.178	0.107	0.208
$LSD_{0.05}$	0.016	0.012	0.009	0.006	0.011

The analysis of the simple correlation between the *P. infestans* spreading rate and the potato tuber yield was analyzed. Pearson's simple correlation coefficients indicate a high positive correlation between the rate of spread of the plague in individual early age groups (r = 0.87 to 1.00), which results from the internal intercorrelation (Table 5). The relationship between the rates of potato late blight spreading and the size of the total, commercial yield of medium-early to late varieties and the yield of very early and early varieties harvested 60 and 75 days after planting turned out to be significantly negative, which means a negative impact of the rate of potato late blight spreading on the yield general and commercial, as well as the yield of early varieties harvested both in the first and second harvest dates. The strongest negative relationship between the rate of spread of potato blight and the potato yield was observed for the commercial yield of tubers (r = −0.62 to −0.71). For very early and early cultivars, the rate of the spread of the blight did not have a significant negative effect, which results from the short vegetation period of these cultivars and their "escape" from infection with *P. infestans*. The shortest time of destruction of 50% of the lichen surface was found in the group of susceptible cultivars (very early and early cultivars with a resistance of 2–4° on a 9° scale).

Table 5. Simple Pearson's correlation coefficients between the rate of spread of potato blight and the yield of tubers.

Specification	X_1	X_2	X_3	X_4	X_5	Y_1	Y_2	Y_3	Y_4
X_1	1.00								
X_2	0.90 **	1.00							
X_3	0.89 **	0.96 **	1.00						
X_4	0.87 **	0.92 **	0.93 **	1.00					
X_5	0.96 **	0.98 **	0.98 **	0.95 **	1.00				
Y_1	−0.67 **	−0.67 **	−0.65 **	−0.59 **	−0.68 **	1.00			
Y_2	−0.68 **	−0.71 **	−0.70 **	−0.62 **	−0.70 **	0.98 **	1.00		
Y_3	−0.40 **	−0.37 **	−0.36 **	−0.40 **	−0.39 **	0.50 **	0.46 **	1.00	
Y_4	−0.15	−0.17	−0.12	−0.19 *	−0.16	0.36 *	0.31 *	0.85 **	1.00

X_1—*P. infestans* spreading rate coefficient of very early varieties; X_2—factor of the spread of late blight of early cultivars; X_3—coefficient of the spreading rate of mid-early cultivars; X_4—coefficient of the spreading rate of medium-late and late cultivars; X_5—the average rate of the spread of the plague; Y_1—total yield of potato tubers; Y_2—marketable yield of tubers; Y_3—yield of early potato cultivars 60 days after planting; Y_4—yield of early potato cultivars after 75 days from planting; *—significant at $p \leq 0.05$. **—significant at $p \leq 0.01$.

Among the assessed groups of potato cultivars, a large variation in the pace of development of *P. infestans* was observed, from the onset of the disease to the cessation of harvesting. The differences resulting from the level of resistance were reflected in the assessment of the destruction of the assimilation area at the end of the growing season. It was found that the destruction of the aerial part decreased with an increasing degree of resistance on *P. infestans*. The lowest values of the destruction of the assimilation area were found in the group of medium-late and late cultivars.

4. Discussion

4.1. Climate Changes and Potato Yield and Productivity

Climate change is having an increasingly strong impact on agriculture, but there is no unanimity in the scientific community and there is still no clarity on the directions of this impact. Earlier forecasts of the impact of climate change on the agricultural economy were more radical and assumed very rapid changes. One of such early forecasts was made in 1991 at the Institute of Soil Science and Plant Cultivation in Puławy on the basis of the General Circulation Model developed by the Goddard Institute for Space Studies. According to this model, climate changes were to be beneficial for agriculture and to bring an increase in the yields of all crops, except for potatoes, after 2020, and to significantly increase the area of maize and soybean cultivation [33]. The last two forecasts were confirmed, the others, unfortunately, were not. It was assumed that the growing season would significantly extend, which would allow for the extension of the assortment of arable crops and the improvement of animal production efficiency. This optimism resulted from the prediction of an increase in the average annual air temperature by 3 °C, an extension of the growing season by about 30–40% and an increase in the average amount of rainfall from 625 to 1100 mm [33]. It is now known that such a scenario is not realistic. Recently, the prevailing view is that, on a general scale, the expected changes in the form of global warming will bring beneficial effects in the agricultural economy of Europe. According to most authors [34–43], the production potential of agriculture was supposed to increase, but unfortunately in the case of potatoes, this scenario did not work. Our research, carried out on the basis of the results of the Central Statistical Office and COBORU research in southeastern Poland show that, in the last 20 years, both the yield, area and harvest of potatoes decreased. The reasons for this condition are manifold. Tomczyk et al. [44] proved that, in Poland, in the last 20 years, there was a significant increase in Tmax in the summer period. This shows, inter alia, for the ten warmest years in the analyzed period, mainly after 2000. The consequence of the increase in Tmax is the increase in the frequency of hot days. A further increase in the number of hot days in Poland is forecast in the coming decades. The smallest changes are predicted for the areas with the most intense changes in Tmax. Agriculture is a sector that is particularly vulnerable to the effects of climate change, and climatic factors are important factors in the success of agricultural production. Therefore, the effects of climate change already have a negative impact on the food security of society [45,46]. The Climate Coalition (CC) report "The impact of climate change on Poland's food security" warns that if Poland does not achieve the goals of the Paris Agreement and stops the increase in the average global temperature below 2 °C, the food security of citizens at the global level, regional and local will be increasingly threatened. Therefore, both the reduction in greenhouse gas emissions and multi-directional adaptation measures in agriculture are necessary.

According to the reports of the Agricultural Drought Monitoring System in Poland [47], in the years 2009–2011, there was a significant risk of drought in potato crops. In the years 1983–2002, the level of precipitation during the potato growing season reached the optimum only in five of the twenty years studied [35,47,48], and the losses of tuber yield in Poland for this reason ranged from 7% to 45% [47]. The high acclimatization capacity of crops may turn out to be a disadvantageous phenomenon, as it was associated with significant energy expenditure on the reconstruction of structures and adapting their functions to stressful conditions, which results in a reduction in agricultural yields [12,34,38,39,49]. In research on plant productivity, the decline in agricultural yield is often used as a measure of the possibilities of the studied cultivars for the stress syndrome occurring during the growing season. Under such assumptions, the yield is the resultant of many different mechanisms responsible for the diverse sensitivity of plants to environmental stresses, and each of these mechanisms may work differently under specific environmental conditions. The approach that takes yield loss as a measure of resistance may be useful for the final evaluation of the effectiveness of breeding treatments and has a greater selection value than the resistance criteria [38]. The research results presented in the paper concern issues related to the reaction of potatoes to the diversified course of meteorological conditions during the potato growing season and are based on numerous figures from the period of 20 years (2000–2019), taken from several sources (Central Statistical Office, Provincial Statistical Office, Institute of Meteorology and Water Management, and Experimental Stations of the Central Research Center for Cultivated Plants). These results prove that climate change in this period had a significant impact on all the studied economic, meteorological and physiological-natural features.

Precipitation is the most sensitive element of climate changes and that changes in time and space. According to Kalbarczyk [39], no permanent trend has taken place over the last 500 years. This period was subject to much thermal continentalism than it is today. According to the calculations of the Sadowski continentalism index [50], it was found that in Eastern Europe, the warmest century was the 11th century, and in Western Europe, it swas the 20th century, while the coldest centuries were in the 15th and 12th centuries, respectively. From the fifteenth century, the degree of continental climate in Poland remained at a high level until the nineteenth century. Average annual air temperatures in winter were much lower, by approximately 1.5–3.0 °C, compared to the present situation, while summer temperatures were higher than today by 0.9–1.5 °C. Over the past 100 years, Earth has warmed by 0.85 °C, and the speed of this process is increasing. In Europe, the temperature has risen by almost 1 °C.

The impact of climate change on world agriculture can be considered in two main aspects: natural and socio-economic. The first is direct, and the second is indirect, usually resulting from the former. The changes in the natural basis of the agri-environmental economy relate primarily to the greenhouse effect associated with an increase in the concentration of carbon dioxide in the atmosphere. The main cause of the rapidly following climate change is the increase in carbon dioxide (CO_2) content in the atmosphere, causing the so-called greenhouse effect. This will enable some cereal crops, such as wheat or rice, as well as potato plants to photosynthesize more intensively and, consequently, result in faster development with higher yields [50–53]. It was assumed that, as a result, increasing plant production may reduce the specter of hunger but only in the case of organized international activities. It is generally known that areas of hunger are concentrated in the poorest countries, where climatic conditions generally pose problems for proper farming. It is primarily the dry zone of Africa and some regions of Asia [54]. According to Kulig [5] and Ziernicka-Wojtaszek [6,7], the optimism related to the increase in plant production as a result of the increase in CO_2 in the atmosphere was, however, premature. These authors propose that the influence of carbon dioxide concentration in the atmosphere on the global agricultural production should be considered in direct and indirect categories. The former concerns the intensification of photosynthesis and the possibility of plant development with lower water resources and their more effective use. Indirect impacts should be seen in the aspect of climate and soil changes, as well as in the development of diseases and pests. In the case of direct impacts, our research indicates a very large differentiation in the increase in yields resulting from the increase in the carbon dioxide content, but under natural conditions. Laboratory experiments confirm that plants absorbing more carbon grow faster and are larger [39]. Moreover, the increased concentration of carbon dioxide increases the efficiency of water use. This applies in particular to plants of the so-called C3 group (e.g., wheat, rice, soybean and potato), which show, under conditions of increased CO_2 content, an increased photosynthesis rate and a moderate decrease in transpiration. On the other hand, plants from the C4 group (maize, sugar cane, sorghum, etc.) show relatively slower photosynthesis (slower biomass growth) under these conditions [5,49].

Repeated droughts and desertification on almost all continents already threaten the livelihoods of some 1.2 billion people [4]. For example, an increase in CO_2 from 330 to 660 ppm (parts per million—gas particles per million air particles per unit volume) resulted, under optimal conditions, in an increase in cotton yield due to the concentration of carbon dioxide. The main conclusion of physiological studies [49] is the fact that the positive effect of an increase in CO_2 than twice lower than other important environmental factors (humidity and thermal conditions, the content of mineral nutrients and another) can counteract this influence. The concentration in laboratory conditions on plant production is not confirmed under the conditions of natural plant cultivation.

The research of many authors [5–7,35–37,51–53] shows that the most likely scenario will be a slow temperature increase resulting from an increase in the concentration of carbon dioxide in the atmosphere. The result will be a shrinkage of cool climate zones and an expansion of hot climate zones. The effects of the temperature rise will be more pronounced in areas near the poles than in the equatorial areas. Therefore, the shift in climatic zones will be more marked in higher latitudes. In regions with a temperate climate, such as Central and Eastern Europe, the shift by 1 °C will be from 200 to 300 km [54]. This will have a direct impact on the extension of the range of some crops, including sweet potatoes [12].

The production potential will increase mainly in the temperate climate zone. Global changes in world agriculture under the influence of climatic changes will cause many processes in the natural environment that are still difficult to identify. They will also shape socio-economic processes and phenomena. The considerations to date show that the greatest increase in production possibilities will take place in the most economically developed countries, where problems with food overproduction are observed. On the other hand, in the poor countries of Africa and Asia, where there are hunger zones, there may be growing food problems resulting from reduced production possibilities [54]. The increase in the production potential of agriculture in rich countries will result in an increase in the average global agricultural production per unit area. There is still a problem of food overproduction in the European Union countries, which the Common Agricultural Policy is trying to mitigate and eliminate. One of the directions of activities is the widely understood extensification of agricultural production, including the reduction in the area of land developed by agriculture [7,50].

Europe and North America have relatively ample room to adapt to the effects of climate change. The research conducted so far shows that there will be rather favorable changes in terms of agricultural production possibilities. However, in subtropical regions (e.g., in southern Europe), large areas may be exposed to drought, while on the continents of the Americas, the risk of extreme phenomena will increase: floods, droughts and cyclones. The warming of the climate will cause the extension of dry areas also to the areas of Southern Europe and the necessity to take decisive measures in the field of water retention and the irrigation of farmland [45,50,54].

Global climate change poses a serious challenge to global food security. The sensitivity of germs, potentially toxin-producing microorganisms and other pests to climate factors, shows that climate change can affect the incidence and intensity of certain food-borne diseases.

4.2. Climate Change and the Spread of Pathogens

Climate change and changes in ecological conditions can promote the spread of pathogens, parasites and diseases, with the potential to seriously affect human health, agriculture and the environment [54–57]. They are one of the important stressors that can contribute to the extinction of many species. The IPCC estimates that 20–30% of plant and animal species assessed so far in climate change studies are at risk of extinction if temperatures reach levels projected at the end of this century [50].

Poland is one of the largest potato producers in the world, and it is one of the key edible and industrial plants. Unfortunately, its yields are almost half of those obtained in other EU countries. The main reasons are as follows: high susceptibility of genetically homogeneous cultivars to *P. infestans* and imperfect protection of plantations against pathogen and climate variability, which is favorable for the development and spread of this pathogen [11,56,57] and the ability to carry infectious material. Andrivon et al. [58] believed that the spores of Ph. infestans can spread to a distance of 70–80 km from the site of infection. Aylor et al. [59] state that the spores of the blight at wind speeds of 20–40 km h^{-1}. They can spread from the site of infection to 80–160 km in 4 h. Moreover, the short infection cycle and the possibility of producing a large amount of infectious material creates favorable conditions for the development of blight, and 100,000 spores can be formed from one pathological lesion on a leaf. Fry [56] estimates the number of sporangia at 300,000 for 3 days, and sporulation may begin as early as one or two days after the onset of symptoms [58]. The amount of yield losses ranges from 30 to 60% [60], 70% [61,62] and 100% [57], and the growth of this pathogen, in the absence of protection, amounts to about 10–50% due to the premature destruction of the tops and 0–40% due to the destruction of tubers [12–14]. Haverkort et al. [63] estimate that, in Europe, the annual expenditure on combating potato blight amounts to EUR 900 million. In the USA, the amount of expenditure on protection against the plague is estimated at USD 3 billion per year. Global conservation costs and crop losses are estimated at USD 6.7 billion [5].

In the conducted research, the group of early varieties related to the resistance of cultivars to this pathogen had the greatest influence on the pace of spreading *P. infestans*. The influence of cultivar-related resistance on these plant health traits is confirmed by the studies of Croxall and Smith [64], Kapsa [61,62], Osowski [15] and Sawicka [12–14]. In Poland, where protection against late blight is carried out only on about 40% of potato plantations, the average yield loss was 20–25%. Losses on unprotected plantations are estimated at 70–80%. There are two phases in the development of late blight: early (hidden) and epidemic. During the first stage, the fungus multiplies, leading to local infections and the growth of primary infectious foci depending on the following: density and location of primary foci, susceptibility of the cultivars to late blight, plant physiological condition, weather conditions and changes in microclimate and ecoclimate [12,14].

The date of the outbreak in Polish conditions is usually June or July and depends mainly on the air temperature and precipitation patterns during the growing season. The development of the disease after reaching the epidemic stage is usually rapid, and usually, after a few or several days, the plants are almost completely destroyed by the plague. The date of the outbreak and the pace of its development determine the reduction in potential potato yields [11,13,15]. Over the past two decades, an increase in the infectivity of *P. infestans* has been observed, which results from changes in the population of this pathogen. The result of these changes is an earlier onset of more rapid disease development, increased pathogenicity of the fungus, changes in epiphytotic the development of primary *P. infestans* infections and the breaking of genetic resistance of many potato cultivars and the ineffectiveness of traditional methods of plantation protection against the pathogen [17,18]. The increased severity of plague and the losses it causes justify the need to combat it. However, the goal of most potato blight control systems, due to their commercial nature, is a comprehensive plant protection strategy with the recommendation of specific chemical preparations and their dosage, while predicting the timing of an outbreak of *P. infestans* is only auxiliary or even switched off or is established only in case of occurrence through linear models. Almost all decision-making programs were developed on the basis of observations of the development of potato blight in Western Europe, which makes them completely unadjusted to other climatic conditions [2]. Therefore, there is an urgent need to monitor and predict the timing of an outbreak based on meteorological data and/or potato development phases.

Using the method of determining the upper and lower limits of the trait value for groups of varieties significantly different from each other, the cultivars tested were divided into three groups: cultivars with the lowest share of plants with symptoms of potato blight, in which the infection was less than 4%; cultivars with an average share of individuals with symptoms, where the paralysis ranges from 4.1 to 8.0%; and cultivars with a high proportion of plants infected with this pathogen, with an infection rate of > 8%. The late varieties showed the highest resistance to late blight; the lowest resistance was observed in very early and early varieties. It can be assumed that specific defense substances (so-called phytoalexins), which are activators of defense reactions to pathogenic factors, can trigger the trigger mechanism of resistance to plant infection by *P. infestans* in plants. Stark et al. [49] states that the compounds of this type can act as effectors of the expression of the plant resistance genome, as well as activate enzymes, transfer physiological stimuli from membrane receptors to the genome, etc., and, therefore, fulfill the function of the first informants in the pathogenesis and resistance of plants in establishing parasitic contact. The plant tries to preserve the species in this targeted way.

4.3. Ensuring Food Security

Providing a food base is the basis of food security for each country. The definition of food security adopted at the Food Safety Summit covers four aspects: food availability, access to food, food use and stability [1,2,11]. Food security can be analyzed on several levels (security dimensions): individual or family security, also referred to as food security of households, national food security and international food security [2,65]. Economic globalization justifies the use of the concept of global food security [66]. Particular attention is now paid to the food security of households. The Food and Agriculture Organization of the United Nations (FAO) defines them in such a manner that all households have physical and economic access to sufficient food for all members, and there is no risk that they will lose this access [2,65].

In the legal sense of "food security", it is said that food security is an optimal state assumed by the legislator. The means that the relevant provisions of national and EU law and, above all, the provisions of international law are required to achieve food security. According to the FAO/WHO (Food and Agriculture Organization of the United Nations), food security means a state in which all people have access to food that is safe for health and value. In legal terms, the concept of food security is related to the human right to food and the need to combat hunger [2,4].

In turn, from a philosophical point of view, "food security" is expressed in the fact that society should not allow any of its members to starve. In this approach, the opposite of food security is hunger related to the body, which can result in the loss of human dignity [4].

Food safety is considered in a production context, especially potato production. Food safety mainly depends on the health quality of food. The Act of August 25, 2006, as amended (2020/2021) on food and nutrition safety, specifies the requirements and procedures necessary to ensure food and nutrition safety in accordance with the provisions of Regulation (EC) No. 178/2002 of the European Parliament and of the Council of 2002 establishing the general principles and requirements of food law, establishing the European Food Safety Authority and laying down procedures for food safety (Journal of Laws EC L 31 of 01.02.2002, p. 1; EU Journal of Laws), called "Regulation No 178/2002", "by regulation No. 1935/2004" [67–70]:

(1) Food health requirements—in the scope not regulated in the regulations of the European Union.
(2) Requirements for compliance with hygiene rules:
    (a) Food—within the scope not regulated in Regulation (EC) No. 852/2004 of the European Parliament (Journal of Laws UE L 139 of 30 April 2004, p. 1),
    (b) Materials and articles intended to come into contact with food—within the scope not regulated in Regulation (EC) No. 1935/2004 of the European Parliament of 27 October 2004 on materials and articles intended to come into contact with food (Journal of Laws UE L 338 from 13 November 2004),
(3) The competence of the authorities to carry out official food controls in accordance with the principles set out in Regulation (EC) No 882/2004 of the European Parliament and of the Council of 29 April 2004 on official controls performed to verify compliance with feed law and food and animal health and animal welfare rules (Journal of Laws UE L 191 of 30 April 2004, p. 1), hereinafter referred to as "Regulation No. 882/2004",
(4) Requirements for the performance of official food controls—within the scope not regulated in Regulation No 882/2004.

The results of the research by Otekunrin et al. [4] carried out in Nigeria revealed that only 12.8% of households had food security, while 87.2% had varying levels of food insecurity. In Europe, food security is at a higher level than in developing countries, but there are large regional and national differences indicated by FAO/WHO [2], Otekunrin et al. [4,54], Yuen [10], Karaczun [45], Scott et al. [70] and Tripathi et al. [71]. The main objectives of the European food safety policy are to protect human health and consumer interests and to support the smooth operation of the European single market. Thus, the European Union supervises the legal status, the establishment and compliance with control standards in such areas as follows: hygiene of feed and food products, animal health, plant health and the prevention of food contamination by external substances. In addition, the farm-to-fork approach in the EU aims to ensure a high level of safety at all stages of the production and distribution process of all food products placed on the EU market, regardless of whether they are produced in the EU or imported from third countries. This piece of legislation is a complex and comprises an integrated system of rules governing the entire food chain. These rules will be further developed in the context of the Commission's Farm to Fork Strategy, which was launched in 2020 as part of the European Green Deal. Nevertheless, the level of food security is clearly differentiated between European countries, and Europe is ranked behind North America according to the Global Food Security Index created in cooperation with the Economist Intelligence Unit [4,72]. The first "20" of the index includes eleven EU countries, with Finland, Ireland and the Netherlands at the fore. Bulgaria, Slovakia and Hungary ranked lower; 44, 40 and 36, respectively, Poland was placed 25th in this ranking. The need to implement more solutions that will provide people in need with access to a wide range of food resources is stressed. This variation in food safety is due to an internal discrepancy. The level of food security in the EU is therefore not uniform, as the results of both Mediterranean and Central and Eastern European countries are much weaker than in Western Europe or Northern Europe. Currently, the EU's food system is dominated by other challenges such as food waste, over-consumption, obesity and the environmental impact of food production [72].

The results presented by the United Nations [3] indicate that age, years of education of the household head, gender, farm size, farm experience, non-farm income, food expenditure and access to advisory services significantly contribute to food insecurity among farms. The state's efforts should, therefore, be directed towards promoting education-related household intervention programs in order to broaden their knowledge of nutrition, which may improve their food security status. In addition, rural infrastructure facilities should be provided, such as water supply, rural gasification, internet services and healthcare services that promote healthy lifestyles and increase agricultural productivity in households.

The relationship between food safety and food security is also important. The food law does not explicitly state the relations between the two concepts in question, although both are legal concepts. However, the literature on the subject indicates that food safety is an element of a wider issue, which is food safety. Food safety in the legal sense seems to correlate with the aspect of "Food availability" as part of food security in economic terms. Food safety is primarily related to its health values, and "food security" in the "Food access" dimension also means the need to provide food that is safe for health. Therefore, both of these terms draw attention to the need to eliminate pathogenic substances that are dangerous to human health from food products. Thus, it can be said that (in the indicated aspects) the terminological scope of food safety is wider than the terminological scope of food safety and covers it.

The research carried out was innovative as it combined the assessment of climate change with food and food safety. This will allow for better planning of the supply of food raw materials, forecasting possible drops in the potato yield based on the monitoring of potato blight and ensuring a healthy raw material, both for direct consumption and food processing in the southeastern part of Poland.

5. Conclusions

Climate change creates, on the one hand, new opportunities for potatoes as an alternative source of human food and animal feed, as well as a raw material for the production of bioethanol and starch, and on the other hand, unfavorable weather conditions may contribute to a decrease in the acreage suitable for potato cultivation as a result of the deepening hydrological drought and the reduction in yielding, in relation to the actual and potential yield.

The effect of climate warming in the southeastern part of Poland is a prolonged growing season, an earlier start of vegetation, no or little snow cover and the development of diseases, including a faster spread of potato blight, which results in a decrease in the total and commercial yield of early and medium-early varieties, medium-late and late as well as a decrease in the yield of early varieties intended for very early and early potato harvesting.

The meteorological conditions exerted a significant influence on the variability of the potato yielding. The positive dependence of tuber yield was demonstrated on the following: air temperature in April–May and June–July and negative temperature in August–September. Along with the increase in the average temperature in April–May and June-July, tuber yields should be expected to increase, assuming that the increase in temperature will be accompanied by an increase in rainfall in this period.

It has been documented that thermal, precipitation and precipitation conditions have a negative effect on potato yields, especially in the early stages of the growing season. Excessive rainfall in April–May may significantly reduce the yield of potatoes. The yield limited by the excess of water was 39% lower than the actual yield and 64% lower than the potential potato yield.

The modeling method used in this study, after the necessary modification, can be used to forecast the yields of other crop species, which in turn can bring about measurable macroeconomic and microeconomic effects.

Climate changes affect food security in the field of basic food raw materials, including potato, as one of the most important food products and raw material for the processing of main food products such as French fries, crisps, dried products, frozen and freeze-dried products, raw material for the production of starch and bioethanol and influence on the state's food security policy.

Author Contributions: Conceptualization, B.S., P.B. and P.P.; methodology, P.P. and P.B.; software F.Y.; validation, D.S., P.B. and B.K.-M.; formal analysis, F.Y.; investigation, P.B.; resources, D.S. and P.P.; data curation, P.P.; writing—original draft preparation, D.S.; writing—review and editing, B.S., P.B. and B.K.-M.; visualization, F.Y. and D.S.; supervision, B.S. and P.P.; project administration, P.B. and P.P.; funding acquisition, B.S. All authors have read and agreed to the published version of the manuscript.

Funding: This research received no external funding.

Institutional Review Board Statement: Not applicable.

Informed Consent Statement: Not applicable.

Data Availability Statement: Not applicable.

Acknowledgments: Thank you for supporting the Central Research Center for Cultivated Plants in Słupia Wielka and for sharing the data.

Conflicts of Interest: The authors declare no conflict of interest.

References

1. FAO. *Hunger Hotspots—FAO-WFP Early Warnings on Acute Food Insecurity: February to May 2022 Outlook*; FAO, WFP: Rome, Italy, 2022; ISBN 978-92-5-135662-3. [CrossRef]
2. FAO; IFAD; UNICEF; WFP; WHO. The State of Food Security and Nutrition in the World 2021. In *Transforming Food Systems for Food Security, Improved Nutrition and Affordable Healthy Diets for All*; FAO: Rome, Italy, 2021.
3. WHO. Draft WHO Global Strategy for Food Safety 2022–2030. Towards Stronger Food Safety Systems and Global Cooperation. 2021, p. 48. Available online: https://cdn.who.int/media/docs/default-source/food-safety/who-global-strategy-food-safety-2022-2030.pdf (accessed on 30 March 2022).
4. Otekunrin, O.A.; Otekunrin, O.A.; Sawicka, B.; Pszczółkowski, P. Assessing Food Insecurity and Its Drivers among Smallholder Farming Households in Rural Oyo State, Nigeria. *HFIAS Approach. Agric.* **2021**, *11*, 1189.
5. Kulig, B.; Skowera, B.; Klimek-Kopyra, A.; Kołodziej, S.; Grygierzec, W. The Use of the WOFOST Model to Simulate Water-Limited Yield of Early Potato Cultivars. *Agronomy* **2020**, *10*, 81. [CrossRef]
6. Ziernicka-Wojtaszek, A. Porównanie wybranych wskaźników oceny suszy atmosferycznej na obszarze województwa Podkarpackiego (1901−2000). *Woda Środowisko Obszary Wiejskie* **2012**, *12*, 365–376. (In Polish)
7. Ziernicka-Wojtaszek, A. Pluviothermic conditions concerning the vegetation of key crops in south-eastern Poland 1901–2010. *Appl. Ecol. Environ. Res.* **2020**, *18*, 839–848. [CrossRef]
8. Yuen, J.E.; Andersson, B. What is the evidence for sexual reproduction of *Phytophthora infestans* in Europe? *Plant Pathol.* **2013**, *62*, 485–491. [CrossRef]
9. Bostancı, S.H.; Yıldırım, S. Sustainable Communities vs. Climate Refugees: Two Opposite Results of Climate Change. In *Handbook of Research on Novel Practices and Current Successes in Achieving the Sustainable Development Goals*; Popescu, C., Raluca, G., Eds.; IGI Global: Hershey, PA, USA, 2021; pp. 298–319. [CrossRef]
10. Mezghani, A.; Parding, K.M.; Dobler, A.; Benestad, R.E.; Haugen, J.E.; Piniewski, M. Projections of Changes in Temperature, Precipitation and Cove in Poland. In *Climate Changes and Their Impact on Selected Sectors in Poland*; Zbigniew, W., Kundzewicz, Ø.H., Okruszko, T., Eds.; Polska Akademia Nauk: Poznań, Poland, 2017; pp. 94–118. (In Polish)
11. Yuen, J. Pathogens which threaten food security: *Phytophthora infestans*, the potato late blight pathogen. *Food Sec.* **2021**, *13*, 247–253. [CrossRef]
12. Sawicka, B.; Michałek, W.; Pszczółkowski, P. The relationship of potato tubers chemical composition with selected physiological indicators. *Zemdirb. Agric.* **2015**, *102*, 41–50. [CrossRef]
13. Sawicka, B.; Kapsa, J. Effect of varietal resistance and chemical protection on the potato late blight (*Phytophthora infestans* [Mont.] de Bary) development. *Potato Res.* **2001**, *44*, 303–304.
14. Sawicka, B.; Pszczółkowski, P.; Krochmal-Marczak, B. Efektywność różnych strategii ochrony ziemniaka przed *Phytophthora infestans*. *Pamiętnik Puławski* **2006**, *142*, 411–428. (In Polish)
15. Osowski, J. The rate of potato late blight development (*Phytophthora infestans*) in 2005–2007 depending on weather conditions and cultivar resistance. *Biuletyn IHAR* **2010**, *255*, 109–118.
16. Kisinga, B. Predicting potato late blight disease pressure using in artificial neural network. *Res. Rep. Math.* **2019**, *42*, 1–46.
17. Kapsa, J. Zmiany stanu zagrożenia i ochrony plantacji przed zarazą (*P. infestans*) w Polsce na tle krajów europejskich. *Prog. Plant Protect* **2004**, *44*, 129–137.
18. Kapsa, J. Effect of climatic conditions on infection pressure of *Phytophthora infestans* and harmfulness of the pathogen to potato crops. *J. Plant Prot. Res.* **2007**, *47*, 357–366.
19. Kaszewski, B.M. Polish climate changes in the works of Polish climatologists. *Geophys. Rev.* **2015**, *3–4*, 217–235. (In Polish)
20. The State of the Environment in the Podkarpackie Province in 1999–2008. VII Soils. Available online: https://wios.rzeszow.pl/cms/upload/edit/file/opracowania/raporty/1999-2008/r7.pdf (accessed on 30 January 2022). (In Polish).

21. *Biuletyn Państwowej Służby Hydrologiczno-Meteorologicznej 2000−2019*; Wydawnictwo IMiGW: Warszawa, Poland, 2019. (In Polish)
22. Stöckle, C.; Donatelli, M.; Nelson, R. CropSyst, a cropping systems simulation model. *Eur. J. Agron.* **2003**, *18*, 289–307. [CrossRef]
23. Skowera, B.; Puła, J. Skrajne warunki pluwiometryczne w okresie wiosennym na obszarze Polski w latach 1971–2000. *Acta Agroph.* **2000**, *3*, 171–177. (In Polish)
24. Skowera, B.; Kopcińska, J.; Ziernicka-Wojtaszek, A.; Wojkowski, J. Niedobory i nadmiary opadów w okresie wegetacji ziemniaka późnego w województwie opolskim (1981–2010). *Acta Sci. Pol. Form. Circumiectus* **2016**, *15*, 137–149. (In Polish)
25. Skowera, B. Changes of hydrothermal conditions in the Polish area (1971−2010). *Fragm. Agron.* **2014**, *31*, 74–87.
26. Van der Plank, J.E. *Plant Disease: Epidemics and Control*; Academic Press: New York, NY, USA, 1963; p. 345.
27. Koronacki, J.; Ćwik, J. *Statystyczne Systemy Uczące Się*; Wydanie, A.L., II, Ed.; Akademicka Oficyna Wydawnicza EXIT Andrzej Lang: Warszawa, Poland, 2005. (In Polish)
28. Muriithi, D.K. Application of Response Surface Methodology for Optimization of Potato Tuber Yield. *Am. J. Theor. Appl. Stat.* **2015**, *4*, 300–304. [CrossRef]
29. GUS. Rocznik Statystyczny Rzeczypospolitej Polskiej 2020. 2020. Available online: https://stat.gov.pl/obszary-tematyczne/roczniki-statystyczne/roczniki-statystyczne/rocznik-statystyczny-rzeczypospolitej-polskiej-2020,2,20.html (accessed on 30 January 2022).
30. WUSP. *Wojewódzki Urząd Statystyczny Podkarpacia 1999–2018—Statistical Yearbook Podkarpackie Voievodship 2000–2019*; Urząd Statystyczny: Rzeszowie, Poland, 2000.
31. WUSP. Rocznik Statystyczny Województwa Podkarpackiego 2020. 2020. Available online: https://rzeszow.stat.gov.pl/publikacje-i-foldery/roczniki-statystyczne/rocznik-statystyczny-wojewodztwa-podkarpackiego-2020,5,17.html (accessed on 30 January 2022).
32. Kranz, J. *Epidemics of Plant Diseases. Mathematical Analysis and Modeling*; Springer: Berlin/Heidelberg, Germany, 1990; Volume 268. [CrossRef]
33. Bis, K. Przewidywane zmiany klimatyczne i ich ekonomiczne konsekwencje dla rolnictwa polskiego. [w:] Rolnictwo w gospodarce rynkowej. In Proceedings of the Międzynarodowa Konferencja Naukowa, Kraków, Poland, 13–15 May 1993; Zeszyty Naukowe AR w Krakowie: Krakow, Poland, 1993; Volume 279, pp. 187–199.
34. Bartoszuk, W. Zniżka plonu powodowana niedoborem wody w czasie wegetacji. *Biuletyn Instytutu Ziemniaka* **1987**, *36*, 43–51. (In Polish)
35. Boguszewska, D. Zmiany klimatyczne Polski w latach 1983–2002, A perspektywy uprawy ziemniaka. *Ziemniak Polski* **2008**, *3*, 19–22. (In Polish)
36. Boguszewska, D.; Grudkowska, M.; Zagdańska, B. Drought-responsie antioxidant enzymes in potato (*Solanum tuberosum* L.). *Potato Res.* **2010**, *53*, 373–382. [CrossRef]
37. Boguszewska-Mańkowska, D. Potato soil drought resistance and assessment methods. *Biul. Inst. Hod. I Aklim. Roślin* **2016**, *279*, 65–75.
38. Kacperska, A. Plant reactions to abiotic stress factors. In *Plant Physiology*; Kopcewicz, J., Lewak, S., Eds.; PWN: Warsaw, Poland, 2005; pp. 612–678. (In Polish)
39. Kalbarczyk, E.; Kalbarczyk, A. *The Course of Phenological Phases of Potato and Its Determination by Multi-Annual Variability of Air Temperature in Poland*; Annales Universitatis Mariae Curie-Skłodowska: Lublin, Poland, 2010; Volume 64.
40. Kalbarczyk, R. Próba prognozowania plonów ziemniaka w Polsce na podstawie danych meteorologicznych. *Folia. Univ. Agric. Stetin.* **2004**, *234*, 145–158. (In Polish)
41. Kalbarczyk, R. Strefy klimatycznego ryzyka uprawy ziemniaka w Polsce. *Folia Univ. Agric. Stetin Agric.* **2005**, *244*, 83–90. (In Polish)
42. Kalbarczyk, R. Wpływ warunków agrometeorologicznych na rozwój ziemniaka średnio późnego w Polsce. *Folia. Univ. Agric. Stetin.* **2003**, *231*, 39–46. (In Polish)
43. Kalbarczyk, R.; Kalbarczyk, E. Needs, and deficiencies of atmospheric precipitation in the cultivation of medium late and late potato in Poland. *Infrastruct. Ecol. Rural. Areas* **2009**, *3*, 129–140.
44. Tomczyk, A.M.; Piniewski, M.; Eini, M.R.; Bednorz, E. Forecasts of changes in the maximum air temperature and hot days in Poland. *Int. J. Climatol.* **2022**, *42*. [CrossRef]
45. Karaczun, Z.M.; Kozyra, J. *The Impact of Climate Change on Poland's Food Security*; SGGW Publishing House: Warsaw, Poland, 2020; p. 120. Available online: http://www.koalicjaklimatyczna.org/uploads/publikacje/Raport_Klimat_bezpieczenstwo_zywienowe_Karaczun_20.03.pdf (accessed on 28 January 2022).
46. Anonymous. SMSR—System Monitoringu Suszy Rolniczej. 2021. Available online: https://susza.iung.pulawy.pl/ (accessed on 30 January 2022).
47. Chmura, K.; Chylińska, E.; Dmowski, Z.; Nowak, L. *Role of the Water Factor in Yield Formation of Chosen Field Crops. Infrastructure and Ecology of Rural Areas. Nr 9/2009*; Polska Akademia Nauk, Oddział w Krakowie, Komisja Technicznej Infrastruktury Wsi: Krakow, Poland, 2009; pp. 33–44. (In Polish)
48. Anonymous. Prognozowanie Plonów. 2015. Available online: http://www.igik.edu.pl/pl/teledetekcjaprognozowanie-plonow (accessed on 18 January 2022). (In Polish).
49. Stark, C.; Love, S.; King, B.; Marshall, J.; Bohl, H.; Salaiz, T. Potato variety response to seasonal drought patterns. *Am. J. Potato Res.* **2013**, *90*, 207–216. [CrossRef]

50. Sadowski, M. Zmiany Klimatu i Ich Skutki Teraz i w Przyszłości. 2020. Available online: https://www.pzh.gov.pl/wp-content/uploads/2020/11/1_-prof-Sadowski.pdf (accessed on 29 January 2022).
51. Rykaczewska, K. Influence of high temperature during the growing season on the potato (*Solanum tuberosum* L.) yield, tuber dormancy period and the yield value of seed potatoes. I. Plant development and yield. *Zesz. Probl. Post. Nauk Roln.* **2004**, *496*, 185–198. (In Polish)
52. Rykaczewska, K. The impact of high temperature during growing season on potato cultivars with different response to environmental stresses. *Am. J. Plant Sci.* **2013**, *4*, 2386–2393. [CrossRef]
53. Rykaczewska, K. The effect of high temperature occurring in subsequent stages of plant development on potato yield and tuber physiological defects. *Am. J. Potato Res.* **2015**, *92*, 339–349. [CrossRef]
54. Otekunrin, O.A.; Otekunrin, O.A. Healthy and Sustainable Diets: Implications for Achieving SDG2. In *Zero Hunger. Encyclopedia of the UN Sustainable Development Goals*; Leal Filho, W., Azul, A., Brandli Özuyar, P., Wall, T., Eds.; Springer: Cham, Switzerland, 2021.
55. Czarnecka, M.; Nidzgorska-Lencewicz, J. The role of meteorological elements in the formation of daily (24 h) variability of the light soil moisture in winter. *Ann. Univ. Mariae Curie-Skłodowska Sect. B Geogr. Geol. Min. Alogia Petrogr.* **2006**, *61*, 125–130.
56. Fry, W.E. *Phytophthora infestans*: The plant (and R gene) destroyer. *Mol. Plant Pathol.* **2008**, *9*, 385–402. [CrossRef]
57. Harrison, J.G. The effect of aerial environment on late blight of potato foliage—A review. *Plant Pathol.* **1992**, *41*, 384–416. [CrossRef]
58. Andrivon, D.; Evenhuis, B.; Schepers, H.; Gaucher, D.; Kapsa, J.; Lebecka, R.; Nielsen, B.; Ruocco, M. Reducing Primary Inoculum Sources of Late Blight. *ENDURE Potato Case Study-Guide* **2008**, *1*, 4. Available online: www.endure-network.eu (accessed on 5 January 2022).
59. Aylor, D.E.; Fry, W.E.; Mayton, H.; Andrade-Piedra, J.L. Quantifying the rate of release and escape of *Phytophthora infestans* sporangia from a potato canopy. *Phytopathology* **2001**, *91*, 1189–1196. [CrossRef]
60. Lung'aho, C.S.K.N.; Nderitu, S.K.N.; Kabira, J.N.; El Bedewy, R.; Olanya, O.M.; Walingo, A. Yield performance and release of four late blight tolerant potato varieties In Kenya. *J. Agron.* **2006**, *5*, 57–61. [CrossRef]
61. Kapsa, J. Zaraza (*Phytophthora infestans*/Mont./de Bary) występująca na łodygach ziemniaka. Monography 11. *IHAR Radzików* **2001**, *11*, 7–10.
62. Kapsa, J.; Mrówczyński, M.; Erlichowski, T.; Gawińska-Urbanowicz, H.; Matysek, K.; Osowski, J.; Pawińska, M.; Urbanowicz, J.; Wróbel, S. Ochrona ziemniaka zgodna z zasadami integrowanej ochrony roślin. Cz. II. Metoda zrównoważonej chemicznej ochrony ziemniaka. *Biul. IHAR* **2014**, *273*, 145–159. (In Polish)
63. Haverkort, A.J.; Boonekamp, P.M.; Hutten, R.; Jacobsen, E.; Lotz, L.A.P.; Kessel, G.J.T.; Visser, R.G.F.; van der Vossen, E.A.G. Societal costs of late blight in potato and prospects of durable resistance through congenic modification. *Potato Res.* **2008**, *51*, 47–57. [CrossRef]
64. Croxall, M.E.; Smith, L.P. The epidemiology of potato blight in the east Midlands. *Ann. Biol.* **1976**, *82*, 451–466. [CrossRef]
65. Małysz, J. Economic interpretation of food security. In *Food Safety in the Era of Globalization*; Kowalczyk, S., Ed.; Wydawnictwo SGH: Warszawa, Poland, 2009. (In Polish)
66. Kraciuk, J. Food security from the perspective of underdeveloped and highly developed countries. *Sci. Ann. Assoc. Agric. Agribus. Econ.* **2015**, *17*, 205–209. (In Polish)
67. ACT of 25 August 2006, on food and nutrition safety. Journal of Laws 2020.2021, i.e., the current act. Version from 1 January 2021, to 3 February 2022. Available online: https://sip.lex.pl/akty-prawne/dzu-dziennik-ustaw/bezpieczenstwo-bezpiecznosci-i-zywien-17302608 (accessed on 25 January 2022).
68. Articles 43, 114 and 168 (1) 4 and art. 169 of the Treaty on the Functioning of the European Union. Available online: https://www.europarl.europa.eu/ftu/pdf/pl/FTU_2.2.6.pdf (accessed on 25 January 2022).
69. Fortuna, G.; Łobodziński, W. What Is the Level of Food Security in the EU? Available online: https://www.euractiv.pl/section/rolnictwowpr/news/zywnosc-unia-europejska-bezpieczenstwo-zywnosciowe-pandemia-koronawirus-lancuchy-dostaw-od-pola-do-stolu-wpr-wojciechowski-srodowisko-susza-klimat-powodz/ (accessed on 25 January 2022).
70. Scott, G.J.; Rosegrant, M.W.; Ringler, C. Global projections for root and tuber crops to the year 2020. *Food Policy* **2000**, *25*, 561–597. [CrossRef]
71. Tripathi, A.; Tripathi, D.K.; Chauhan, D.K.; Kumar, N.; Singh, G.S. Paradigms of climate change impacts on some major food sources of the world: A review on current knowledge and future prospects. *Agric. Environ.* **2016**, *216*, 356–373. [CrossRef]
72. Communication from the Commission to the European Parliament, the European Council, the Council, the European Economic and Social Committee and the Committee of the Regions. The European Green Deal. Brussels, 11 December 2019 COM (2019) 640 Final. Available online: https://ec.europa.eu/info/sites/default/files/european-green-deal-communication_en.pdf (accessed on 30 January 2022).

climate

Article

Water Profitability Analysis to Improve Food Security and Climate Resilience: A Case Study in the Egyptian Nile Delta

Adham Badawy [1,2,*], Amgad Elmahdi [1], Sayed Abd El-Hafez [3] and Ali Ibrahim [3]

1 MENA Office, International Water Management Institute (IWMI), Giza 12661, Egypt; amgad.elmahdi@gmail.com
2 Department of Earth & Environment, Boston University (BU), Boston, MA 02215, USA
3 Soil, Water and Environmental Research Institute (SWERI), Giza 12613, Egypt; s.a.abdelhafez@gmail.com (S.A.E.-H.); dr.ali_66@yahoo.com (A.I.)
* Correspondence: adhamb@bu.edu

Abstract: The food self-sufficiency policy has always featured as an unquestionable policy objective for Egypt. This is understandable when one considers both the high population growth and the social and political vulnerability associated with a dependence on food imports and world market food prices such as wheat. Intensive agriculture has led to a growing subsidy burden for the Egyptian government. In addition, the agricultural fields in Egypt are commonly distributed with relatively small sizes parcels that usually reduce the reliability of the agricultural sector, particularly in the delta region, to meet the national food policy. On top of that, climate change, through changing weather patterns and increased temperatures, is affecting agricultural yields and thus farmers' livelihoods. A water profitability analysis was conducted for three governorates in the Nile Delta in Egypt to establish a baseline and assess the net return per unit of water of the main crops in each of these governorates; this can act as a reference of the water profitability of different crops before they are affected by climate change and other internal and external factors. The analysis was based on extensive in-person surveys in each governorate in addition to workshop discussions with farmers. The study has highlighted the impact of a lack of extension services, which limits farmers' ability to increase their land and water productivity. Farmers with more access to subsidized production inputs managed to achieve higher levels of water profitability even on smaller lands. Finally, we drew from our findings key policy actions to improve water profitability and land productivity for farmers in the Nile Delta to achieve higher levels of food security. This will help build resilient food production systems that are reliable in the face of climate change and other drivers. In addition, an integrated nexus strategy and plan for the inter- and intra-country is recommended to address the challenges related to food and climate security.

Keywords: water profitability; water productivity; Nile Delta; food security; water security

Citation: Badawy, A.; Elmahdi, A.; Abd El-Hafez, S.; Ibrahim, A. Water Profitability Analysis to Improve Food Security and Climate Resilience: A Case Study in the Egyptian Nile Delta. *Climate* **2022**, *10*, 17. https://doi.org/10.3390/cli10020017

Academic Editor: Ying Ouyang

Received: 29 December 2021
Accepted: 28 January 2022
Published: 30 January 2022

1. Introduction

Since the 1950s, the food self-sufficiency policy has always featured as an unquestionable policy objective for Egypt. Climate change can have a severe impact on the agricultural sector and the stability of food security in Egypt. This is understandable when one considers both the high population growth and the social and political vulnerability associated with a dependence on food imports and world market food prices such as that of wheat. Egypt is considered one of the largest importers of wheat and a country where people rely on wheat products for around one-third of their food consumption in terms of calorie intake [1]. It is also expected that the food and water gaps that Egypt is facing will significantly widen by 2050 [2].

Food security, job creation, and limited per-capita land endowment in the Old Lands were always the determining factors for water and agricultural policy and are constantly used as an indisputable rationale for the expansion of irrigation, as illustrated in the Ministry of Agricultural and Land Reclamation's (MALR) sustainable development strategy towards 2030 [3]. Moreover, the responsibility of MALR is to ensure that food production is sufficient for to meet demand and sustainable at the same time, in addition to the monitoring and evaluation of sudden climatic changes and their impact on crop productivity to mitigate climate impacts on the quality and productivity of crops under stress.

Climate change can have a severe impact on the agricultural sector and the stability of food security in Egypt and in the Middle East and North Africa (MENA) region [4]. It is expected that crop production will be affected negatively due to the expected increases in temperature, extreme weather events, drought, plant diseases, and pests. Additionally, land use will be affected due to seawater intrusion and salinization. Water resources will be affected due to global warming and decreases in precipitation. Moreover, crop water requirements are expected to increase [5]. The compound effect of all these components represents the main challenge for researchers; moreover, the current cropping systems must be changed to comply with the future demands of the growing population and the threat of climate change [6]. The negative impacts of climate change on crop production can be reduced by the implementation of integrated farm-level adaptation strategies, starting with adopting changes including different seed varieties, planting dates, rationalizing the use of water and fertilizers, and changing irrigation intervals.

In addition, the sustainable development goals (SDGs) 1, 2, and 6, which are promoting sustainable agricultural practices to end poverty and increase water use efficiency [7], need greater efforts and resources at the country level to ensure the even and equal achievement of targets [8]. Therefore, further efforts are required to face these challenges, including more investments in agricultural and food systems and adapting sustainable alternative crops to the impact of water scarcity and climate change. For the sake of rationalizing the use of resources in the agricultural production system in Egypt, there is a need to understand the agricultural system (crops) and its related costs, returns, and profitability for farmers in terms of both land and water.

Water profitability analysis for policy planning—while still relatively a new concept—has been conducted in multiple regions to assess the net return per unit of water consumed in agriculture for crop production. In the Middle East and North Africa (MENA) region, several of these studies were performed. Oulmane et al. [9] assessed the water productivity and water value of three crops under normal conditions and used water-saving technologies in Algeria. The water value was calculated using gross margin, water costs, and applied water. It highlighted the increase in net returns per cubic meter of water due to the use of water-saving technologies. In Jordan, ref. [10] conducted a multicriteria analysis for water productivity to evaluate the economic value of water under maximum yields for selected crops. The study showed date palm to be the most profitable crop regarding water productivity. In Lebanon, a water profitability analysis was conducted to optimize cropping patterns based on the net revenue per unit of water [11]. In Oman, Al-Said et al. [12] assessed the water productivity of vegetables under modern irrigation methods. They analyzed the income per unit of water for five different crops and showed the increased returns and savings gained using drip irrigation for vegetable production.

Further, economic water productivity has been assessed by scholars following the water footprint concept [13]. Chouchane et al. [14] analyzed the economic water and land productivity of 11 crops in Tunisia. The study highlighted that the highest economic water productivity was reported for tomatoes and potatoes, while the lowest was recorded for olives, which are one of the major export products of the nation. In Pakistan, a study was conducted to compare the water productivity and return per unit of water for different rice types [15].

Yakubu et al. [16] analyzed net farm income per unit of land for four major strategic crops in the Kano River irrigation project in Nigeria. The study highlighted the profitability of maize, rice, and wheat compared to tomatoes. Tashikalma et al. [17] compared the crop profitability per unit of land under both rainfed and irrigated conditions in Nigeria. The study highlighted the major inputs that are costly for farmers, which, if subsidized, could drastically improve their incomes. Khansa [18] analyzed the average farm income under normal cropping patterns and used alternative saving crops. The author showcased the potential increase in farm income and water savings by changing the cropping pattern. Similar studies were conducted in Turkey [19], Peru [20], Bolivia [21], and Mexico [22].

However, the study of water profitability has been scarcely implemented in Egypt and thus there are few data on the baseline for its assessment in the Nile Delta. The only study analyzing water profitability in Egypt was conducted by Hosni et al. [23], who assessed the economic value of water used in irrigation in three governorates. Furthermore, they used linear programming (LP) to optimize the cropping patterns of these governorates to maximize water profitability and water savings. However, Osama et al. [24] studied the net return obtained per unit area (feddan) of all allocated crops for the cropping pattern (2008–2012). They used a linear programming (LP) technique to optimize the area allocated for each crop to achieve an overall increase in net benefits. These studies were conducted using reported data and lacked local farmers' information and voices.

This study attempts to set a baseline for the water profitability of multiple strategic crops in the Egyptian Nile Delta and compares different crops in three different governorates based on primary data collected from farmers' input. These crops were selected due to their importance in terms of the cultivated area, food insecurity, economy, and employment in Egypt. The analysis sheds the light on the main factors contributing to the heterogeneity of the water profitability levels across the Nile Delta and what policy recommendations and actions could be followed to increase and improve water and land profitability and productivity for farmers.

2. Materials and Methods

2.1. Description of the Study Area

This study was conducted in three governorates in the Egyptian Nile Delta, in Sharkia, El-Beheira, and Kafr El-Sheikh governorates, as seen in Figure 1. The three governorates were selected as they are a good representative of the 'old lands' in the Egyptian Delta, where most smallholder farmers are situated. The three governorates also represent the east, west, and middle of the Delta. The study area is representative as many of the strategic crops in Egypt are grown in these governorates.

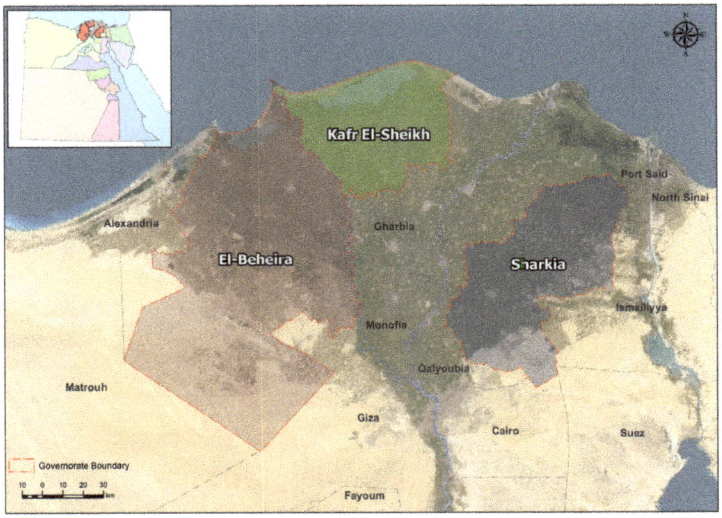

Figure 1. Study area in the Egyptian Nile Delta: Sharkia, El-Beheira, and Kafr El-Sheikh governorates. Source: Google Earth.

2.2. Data Collection

The study operationalizes both quantitative and qualitative data. The data were collected through two approaches. Firstly, structured interviews for farmers through a structured questionnaire were conducted. For each governorate, a minimum of 100 farmers were interviewed during 2020 to establish the baseline before COVID-19. Secondly, farmer consultation workshops were conducted in each governorate to collect the information and data required but also to validate the data collected during the individual interviews. Three workshops were conducted, one at every governorate with a minimum of 20 farmers. Farmers were randomly selected from each governorate to represent smallholder farmers that have farms ranging from less than one feddan up to five feddans. Table 1 shows the number of farmers in each data collection approach.

Table 1. Number of forms collected and farmers attending the workshops.

Governorate	Number of Forms Collected	Number of Participants in the Workshop
Sharkia	102	22
El-Beheira	110	20
Kafr El-Sheikh	120	20

The questionnaire used for data collection included both qualitative and quantitative data [25]. It included socio-economic data, farmer family structure, a detailed breakdown of all production inputs and their cost, water consumption, agricultural yields and productivity, self-consumption, market access, and selling prices.

The production input costs included:

- Land preparation;
- Seeding and planting;
- Irrigation;
- Fertilization;
- Weeding;
- Pest Control;
- Harvesting;
- Transportation;
- Other Expenses.

2.3. Analytical Methodology

Assessing water profitability in this study was achieved using the economic water productivity analytical method [12,23]. This analytical method was chosen due to several reasons. Firstly, the data collected and questionnaires were designed to follow the same structure of production input classification as the agricultural statistics bulletin. Moreover, the data collected allowed for collecting actual water applied by farmers. Finally, the chosen analytical method can accommodate the nuances and differences between farmers regarding access to agricultural inputs and water application compared to statistical averages and experts' estimations. The following steps describe the calculation method:

- **Total Costs (TC):** This is the summation of all the production costs.

$$\text{Total Costs} = \sum(\text{Land Preparation} + \text{Seeding and Planting} + \ldots + \text{Other Expenses.}) \quad (1)$$

- **Total Revenue (TR):** the yield per unit area multiplied by the selling price of all crops on land including primary and secondary crops.

$$\text{Total Revenue} = \text{Yield} \times \text{Crop Price} \quad (2)$$

- **Net Return (NR):** This is the difference between the total revenue and total costs.

$$\text{Net Return} = \text{TR} - \text{TC} \tag{3}$$

- **Water Applied (WA):** the amount of water applied per unit area for that crop production per season.
- **Water Profitability (WP):** the net return per unit of water applied for that crop's production.

$$\text{Water Profitability} = \text{NR}/\text{WA}$$

Units for results:

- EGP (Egyptian pound);
- Feddan, unit of area commonly used in Egypt, equivalent to 4200 m^2.

3. Results

This section presents the results of the collected data and the water profitability analysis conducted for the crops identified by sampled farmers as crops they had planted As the smallholder farmers were randomly selected, data for some crops are not available for any of the three governorates.

3.1. Sharkia

Table 2 depicts the water profitability analysis conducted for Sharkia from the collected forms in the governorate. The analysis was conducted for five main crops: wheat, sugar beet, clover, rice, and maize. Regarding total costs per feddan, sugar beet was the highest followed by rice, and the lowest was clover. Total revenue was highest for sugar beet and clover and lowest for maize. For the net return per feddan, clover was the highest 17,480 EGP/feddan, and the lowest value was found for maize at 723 L.E/feddan. Furthermore, rice is considered to be the most water-intensive crop, requiring about 6480 m^3/feddan, which is nearly double the amount required for the other crops, and the least water-intensive is wheat, using 2160 m^3/feddan. Finally, regarding water profitability, clover was the most profitable at 5.2 EGP/m^3 followed by sugar beet and wheat, while maize was the least water-profitable crop at 0.22 EGP/m^3.

Table 2. Water profitability analysis for Sharkia governorate in 2020, source: field data collected from farmers and verified by workshops.

Production Inputs	Wheat	Sugar Beet	Clover	Rice	Maize
Land Preparation (EGP)	783	1320	736	1097	959
Seeding and Planting (EGP)	899	1575	466	1634	1174
Irrigation (EGP)	421	545	589	1138	608
Fertilization (EGP)	949	2310	938	1101	1472
Weeding (EGP)	200	1700	-	521	715
Pest Control (EGP)	261	900	-	360	399
Harvesting (EGP)	2224	2750	920	2537	1414
Transportation (EGP)	587	700	854	634	680
Other Expenses (EGP)	231	200	217	237	239
Total Cost Without Rent (EGP)	6555	12,000	4720	9257	7661
Productivity (Ton/feddan)	2.82	45	37	3.675	2.62
Price (EGP/Ton)	4400	500	600	3500	3200
Revenue (EGP/feddan)	12,408	22,500	22,200	12,863	8384
Net Return (EGP/feddan)	5853	10,500	17,480	3605	723
Water Applied (m^3/feddan)	2160	3520	3360	6480	3240
Water Profitability (EGP/m^3)	**2.71**	**2.98**	5.2	**0.56**	**0.22**

Figure 2 shows the water profitability for the selected crops in Sharkia Governorate calculated using the Ministry of Agriculture and Land Reclamation–Economic Affairs Sector's Agricultural Statistics Bulletin for 2019 and data from the Land and Water Research Institute of the Water Standards Department. The figure also shows that clover is the most water-profitable crop, followed by sugar beet and wheat. The least water-profitable crops are rice and maize, which aligns with the data collected from the surveys and workshops. However, the collected data reflect the real value on the ground that show the sugar beet and wheat values were almost double the published value. These on-ground data need to be reflected in the policy planning and recommendations.

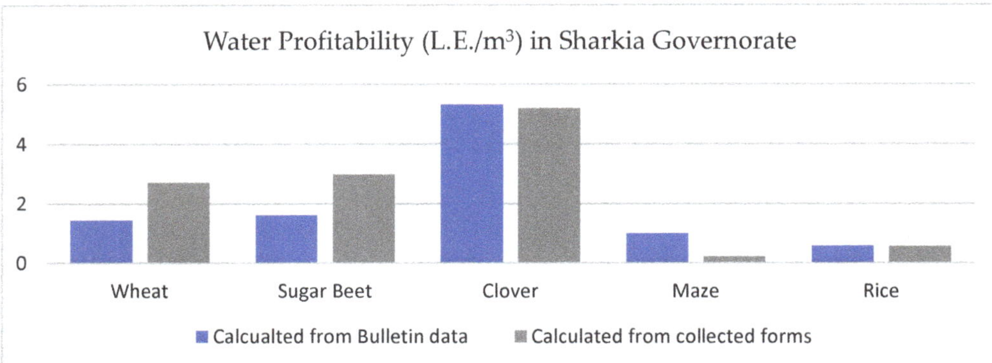

Figure 2. Water profitability of Sharkia governorate from Bulletin data compared to collected forms. Source: In blue, Ministry of Agriculture and Land Reclamation–Economic Affairs Sector's Agricultural Statistics Bulletin for 2019 and Land and Water Research Institute's Water Standards Department's unpublished data. In gray are the in-person forms collected in 2020.

3.2. El-Beheira

Table 3 depicts the water profitability analysis conducted for El-Beheira from the collected forms in the governorate. The analysis was conducted for nine main crops: wheat, sugar beet, broad bean, clover, rice, maize, watermelon pulp, tomato, and cotton. Regarding total costs per feddan, tomato was the highest followed by cotton, and the lowest was broad bean. Total revenue was highest for cotton and tomato and lowest for maize. For the net return per feddan, watermelon pulp was the highest 19,391 EGP/feddan and the lowest was maize at 10,799 L.E/feddan. Furthermore, rice was found to be the most water-intensive crop, followed by maize. Finally, for water profitability, watermelon pulp was the most profitable at 13.47 EGP/m^3 followed by broad bean at 12.96 EGP/m^3, while maize was the least water-profitable crop at 1.71 EGP/m^3.

Figure 3 shows the water profitability for the selected crops in the El-Beheira governorate calculated using the Ministry of Agriculture and Land Reclamation–Economic Affairs Sector's Agricultural Statistics Bulletin for 2019 and data from the Land and Water Research Institute's Water Standards Department compared to results calculated from the collected forms/study. Moreover, the water profitability values gathered from the collected forms are significantly higher than those from the Bulletin and Water Standards Department. This is reflected in ground farmers' information and local conditions. This analysis shows the potential of cotton and wheat in El-Beheira.

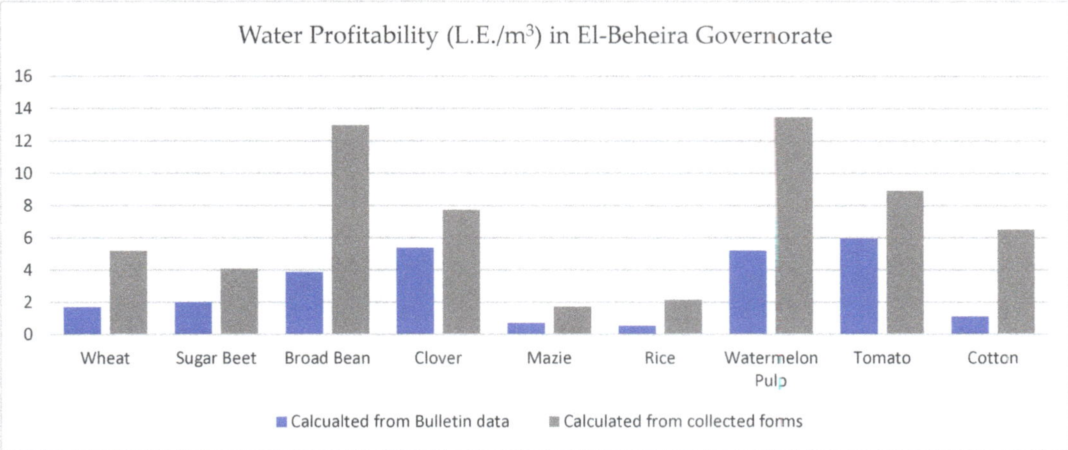

Figure 3. Water profitability for El-Beheira Governorate from Bulletin data compared to collected forms. Source: In blue, Ministry of Agriculture and Land Reclamation–Economic Affairs Sector's Agricultural Statistics Bulletin for 2019 and Land and Water Research Institute–Water Standards Department's unpublished data. In gray are the in-person forms collected in 2020.

Table 3. Water Profitability Analysis for El-Beheira Governorate in 2020, source: field data collected from farmers and verified by workshops.

Production Inputs	Wheat	Sugar Beet	Broad Bean	Clover	Rice	Maize	Watermelon Pulp	Tomato	Cotton
Land Preparation (EGP)	416	368	294	314	483	548	413	1280	690
Seeding and Planting (EGP)	656	341	744	597	973	782	370	1380	730
Irrigation (EGP)	396	600	391	494	682	564	536	360	340
Fertilization (EGP)	858	968	596	624	870	1018	1023	2270	1060
Weeding (EGP)	414	573	175	607	605	506	526	1160	1000
Pest Control (EGP)	350	386	225	864	534	474	735	4700	1540
Harvesting (EGP)	1001	1000	763	1076	879	838	657	1340	2140
Transportation (EGP)	269	473	475	166	292	303	130	500	250
Other Expenses (EGP)	23	67	-	21	30	56	21	200	-
Total Cost Without Rent (EGP)	4371	4764	3661	4758	5333	5060	4409	13,190	7750
Productivity (Ton/feddan)	3	22	1.86	35	4	3.36	0.7	25	1.575
Price (EGP/Ton)	4467	650	12,000	680	3700	3214	34,000	1000	16,825
Revenue (EGP/feddan)	13,401	14,300	22,320	23,800	14,800	10,799	23,800	26,000	26,499
Net Return (EGP/feddan)	9030	9536	18,659	19,042	9467	5739	19,391	12,810	18,749
Water Applied (m^3/feddan)	1740	2340	1440	2460	4440	3360	1440	1440	2880
Water Profitability (EGP/m^3)	**5.19**	**4.08**	**12.96**	**7.74**	**2.13**	**1.71**	**13.47**	**8.9**	**6.51**

3.3. *Kafr El-Sheikh*

Table 4 depicts the water profitability analysis conducted for Kafr El-Sheikh from the collected forms in the governorate. The analysis was conducted for ten crops produced in the area: wheat, sugar beet, broad bean, maize, watermelon pulp, clover, cotton, dry peas, and onion. Regarding total costs per feddan, cotton was the highest followed by sugar beet, and the lowest was clover. Total revenue was highest for dry peas followed by onion and cotton and lowest for maize and then clover. For the net return per feddan, dry peas were the highest at 25,000 EGP/feddan followed by onion at 22,544 EGP/feddan, and the lowest was maize at 1697 L.E/feddan. Furthermore, rice was found to be the most water-intensive crop, and the least intensive was dry peas using 1400 m^3/feddan. Finally, for water profitability, dry peas were the most water profitable at 17.86 EGP/m^3 followed by onion and broad bean, while maize was the least water-profitable crop at 0.49 EGP/m^3.

Table 4. Water profitability analysis for Kafr El-Sheikh Governorate in 2020, source: field data collected from farmers and verified by workshops.

Production Inputs	Wheat	Sugar Beet	Broad Bean	Rice	Maize	Watermelon Pulp	Clover	Cotton	Dry Peas	Onion
Land Preparation (EGP)	616	726	400	691	638	648	389	795	700	700
Seeding and Planting (EGP)	828	823	1029	1178	901	841	417	985	375	1000
Irrigation (EGP)	494	588	236	972	508	371	439	720	425	800
Fertilization (EGP)	942	1538	414	1122	1051	1064	526	1520	800	1500
Weeding (EGP)	355	859	643	545	596	580	136	999	625	300
Pest Control (EGP)	424	825	664	747	360	964	231	1703	850	800
Harvesting (EGP)	850	1054	1007	809	644	757	217	2663	1000	800
Transportation (EGP)	261	573	314	270	306	173	103	248	225	200
Other Expenses (EGP)	204	300	175	231	200	200	0	250	0	0
Total Cost Without Rent (EGP)	4974	7286	4882	6566	5204	5596	2459	9884	5000	6100
Productivity (Ton/feddan)	2.85	27	1.395	3.25	2.1	0.7	25.5	1.339	2	14
Price (EGP/Ton)	4467	625	12,187	3560	3286	35,000	400	19,810	15,000	2046
Revenue (EGP/feddan)	12,731	16,875	17,001	11,570	6901	24,500	10,200	26,526	30,000	28,644
Net Return (EGP/feddan)	7757	9589	12,119	5004	1697	18,904	7741	16,641	25,000	22,544
Water Applied (m^3/feddan)	2088	1740	1218	5568	3480	2320	2262	3712	1400	1740
Water Profitability (EGP/m^3)	**3.72**	**5.51**	**9.95**	**0.9**	**0.49**	**8.15**	**3.42**	**4.48**	**17.86**	**12.96**

Figure 4 shows the water profitability for the selected crops in Kafr El-Sheikh calculated using the Ministry of Agriculture and Land Reclamation–Economic Affairs Sector's Agricultural Statistics Bulletin for 2019 and data from the Land and Water Research Institute–Water Standards Department. The figure also shows that dry peas were the most water-profitable crop, followed by onion, while the least-water profitable crops were rice and maize, which aligns with the data collected from the surveys and workshops. Again, the results indicate the potential consideration of cotton and sugar beet in Kafr El-Sheikh.

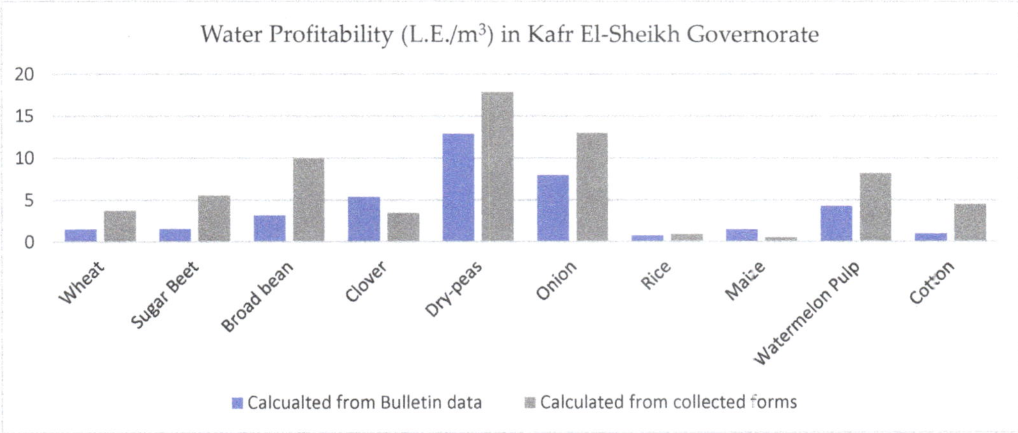

Figure 4. Water Profitability for Kafr El-Sheikh Governorate from Bulletin data compared to collected forms. Source: In blue, Ministry of Agriculture and Land Reclamation–Economic Affairs Sector's Agricultural Statistics Bulletin for 2019 and Land and Water Research Institute–Water Standards Department's unpublished data. In gray are the in-person forms collected in 2020.

3.4. Cross-Governorate Comparison

From a socio-economic perspective, the results show that the majority of farmers are over 50 years old, as their percentage reached 61% in Sharkia Governorate, about 67% in El-Beheira Governorate, and about 73% in Kafr El-Sheikh Governorate. Additionally, most farmers had received a formal education, as the percentage of educated people reached about 89.22% in Sharkia Governorate, about 66.96% in the El-Beheira Governorate, and about 51% in Kafr El-Sheikh Governorate. The study also showed that agricultural incomes varied from one governorate to another, as the annual income from agriculture reached about 28,000 EGP per feddan in Sharkia Governorate, about EGP 13,000 per feddan in El-Beheira Governorate and about EGP 12,000 per feddan in Kafr El-Sheikh Governorate, and the majority of farmers had incomes other than agriculture.

It was found from the farmers' responses that the most important reason for the low productivity per feddan was the lack of fresh water in the water channels, which forces some farmers to supplement their irrigation needs with water from agricultural drainages such as in Kafr El-Sheikh or well water. Furthermore, across the three governorates, the lack of production requirements such as fertilizers, seeds, pesticides, and machinery needed for cultivation and harvesting operations at appropriate prices and times are increasing the costs and limiting productivity. Furthermore, low market prices are affecting the net returns per feddan and unit of water. In addition, low soil fertility and deteriorating water quality are all contributing to a reduction in productivity and yields.

Figure 5 depicts the water profitability of different crops in each of the three governorates. The highest overall profitable crop was dry peas, which were only found in Kafr El-Sheikh at nearly 18 EGP/m^3, followed by watermelon pulp, broad bean, and onion. The least water-profitable crops were rice and maize. However, in El-Beheira wheat was more water profitable than in Sharkia and Kafr El-Sheikh. Furthermore, clover's water profitability value in Sharkia was nearly double that of Kafr El-Sheikh. Moreover, sugar beet in Kafr El-Sheikh was more water profitable compared to the other two governorates. Finally, water profitability was mostly higher for the same crops in Beheira and then Kafr El-Sheikh and lastly Sharkia. These results could be used to reflect on the suitability of the crops per governorate including land and water as well as the socio-economic data of the farmers and accessibility to the market.

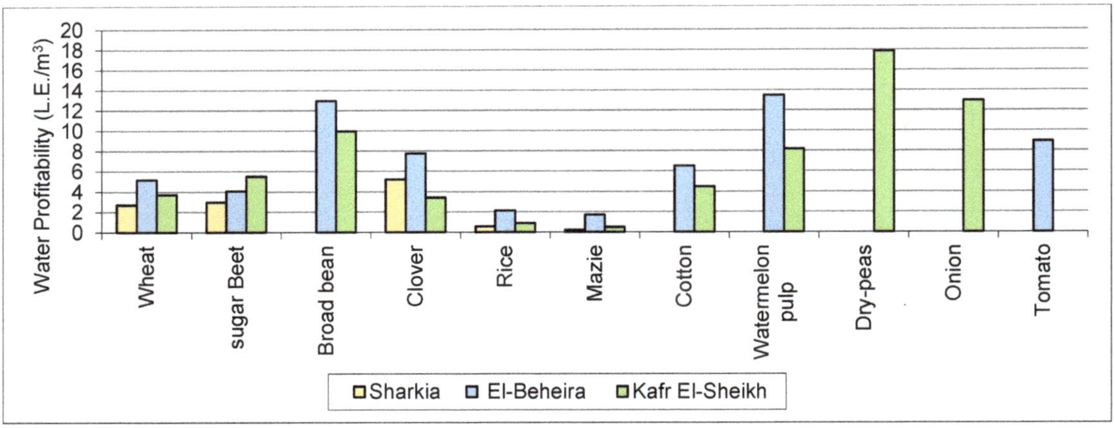

Figure 5. Cross-governorate water profitability comparison for 2020, source: field data collected from farmers and verified by workshops.

The collected data also point to the fact that experience had a great impact on increasing water and land productivity and determining the dates of harvest and the best harvesting technique. The level of education also affected production levels and the change of crops from one year to another. The high costs of production requirements had a significant impact on the profitability potential as many farmers did not have the cash flow required to cover the initial costs of the production of the more profitable crop.

The results show that the economic variables vary from one governorate to another, although the data of Sharkia Governorate show higher values than those of other governorates. Thus, the total costs for the same crop are higher in Sharkia compared to El-Beheira and Kafr El-Sheikh. This is one of the main reasons why the net return per feddan and water profitability are lower in Sharkia compared to the other two governorates. Moreover, in Sharkia Governorate the farm sizes are relatively smaller and more fragmented than the other two governorates. This makes the costs relatively higher and the net return per feddan significantly lower. Finally, for smaller farm sizes the options for profitable crops are limited, which compels farmers to select crops based on considerations other than per feddan net returns and profitability.

4. Discussion

4.1. Discrepancy and Similarity between Governorates

The discrepancy between the water profitability values calculated from the face-to-face forms and the bulletin is due to several reasons. Firstly, the bulletin data were published for the year 2019, which is based on 2018 estimates and numbers, while the forms were collected in 2020. This could affect water profitability through a multitude of ways, such as different market prices and the availability of production inputs and their prices. Moreover, the bulletin data were based on averages of market prices, yields, and experts' estimations of the production inputs and thus the return per feddan. On the contrary, the value of the estimates in this study are calculated from primary collected data from farmers, thus considering the different challenges farmers may face in acquiring certain production inputs such as fertilizers and pesticides.

The estimates of water profitability from the collected are forms are higher compared to the estimates by Hosni et al. in 2013 [23]. For Sharkia Governorate their estimates of water profitability for rice were 0.32 EGP/m^3 compared to our estimates of 0.56 EGP/m^3. Sugar beet water profitability in their study was estimated at 1.68 EGP/m^3 compared to our estimates of 2.98 EGP/m^3 for the same governorate. It is also important to point out that the estimates by Hosni et al. (ibid.) were calculated using the bulletin data and not from primary collected data from farmers.

4.2. Rationale for Crop Selection

Despite maize and rice having relatively lower water profitability, these crops have significant value for farmers. Maize is a major source of animal nutrition, and it provides important supplementary nutrition for livestock. The absence of these nutritious elements leads to a decrease in the production of meat, which affects Egyptian food security. Rice plays an important role in soil management; farmers plant rice to wash their soils and improve their fertility. Rice is also considered one of the most important staple foods and a source of foreign currency when exported. The shortage of this crop affects the volume of agricultural exports. Rice has another significant value as it contributes to the protection of the northern areas of the Nile Delta from seawater intrusion.

Even though cotton has also relatively lower water profitability, it is a significant Egyptian crop due to several reasons; it is an important strategic crop for the textile industry, as well as for exports [26]. Egyptian cotton is world-famous for its quality and has great export potential [27]. Many farmers responded that they continued planting cotton as they inherited the practice from their fathers and grandfathers. They also mentioned that for them there is no convincing alternative.

Wheat is an essential crop for Egyptian food security even if it has lower water profitability compared to other high-value crops [28]. Egypt is considered one of the biggest wheat importers globally; this is due to the high consumption of bread in the Egyptian diet [1,29]. Thus, many farmers in Egypt tend to grow wheat for self or home consumption. Farmers listed several other reasons for growing wheat, for example, the low amount of labor, easiness of growing the crop, having the accumulated experience and knowledge to grow it, and its usefulness as feed livestock.

Regarding the most water-profitable crops as seen in Figure 4, dry peas were the most profitable followed by watermelon pulp. Farmers justified the plantation of dry peas in Kafr El-Sheikh as it has very high net returns per feddan and relatively low costs compared to the profit. Farmers selected watermelon pulp cultivation due to the high return it generates, easiness of cultivation, and the fact that it has a short cycle so does not stay in the ground for a long period. Moreover, farmers chose broad beans because they reduce soil stress and increase its fertility, and its straw is used as fodder for livestock. These represented the secondary values that were often underestimated. Sugar beet was selected by farmers as it has high returns per feddan, it thrives in the soil in Kafr El-Sheikh, and it has a relatively stable selling price when sold to sugar factories. Finally, sugar beet can withstand salinity, which reduces the risk of growing it.

Net return and profitability are not the only factors that impact farmers' crop selection. The smaller the farm size, the fewer the options for profitable crops. However, farmers therefore tend to grow livestock on those lands and grow crops that can be used as fodder such as clover, maize, and crops that have a side product that can be used as fodder such as wheat, broad beans, and sugar beet. The net revenue and profitability of these products are relatively low, but their contribution in the value chain for farmers is high and satisfies the need for fodder for livestock, which would be expensive if purchased from the market.

4.3. Recommendations

The above analysis revealed the need for a new paradigm shift in the Egyptian water and food sectors in an effort to address these challenges and mitigate the risks. This paradigm has three main directions in which Egypt's water sector and food sector can transform to be able to accommodate and deal with its challenges and meet future needs, including the socio-economic development ambitions. The first dimension is the digital transformation of the agricultural sector. The second dimension is the investment in the agricultural sector and focus on its development. The third dimension is to adopt more bottom-up planning and implementation to improve equity in water access and use with respect to agricultural water investments, which are part of a bigger picture of system management, as they are efficient from economic, social, and environmental perspectives. This entails the concept of nexus governance, requiring policy actors to engage across policy domains and the public and private spheres, and by extension, strengthening human capital and institutions for policy coherence and participatory mechanisms.

The role of education and extensions services is clear in improving land and water productivity [30]. Investment in strong extension services and awareness campaigns for farmers can significantly increase water profitability and contribute to increased levels of food security. One dimension of this could be achieved through the use of digital innovations and information systems [31]. These tools can provide farmers with accurate information and viable interventions at the right time.

The prices and availability of agricultural inputs affect the net returns per feddan and water profitability as has been found in the three governorates. Increasing the allocated quantities of seeds, fertilizers, and pesticides at the agricultural associations in each governorate and increasing the subsidies allocated to these items would positively impact water profitability for farmers, in particular the smallholders. In addition, providing machinery for farming and harvesting different crops at subsidized rates or through establishing farmers' associations could positively impact the profitability and respond to the lack of manpower and its high cost.

The next step required to better understand the agricultural system in the Nile Delta is to assess the water profitability of cropping sequences, not just single crops. Assessing common cropping rotations in the three governorates will paint a clearer picture of the small farmers in the region. Common rotations are the plantation of rice and sugar beet followed by cotton and then wheat, or starting with cotton and then wheat followed by rice and ending with wheat again [32]. Moreover, analyzing water profitability over a year, thus including every season, would take into account the same temporal scale for analyzing net return for farmers.

Assessing the water- and soil-quality effects on water profitability is essential, as it would open the door for understanding the links between the soil characteristics, land productivity, yield, and production inputs and costs. In addition, some crops are selected by farmers to improve soil fertility and to protect the land from deterioration. These links and benefits should be considered when analyzing water profitability.

Conservation agriculture is key to addressing the challenges related to food insecurity and climate change. Transformation of agricultural systems by adopting climate-smart agriculture practices can increase resilience while increasing productivity.

Finally, having a baseline of water profitability for different crops before COVID-19 could be the first step to evaluate potential new crops that have higher water profitability and can contribute directly or indirectly to improving food security in Egypt. Hence, to improve food security in Egypt, more information on crops' water profitability and their values in comparison to the world (similar countries) and region practices are essential to inform policymakers in deciding strategies regarding cropping patterns. This would create a backdrop based on which future patterns can be assessed and evaluated taking into account the pressures of climate change and economic development ambitions.

5. Conclusions

This study on water profitability analysis was conducted for the major crops in three governorates in Egypt. The analysis was conducted for the Sharkia, El-Beheira, and Kafr El-Sheikh governorates, situated in the Egyptian Nile Delta. The study shed light on the water profitability of different crops in the study areas based on field primary data collected from farmers in each of those governorates and verified these data through consultation workshops. This study approach has not been implemented in the Egyptian Delta before and thus reveals the actual water profitability of different crops produced by smallholder farmers. This study provides insights into the different difficulties farmers face that affect their land and water profitability and shows how these problems could be addressed to improve food and climate security. The analysis showed the differences in water profitability among the three governorates even for the same crop and the contributing factors that affect it. Furthermore, the limiting factors for improving water profitability were identified, such as limited extension services, deteriorating water and soil quality, and inaccessibility of production inputs. Such an assessment can set the baseline for the water profitability of different crops and allow more climate-resilient cropping patterns to be planned accordingly as well as act as a guide for future policies. Monitoring the change of water profitability over time can deepen our understating of the factors that impact it along the production chain and highlight opportunities to improve it. Consequently, analyzing the water profitability of crops downstream the supply chain can paint a clearer picture of their contribution to GDP and national growth. Taking the analysis one step further and analyzing the number of family members benefiting from the generated profits could provide fresh insights into water profitability social distribution and the number of beneficiaries. Finally, we provided policy actions and recommendations for improving water profitability for farmers and future pathways for a deeper understanding of the water profitability of the farmers in the Nile Delta and how this knowledge could improve Egyptian food and climate security.

Author Contributions: The field research conceptualization, methodological design, and data collection was overseen and conducted by A.B., A.E., S.A.E.-H. and A.I. The data collected were analyzed by A.I. The initial figures and tables were generated by A.I. and A.B. and subsequently finalized by A.B. The paper was drafted by A.B. under the supervision of A.E., S.A.E.-H. and A.I., with final drafts generated by A.B. and finalized and edited by A.B. and A.E. Funding acquisition was carried out by A.E. All authors have read and agreed to the published version of the manuscript.

Funding: This project is part of the CGIAR Research Program on Water, Land and Ecosystems (WEGP) and supported by CGIAR fund donors.

Institutional Review Board Statement: Not applicable.

Informed Consent Statement: Not applicable.

Data Availability Statement: The data supporting the conclusions of this study are available upon reasonable request from the authors, with the exception of data that identify the personal information of the research participants.

Acknowledgments: This project is part of the CGIAR Research Program on Water, Land and Ecosystems (WEGP) and supported by CGIAR fund donors. The research was conducted by a team of scientists based at the International Water Management Institute–IWMI and the Soil, Water and Environmental Research Institute–SWERI, Egypt.

Conflicts of Interest: The authors declare no conflict of interest.

References

1. Tanaka, T. Agricultural Self-Sufficiency and Market Stability: A Revenue-Neutral Approach to Wheat Sector in Egypt. *J. Food Secur.* **2018**, *6*, 31–41. [CrossRef]
2. Abdelkader, A.; Elshorbagy, A.; Tuninetti, M.; Laio, F.; Ridolfi, L.; Fahmy, H.; Hoekstra, A.Y. National Water, Food, and Trade Modeling Framework: The Case of Egypt. *Sci. Total Environ.* **2018**, *639*, 485–496. [CrossRef] [PubMed]
3. Arab Republic of Egypt. *Sustainable Agricultural Development Strategy Towards 2030*; Ministry of Agriculture and Land Reclamation: Cairo, Egypt, 2009.
4. Belhaj, F.; Soliman, A. MENA Has a Food Security Problem, but There are Ways to Address It. Available online: https://www.worldbank.org/en/news/opinion/2021/09/24/mena-has-a-food-security-problem-but-there-are-ways-to-address-it (accessed on 4 January 2022).
5. Ouda, S.A.H.; Zohry, A.E.-H.; Alkitkat, H.; Morsy, M.; Sayad, T.; Kamel, A. *Future of Food Gaps in Egypt*; Springer International Publishing: Cham, Switzerland, 2017. [CrossRef]
6. Mahmoud, M.A. Impact of Climate Change on the Agricultural Sector in Egypt. In *Handbook of Environmental Chemistry*; Springer: Cham, Switzerland, 2019; Volume 74, pp. 213–227. [CrossRef]
7. United Nations (UN). *The Sustainable Development Goals Report 2020*; United Nations Department of Economic and Social Affairs: New York, NY, USA, 2020.

8. Elmahdi, A. Road Maps for Water Accounting Designing and Institutionalizing for Sustainable Water Management in MENA Region. *Int. J. Environ. Sci. Nat. Resour.* **2019**, *22*, 1–3. [CrossRef]
9. Oulmane, A.; Chebil, A.; Frija, A.; Benmehaia, M.A. Water-Saving Technologies and Total Factor Productivity Growth in Small Horticultural Farms in Algeria. *Agric. Res.* **2020**, *9*, 585–591. [CrossRef]
10. Al-Zu'bi, Y.A.; Al-Kharabsheh, A. Multicriteria Analysis For Water Productivity in the Jordan Valley. *Water Int.* **2003**, *28*, 501–511. [CrossRef]
11. ElAwar, F.A.; Darwish, M.R.; Mteirik, R.M.; Nimah, M.N. Optimal cropping pattern for limited water supply: A case study in lebanon. *Appl. Eng. Agric.* **2001**, *17*, 391. [CrossRef]
12. Al-Said, F.A.; Ashfaq, M.; Al-Barhi, M.; Hanjra, M.A.; Khan, I.A. Water Productivity of Vegetables under Modern Irrigation Methods in Oman. *Irrig. Drain.* **2012**, *61*, 477–489. [CrossRef]
13. Hoekstra, A. Virtual Water Trade. In Proceedings of the International Expert Meeting on Virtual Water Trade, Delft, The Netherlands, 12–13 December 2002; pp. 1–244.
14. Chouchane, H.; Hoekstra, A.Y.; Krol, M.S.; Mekonnen, M.M. The Water Footprint of Tunisia from an Economic Perspective. *Ecol. Indic.* **2015**, *52*, 311–319. [CrossRef]
15. Naseer, M.A.U.R.; Ashfaq, M.; Razzaq, A.; Ali, Q. Comparison of Water Use Efficiency, Profitability and Consumer Preferences of Different Rice Varieties in Punjab, Pakistan. *Paddy Water Environ.* **2020**, *18*, 273–282. [CrossRef]
16. Yakubu, A.A.; Baba, K.M.; Mohammed, I. Profitability of Crop Production under the Kano River Irrigation Project, Northern, Nigeria. *J. Adv. Agric.* **2016**, *6*, 963–972. [CrossRef]
17. Tashikalma, A.; Sani, R.; Giroh, D. Comparative Profitability Analysis of Selected Rainfed and Irrigated Food Crops in Adamawa State, Nigeria. *Glob. J. Pure Appl. Sci.* **2014**, *20*, 77. [CrossRef]
18. El Khansa, M. Finding Pathways for Enhancing Irrigated Farming Systems in Lebanon. Ph.D. Thesis, Universiteé Montpellier, Montpellier, France, 2017.
19. Kuşçu, H.; Çetin, B.; Turhan, A. Yield and Economic Return of Drip-Irrigated Vegetable Production in Turkey. *N. Z. J. Crop Hortic. Sci.* **2009**, *37*, 51–59. [CrossRef]
20. Gomez–Pando, L.; Castellanos, E.A.; Ibañez–Tremolada, M.; Falconi–Palomino, J.; Gonzales, K.A. *Introducing Quinoa Mutant Varieties with High Water and Nutrient Use Efficiency to the Peruvian Highlands*; International Atomic Energy Agency (IAEA): Vienna, Austria, 2017.
21. Giuliani, A.; Hintermann, F.; Rojas, W.; Padulosi, S. *Biodiversity of Andean Grains: Balancing Market Potential and Sustainable Livelihoods*; Bioversity International: Rome, Italy, 2012.
22. Scott, C.A.; Flores-López, F.; Gastélum, J.R. Appropriation of Río San Juan Water by Monterrey City, Mexico: Implications for Agriculture and Basin Water Sharing. *Paddy Water Environ.* **2007**, *5*, 253–262. [CrossRef]
23. Hosni, H.; El-Gafy, I.K.; Ibrahim, A.H.; Abowarda, A.S. Maximizing the Economic Value of Irrigation Water and Achieving Self Sufficiency of Main Crops. *Ain Shams Eng. J.* **2014**, *5*, 1005–1017. [CrossRef]
24. Osama, S.; Elkholy, M.; Kansoh, R.M. Optimization of the Cropping Pattern in Egypt. *Alex. Eng. J.* **2017**, *56*, 557–566. [CrossRef]
25. Tobergte, D.R.; Curtis, S. Social Research Methods. *J. Chem. Inf. Model.* **2013**, *53*, 1689–1699. [CrossRef]
26. Dhehibi, B.; Ibrahim, A.A.; El-Shahat, A.; Hassan, A.A. Impacts of Irrigation on Agricultural Productivity in Egypt. *Ann. Arid. Zone* **2016**, *55*, 67–78.
27. Al-Sayyad, Y. The End of Egyptian Cotton | The New Yorker. Available online: https://www.newyorker.com/news/news-desk/the-end-of-egyptian-cotton (accessed on 27 December 2021).
28. Khalil, A.A.; Essa, Y.H.; Hashem, F.A.; Refaie, K.M. Water Consumption Variability for Major Crops in Egypt under Climate Change Conditions. *Int. J. Adv. Res.* **2016**, *4*, 1157–1168. [CrossRef]
29. McGill, J.; Prikhodko, D.; Sterk, B.; Talks, P. *Egypt: Wheat Sector Review*; Country Highlights (FAO) Eng No. 21; FAO Investment Centre: Rome, Italy, 2015.
30. ElMahdi, A.; Badawy, A. Alternative Policy Solutions | Equity and Egypt's Water Paradigm Shift. Available online: https://aps.aucegypt.edu/en/articles/700/equity-and-egypts-water-paradigm-shift (accessed on 27 December 2021).
31. Elmahdi, A.; Nassif, M.-H.; Abi Saab, M.T. Phone App Gives Opportunity to Improve Water Productivity in Lebanon: IWMI. Available online: https://www.iwmi.cgiar.org/2020/04/phone-app-gives-opportunity-to-improve-water-productivity-in-lebanon/ (accessed on 4 January 2022).
32. El-hafez, S.A.A.; Mahmoud, M.A.; Oud, S.A.H. A New Conceptual Framework for Water Conservation Based on Addressing Water Balance, Crop Rotation and Economics. *Int. J. Water Resour. Arid. Environ.* **2014**, *4*, 35–42.

climate

MDPI

Article

Is Drought Increasing in Maine and Hurting Wild Blueberry Production?

Kallol Barai [1], Rafa Tasnim [1], Bruce Hall [2], Parinaz Rahimzadeh-Bajgiran [3] and Yong-Jiang Zhang [1,*]

[1] School of Biology & Ecology, University of Maine, Orono, ME 04469, USA; kallol.barai@maine.edu (K.B.); rafa.tasnim@maine.edu (R.T.)
[2] Jasper Wyman & Son, Milbridge, ME 04658, USA; bhall@wymans.com
[3] School of Forest Resources, University of Maine, Orono, ME 04469, USA; parinaz.rahimzadeh@maine.edu
* Correspondence: yongjiang.zhang@maine.edu

Citation: Barai, K.; Tasnim, R.; Hall, B.; Rahimzadeh-Bajgiran, P.; Zhang, Y.-J. Is Drought Increasing in Maine and Hurting Wild Blueberry Production? *Climate* **2021**, 9, 178. https://doi.org/10.3390/cli9120178

Academic Editors: Christopher Robin Bryant, Andrea Vitali and Azzeddine Madani

Received: 30 October 2021
Accepted: 5 December 2021
Published: 8 December 2021

Abstract: A few severe drought events occurred in the Northeast (NE) USA in recent decades and caused significant economic losses, but the temporal pattern of drought incidents and their impacts on agricultural systems have not been well assessed. Here, we analyzed historical changes and patterns of drought using a drought index (standardized precipitation-evapotranspiration index (SPEI)), and assessed drought impacts on remotely sensed vegetation indices (enhanced vegetation index (EVI) and normalized difference vegetation index (NDVI)) and production (yield) of the wild blueberry fields in Maine, USA. We also analyzed the impact of short- and long-term water conditions of the growing season on the wild blueberry vegetation condition and production. No significant changes in the SPEI were found in the past 71 years, despite a significant warming pattern. There was also a significant relationship between the relatively long-term SPEI and the vegetation indices (EVI and NDVI), but not the short-term SPEI (one year). This suggests that the crop vigor of wild blueberries is probably determined by water conditions over a relatively long term. There were also significant relationships between 1-year water conditions (SPEI) and yield for a non-irrigated field, and between 4-year-average SPEI and the yield of all fields in Maine. The vegetation indices (EVI and NDVI) are not good predictors of wild blueberry yield, possibly because wild blueberry yield does not only depend on crop vigor, but also on other important variables such as pollination. We also compared an irrigated and a non-irrigated wild blueberry field at the same location (Deblois, Maine) where we found that irrigation decoupled the relationship between the SPEI and NDVI or EVI.

Keywords: wild blueberry barrens; temperature; precipitation; drought index (SPEI); drought impact; normalized difference vegetation index (NDVI); enhanced vegetation index (EVI)

1. Introduction

Elevated atmospheric temperatures, increased rainfall variabilities, and more frequent extreme drought events associated with anthropogenic climate change have been significantly damaging agricultural systems and crop production globally [1,2]. Additionally, local and microclimate changes could be more intense and significantly different compared to the reported average global or regional climate changes in terms of temperatures and precipitation [3–5]. For instance, a recent study on wild lowbush blueberry crops in Downeast, Maine (ME), USA has revealed that the summer temperatures of wild blueberry fields have been increasing significantly faster in the past forty years compared to that of the region (the state of Maine, Northeast (NE) USA) [6]. Such higher increasing temperature patterns in agricultural lands will exacerbate the impacts of drought events due to increased water loss [7,8]. Severe drought incidents have been reported in recent decades in NE USA and caused significant economic losses [9–11]. However, the historical trends of drought and their impacts on agricultural systems in this region have not been carefully assessed.

The wild blueberry (*Vaccinium angustifolium* Aiton) is one of the culturally and economically valuable crops in NE USA, which has been growing naturally for hundreds of years at the coastal barrens of the state of Maine in the USA, Atlantic Canada, and the province of Quebec in Canada. It is quite a unique agricultural system, as ~1500 genetically distinct wild blueberry plants can be found in a large field (~5 ha) [12]. This crop grows in a two-year production cycle where the stems, leaves, and buds develop during the first year, referred to as a prune year, and those plants bloom and produce fruits during the second year, referred to as a crop year [12]. After harvesting the berries from the end of July to early August in a crop year, plants are pruned to the ground by mowing or burning and the crop cycle starts again the following prune year. It is still unknown how this unique wild agricultural system will respond to the unprecedented changes in rainfall patterns and decreasing soil moisture in this region [6,7]. In fact, the summer temperatures and, hence, potential evapotranspiration of the wild blueberry fields at Downeast, Maine have been increasing significantly in the past decades [6]. Yet, we do not know whether wild blueberry fields experienced drought stress over the years, not to mention we do not have any scientific evidence of how this crop has been responding to drought incidents over the years. Although some controlled drought experiments revealed that wild blueberry plants are drought tolerant based on one growing cycle (2 years) [13,14], we still do not know whether drought has short-term and long-term effects on the vigor and production of this crop. Hence, the historical drought patterns that the wild blueberry barrens have been experiencing, and their impacts on crop vigor and production, need to be analyzed to guide management practices in the future.

Drought has been a great threat to the agricultural systems worldwide, and has been extensively studied in different regions on varieties of crops [15,16]. Droughts in agricultural lands are related to a lack of precipitation and inadequate water supply to crops. In order to analyze drought severity, several meteorological drought indices have been developed based on different combinations of precipitation, temperature, soil moisture availability, and vegetation conditions. Widely used drought indices include the Palmer Drought Severity Index (PDSI) [17], the standardized precipitation index (SPI) [18], and the standardized precipitation-evapotranspiration index (SPEI) [19]. Here, in order to analyze historical drought patterns for wild blueberry fields in Maine, we adopted the SPEI over other drought indices as it is determined based on precipitation, temperature, and potential evapotranspiration [19]. Moreover, the SPEI would be the most useful index to determine the water conditions (dry/wet) and drought severity of an agricultural system during both the short period and long period [19,20]. This is because the SPEI's multi-scalar character enables it to detect, monitor, and analyze droughts more effectively, as it can quantify the water conditions (dry/wet) and the drought severity according to its duration and intensity [19]. The SPEI also allows the comparison of drought severity through time and space, since it can be calculated over a wide range of climates.

Besides determining the historical drought patterns for the wild blueberry barrens, we also assessed the impacts of drought and water conditions on the crop vigor and production of wild blueberries over the years. In order to determine the crop vigor, which could be indicated by their greenness and biomass, widely used remotely sensed vegetation indices such as the normalized difference vegetation index (NDVI) and the enhanced vegetation index (EVI) were used to analyze the vegetation condition of wild blueberry fields and their response to historical drought incidents. The NDVI and EVI are widely used to assess vegetation health and biomass production, as they represent a composite property of canopy cover, canopy structure, canopy greenness, leaf area, and chlorophyll content [6,21–24]. Moreover, we analyzed the effects of short- and long-term water conditions (or water deficits; indicated by the SPEI) on the wild blueberry crop vegetation status and production in Maine, and the impacts of irrigation in alleviating drought effects and securing production. Our study will provide a complete understanding of how this crop has been affected by frequent erratic changes in rainfall and drought.

Therefore, the objectives of this study were:

1. To test whether drought incidents and severity were increasing in the past 71 years by analyzing the historical changes in temperature, precipitation, and drought index (SPEI),

as well as the changing pattern of the EVI and fruit production of wild blueberry fields in the past two decades in the study sites of the major wild blueberry production region of Maine, USA.

2. To determine the impacts of drought on the vegetation condition and production of the wild blueberry crops in Maine by establishing relationships among the drought index (SPEI), vegetation indices (NDVI and EVI), and yield.

3. To test whether irrigation alleviated the impacts of drought on the vegetation health and production of the wild blueberry crops in Maine by comparing nearby irrigated and non-irrigated fields.

2. Materials and Methods

2.1. Study Area

We analyzed the wild blueberry fields in Maine, USA as a whole, and two specific fields with a good record of yield data (Figure 1). The wild blueberry fields in the major production region are located in the Washington and Hancock counties of Maine (referred to as "Maine WB Fields", Figure 1a,b). The two specific fields selected were the Airport wild blueberry field at Deblois, Maine (referred to as "Airport", Figure 1c), and the Baxter wild blueberry field at Deblois, Maine (referred to as "Baxter", Figure 1c). The soil of the wild blueberry fields in Maine is well-drained sandy loam acidic soil [25]. The studied region has a four-season climate with an average annual minimum temperature of −10.6 °C and a maximum temperature of 24.2 °C, and monthly average precipitation as low as 85.1 mm and as high as 136.4 mm [26].

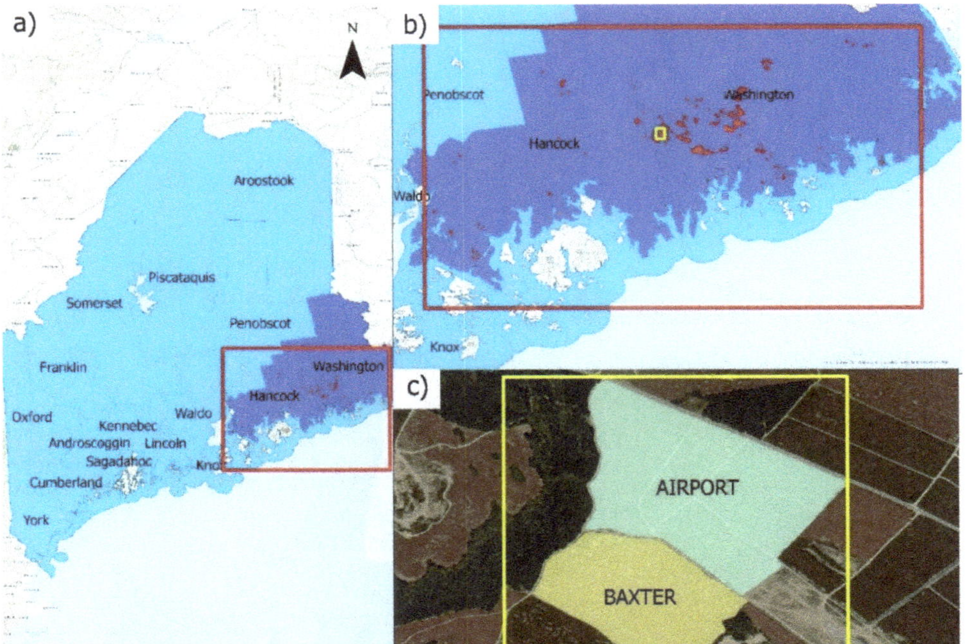

Figure 1. Location of the study sites: (**a**) a map of the state of Maine (light blue color), USA showing the location of the major wild blueberry production region (Washington and Hancock County) in dark blue color; (**b**) a map of the major wild blueberry production region in Maine showing 89 wild blueberry fields in red polygons for this study with an area of 0.06 km^2 (250 m × 250 m) or larger; (**c**) Airport and Baxter wild blueberry fields of Wyman's in Deblois, Washington County, ME, USA.

2.1.1. Wild Blueberry Fields in Major Wild Blueberry Production Region of Maine, USA

Washington and Hancock counties together are the largest producer (~90%) of wild blueberries in Maine [27]. In this study, a total of 89 wild blueberry fields were considered, among which 69 fields were in Washington County, and 20 fields were in Hancock County (Figure 1a,b). Among all the wild blueberry fields of that region, 89 fields with 0.06 km^2 (250 m $\times$ 250 m) or larger areas were selected for this study. The land area threshold of 0.06 km^2 was used because the remote sensing data products that we used had a spatial resolution of 250 m. The selected wild blueberry fields are represented in red color in Figure 1a,b.

2.1.2. Airport and Baxter Wild Blueberry Fields in Deblois, Maine

The Airport and Baxter fields are adjacent to each other (Figure 1c). These two fields are part of the commercial blueberry fields owned by Jasper Wyman and Son in Deblois (longitude: −68.0001° N, latitude: 44.7350° W), Washington County, Maine, USA. In terms of agricultural management, these two fields are historically treated equally except for irrigation. The Airport field is irrigated during the growing season, whereas the Baxter field is non-irrigated. During the growing season from May to September, the Airport field was irrigated when needed with Nelson Full-Circle Impact sprinklers (Walla Walla, WA, USA) uniformly across the field. The irrigation system was set to ensure 0.5 to 1.0 inches of water supply per week by compensating natural precipitation. The area of the Airport field is 77 acres (0.31 km^2), and the Baxter field is 39 acres (0.16 km^2). These two fields were selected to understand and differentiate the effectiveness of current uniform irrigation practices in a single production region with the same climatic conditions.

2.2. Data Acquisition and Methodology

A Keyhole Markup Language Zipped (KMZ) file locating the wild blueberry fields of Maine was produced based on a field survey carried out by the University of Maine Cooperative Extension. The polygons of the 89 wild blueberry fields (area > 0.06 km^2) in the major wild blueberry production region, Maine (Figure 1a,b) were acquired from the KMZ file. The Airport and Baxter field polygons were acquired from a KMZ file based on a field survey by Jasper Wyman & Son, Deblois, Maine.

In this study, the SPEI was used as a drought index, which is calculated based on precipitation, temperature, and potential evapotranspiration data. The SPEI was chosen over another popular drought index, the PDSI, because the PDSI has a fixed temporal scale (between 9–12 months), which prevents the understanding of drought severity on different temporal scales [28]. In addition, the SPEI was chosen over the widely used SPI because the SPI only considers precipitation data and does not include air temperature and evapotranspiration data, which could also significantly influence the understanding of drought impacts on agriculture [19]. The SPEI's multi-scalar character enables it to detect, monitor, and analyze droughts more effectively, as it can quantify the drought severity according to its duration and intensity [19]. The SPEI allows the comparison of drought severity through time and space, since it can be calculated over a wide range of climates. In this study, the SPEI data were collected from the readily available open access database "SPEI Global Drought Monitor" (https://spei.csic.es/map/maps.html, accessed on 10 July 2020) in netcdf format. The netcdf files containing the SPEI data were transferred to ArcGIS Pro Version 2.7 (ESRI, Redlands, CA, USA) to acquire the SPEI data for the study sites (Airport, Baxter, and major wild blueberry region of Maine) over 71 years from 1950 to 2020, using the zonal statistics tool in ArcGIS Pro (Figure 2). The SPEI data were acquired on different temporal scales ranging from 1 month (SPEI_1) to 48 months (SPEI_48). These data were provided on a per-pixel basis at a 4 km spatial resolution. The SPEI (SPEI_6 of September) of only the growing season (April–September) was considered in this study. SPEI_6 of September represents the water conditions of a growing season (April–September). To understand the long-term (multi-year) impact of water conditions on the vegetation indices (EVI and NDVI) and yield of wild blueberry crops, the average

SPEI (SPEI_6 of September) of two, three, and four consecutive years was also calculated. A positive SPEI value represents wet conditions, whereas a negative SPEI value indicates dry conditions.

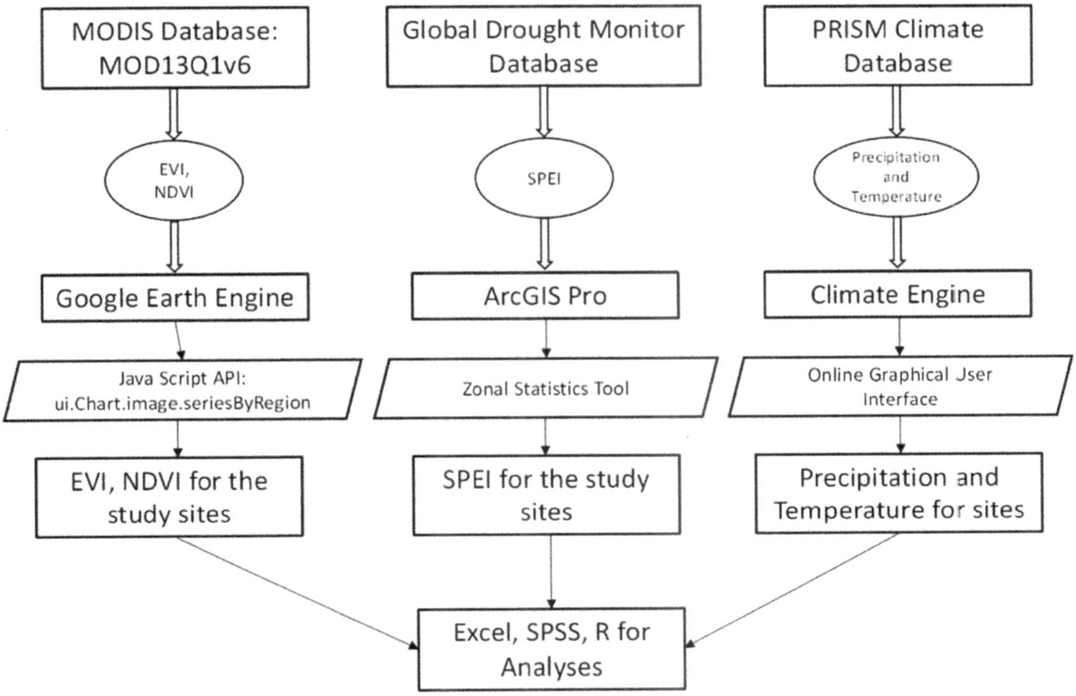

Figure 2. A flowchart showing the steps of data acquisition and analyses for this study.

The dataset of climate variables, such as total precipitation and mean temperature, during the growing season (May to September) over 71 years from 1950 to 2020 for the study sites (Airport/Baxter and the major wild blueberry region of Maine) was acquired from the online tool "Climate Engine" (https://clim-engine.appspot.com/climateEngine, accessed on 17 July 2021) of the Desert Research Institute, University of California, USA. Here, total precipitation refers to an average of monthly total precipitation (mm), and mean temperature refers to an average of the monthly mean air temperature at 2 m from the ground surface for the growing season. The original data sources for the climate variables were obtained from the AN81 m dataset of the PRISM Climate Group (https://prism.oregon-state.edu/explorer/, accessed on 17 July 2021). These data were provided on a per-pixel basis at a 4 km spatial resolution for the conterminous United States with a temporal resolution of one month (daily mean temperature and total precipitation were averaged monthly). This AN81 m dataset is available from January 1895. The extracted data were transferred into Microsoft Excel (Microsoft, Redmond, WA, USA) to calculate the average total precipitation and the mean temperature of the summer months (May to September) of each year (Figure 2).

In order to quantify vegetation responses to drought, satellite-based remotely sensed EVI and NDVI data for 21 years (2000 to 2020) of the studied wild blueberry fields were acquired from Google Earth Engine. These data were originally obtained from the Moderate Resolution Imaging Spectroradiometer (MODIS) product MOD13Q1 (https://lpdaac.usgs.gov/products/mod13q1v006, accessed on 25 July 2021). The MOD13Q1 dataset is preprocessed as well as readily and freely available. The MOD13Q1 Version 6 data have a spatial resolution of 250 m, generated every 16 days. For the development of the MOD13Q1 product, an algorithm was used to select pixels with low clouds, low view angle, and maximum index value to obtain the best available pixels over the 16-day-period image acquisitions [29]. The MOD13Q1 product has two vegetation layers: NDVI and EVI. The NDVI is the most common one used for characterizing canopy leaf chlorophyll content based on the reflectance contrast between the red and the near-infrared (NIR) wavebands [30]. However, the NDVI has some limitations, such as (1) it saturates in dense vegetation, (2) it does not consider the canopy background noise, and (3) its ratioing properties to eliminate noise [31]. These limitations were improved in the EVI to some extent, and thus the EVI has several advantages over the NDVI as it has improved sensitivity over high biomass regions [31]. This dataset is readily available in the Google Earth Engine. The vegetation indices values over the summer months were extracted for the study sites using a JavaScript-based API in the Google Earth Engine (https://code.earthengine.google.com/, accessed on 25 July 2021) using the extraction command "ui.Chart.image.seriesByRegion". The extracted data were transferred into Microsoft Excel (Microsoft, Redmond, WA, USA) to calculate the average EVI and NDVI of the summer months (May–September) of each year (Figure 2).

The historical yield data of Maine were collected from the United States Department of Agriculture (USDA), National Agricultural Statistics Service using a Quick Stats Ad-hoc Query Tool (https://quickstats.nass.usda.gov/, accessed on 26 July 2021). Historical yield data of the entire state of Maine (million lbs.) were available from 1924 to 2020, but the yield per production area data (lbs./acre) were only available from 2012 to 2020 (except 2013). It should be noted that the yield data were considered from all over the state of Maine, where ~90% of the yield was typically from the major wild blueberry production region (Washington and Hancock counties). The historical yield data of the Airport and Baxter fields at Deblois, Maine were provided by Jasper Wyman & Son, Maine. The yield (lbs./acre) data for the Airport and Baxter fields were available from 1993–2019 for every alternate year (except 2001 for the Baxter field).

2.3. Statistical Analysis

In this study, SPSS v23 (IBM Corp., Armonk, NY, USA), and RStudio software (RStudio, PBC, Vienna, Austria) were used for statistical analysis. Trend analyses of the climate variables (SPEI, Precipitation, Temperature) over the last 71 years (1950–2020) at the studied wild blueberry fields (Airport/Baxter and Maine) were conducted using a Mann–Kendall trend test and Sen's slope estimator using the XRealStats (Addinsoft, New York, NY, USA) add-on in Microsoft Excel. The "pheno" package in RStudio was used to analyze the forward (UF) and backward (UB) curves of the sequential Mann–Kendall test statistics. Trend analyses of the EVI and yield over the last 21 years (2000–2020) at the studied wild blueberry fields (Airport/Baxter and Maine) were conducted using a Mann–Kendall trend test and Sen's slope estimator using the XRealStats tool. To assess the statistical significance of the Mann-Kendall trend analysis, the significance level (α) was set to 0.05.

A Pearson correlation analysis was conducted between different temporal scales of SPEI and EVI and yield to understand the drought impact on vegetation (EVI) and yield at different temporal scales using SPSS v23. To assess the statistical significance of the Pearson correlation analysis, the significance level (α) was set to 0.05.

To understand the effects of short- to long-term water conditions (SPEI_1_Year to SPEI_4_Year) on the EVI and NDVI (average of the growing season: May–September) for the studied wild blueberry fields over 21 years (2000–2020), linear (in the form of a + bx) and non-linear (in the form a + bx + cx ^{2}) regression analyses were also conducted using SPSS v23. A similar analysis was conducted to understand the short- to long-term impact of water conditions (indicated by SPEI_1_Year to SPEI_4_Year) on the yield of the studied wild blueberry fields. We determined the statistical significance of the relationship using the coefficient of determination and its significance (α) at $p < 0.05$.

3. Results

3.1. Historical Changes in SPEI, Climate Variables, EVI, and Productivity of Wild Blueberry Systems in Maine, USA

During the last 71 years (1950–2020), the drought index (SPEI, Figure 3a,b) and precipitation (Figure 3c,d) tended to increase marginally (Figure 4a–d; Table 1) in the studied wild blueberry fields in Maine (Figure 3a,c and Figure 4a,c), as well as in two specific fields (Airport/Baxter) at Deblois, Maine (Figure 3b,d and Figure 4b,d). However, the mean atmospheric temperature increased significantly in the wild blueberry fields in Maine overall (Figure 3e,f; Table 1), and in the two fields in Deblois, ME. These patterns were also supported graphically by the upward UF curve (forward trend) mostly being >0.0 and UB (backward trend) curve mostly being <0.0 (Figure 4e,f).

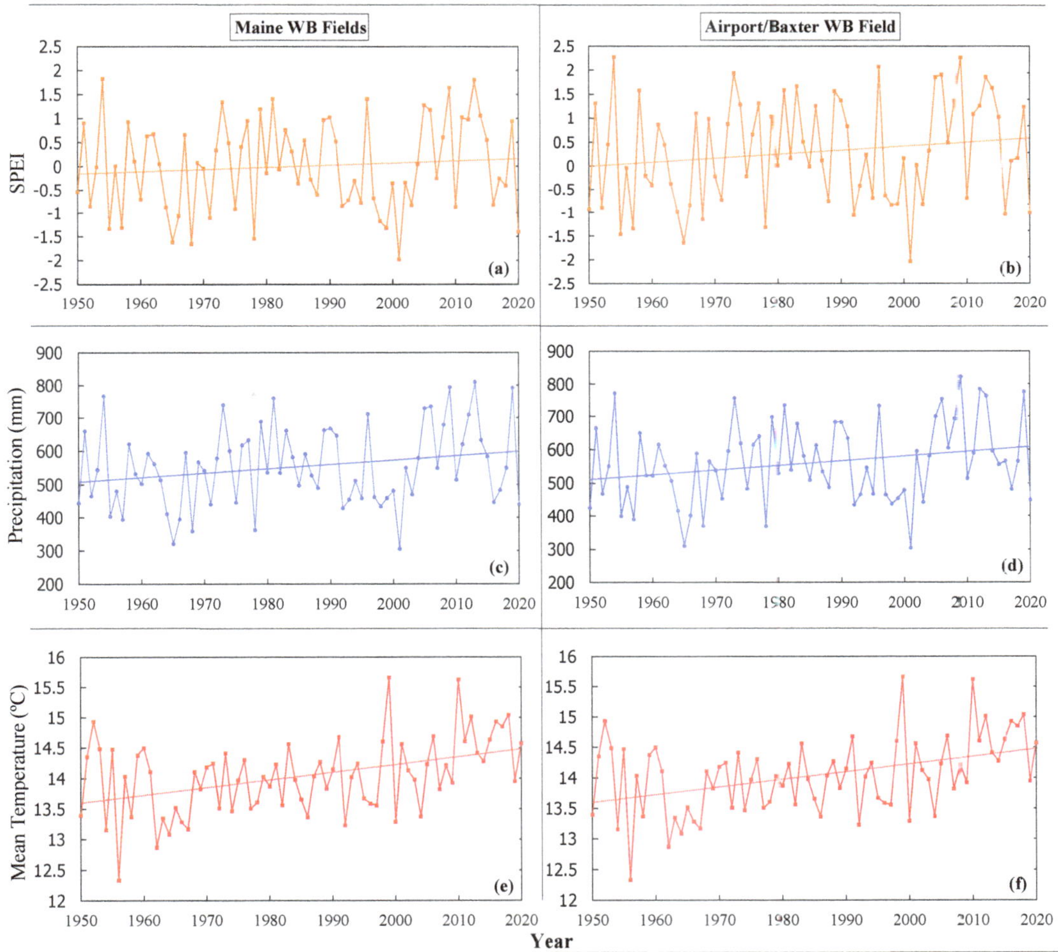

Figure 3. Historical (1950 to 2020) patterns of the (**a**,**b**) SPEI_6 of September; (**c**,**d**) mean precipitation (average of May–September); (**e**,**f**) average temperature (average of May–September) throughout the major wild blueberry production region in Maine as well as at the Airport/Baxter wild blueberry fields in Deblois, ME. A positive SPEI value represents wet conditions, while a negative SPEI value indicates dry conditions. Here, mean precipitation refers to an average of the monthly total precipitation (mm), and mean temperature refers to an average of the monthly air temperature at 2 m from the surface for the growing season.

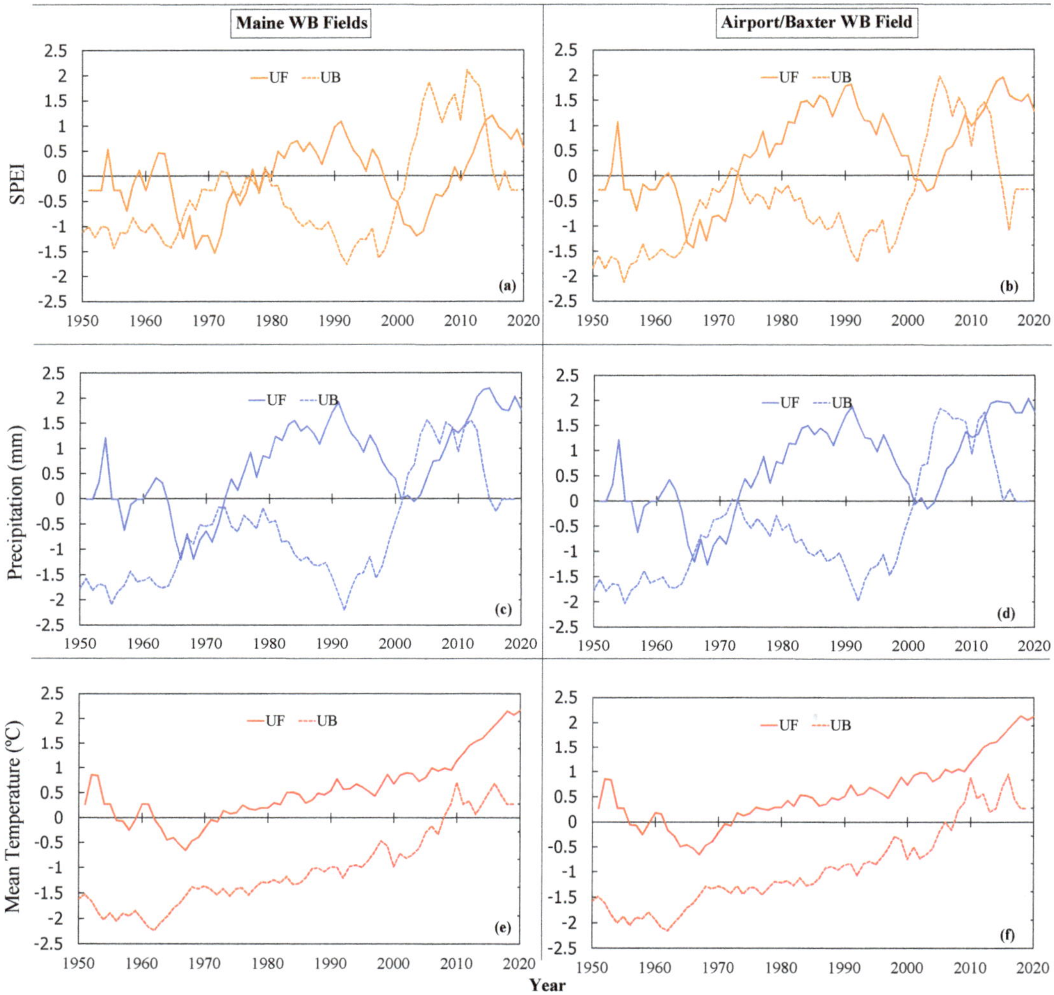

Figure 4. Sequential Mann–Kendall test statistics (UF and UB values) calculated from the (**a**,**b**) SPEI_6 of September; (**c**,**d**) mean precipitation (average of May–September); (**e**,**f**) average air temperature (average of May–September) throughout the major wild blueberry production region in Maine, as well as at the Airport/Baxter wild blueberry fields in Deblois, ME. Here, the upward UF curve (forward trend) mostly being > 0.0, UB (backward trend) curve mostly being < 0.0, and UF and UB not intersecting with each other indicate the significant increasing trends of the mean temperature. The intersections of the curves with the 0.0 line as well as with each other represent the non-significant changing (increasing/decreasing) trends of the SPEI and precipitation.

Based on the yield data from the crop years (every alternate year from 1993–2019), the wild blueberry yield of the Airport field (irrigated) had no significant change (; Figure 5a; Table 2). No significant changes in the EVI during the growing season (April–September) of the Airport field were detected over the last 21 years (2000–2020) (Figure 5d; Table 2). In contrast, both the yield (Figure 5b) and the EVI (Figure 5e) showed significant increments in the Baxter field (non-irrigated; Table 2). No significant changes in yield were observed from the studied wild blueberry fields of Maine over the last 21 years (2000–2020) (Figure 5c). A significant increase in the EVI during the growing season (April–September) was observed over the last 21 years for fields of Maine as a whole (Figure 5f; Table 2).

Table 1. Sequential Mann–Kendall trend analysis of the standardized precipitation-evapotranspiration index (SPEI), precipitation and mean temperature (T_{mean}) at different wild blueberry study zones: Airport/Baxter wild blueberry fields (Deblois, ME), and Maine wild blueberry fields (Washington and Hancock counties, ME). Here, the SPEI refers to SPEI_6 of September. It represents the SPEI of the growing season (April–September) and indicates water conditions and drought severity. T_{mean} represents the average air temperatures during the growing period (May–September).

Mann–Kendall Test	Maine WB Fields			Airport/Baxter, Deblois, ME Irrigated/Non-Irrigated		
	SPEI	**Precipitation**	T_{mean}	**SPEI**	**Precipitation**	T_{mean}
Kendall's Tau	0.062	0.144	0.276	0.114	0.144	0.270
Mann–Kendall Stat (S)	153.000	357.000	687.000	283.000	359.000	671.000
Var (S)	40,588.33	40,588.33	40,588.33	40,588.33	40,588.33	40,588.33
p-value (two-tailed)	0.45	0.07	0.001	0.16	0.07	0.001
Alpha	0.05	0.05	0.05	0.05	0.05	0.05
Trend	Increasing (Non-significant)	Increasing (Non-significant)	Increasing (Significant)	Increasing (Non-significant)	Increasing (Non-significant)	Increasing (Significant)
Sen's Slope Q	0.005	1.344	0.013	0.008	1.304	0.012

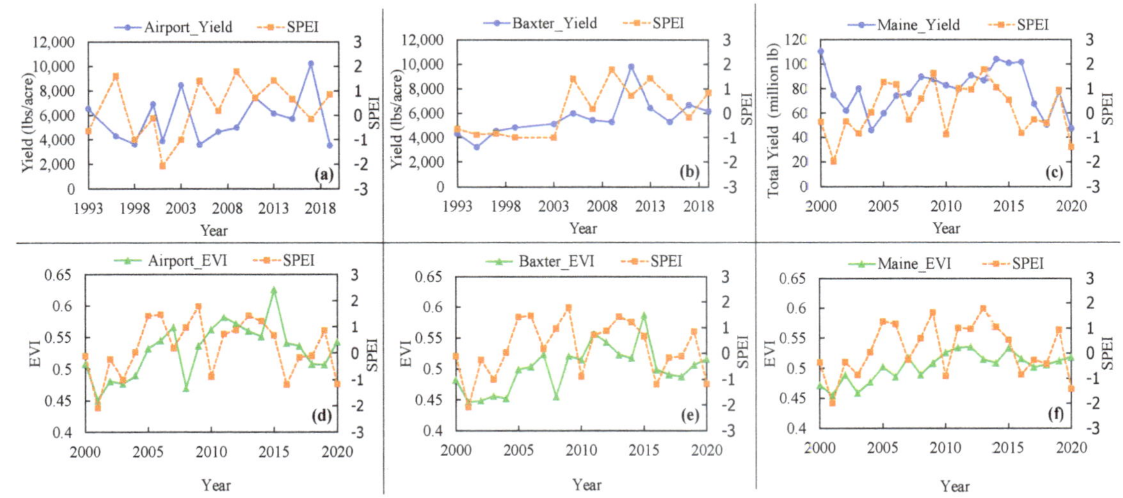

Figure 5. (**a**,**d**) Historical values of yield and SPEI ((**a**): 1993–2019), and of the EVI and SPEI ((**d**): 2000–2020) for the Airport (irrigated) field, Deblois, ME. (**b**,**e**) Historical values of yield and SPEI ((**b**): 1993–2019), and in the EVI and SPEI ((**e**): 2000–2020) for the Baxter (non-irrigated) field, Deblois, ME. (**c**,**f**) Historical values of yield and SPEI ((**c**): 2000–2020), and in the EVI and SPEI ((**f**): 2000–2020) for the major wild blueberry production region in Maine. Here, orange dashed lines indicate SPEI, blue solid lines indicate yield, green solid lines indicate EVI. Here, SPEI refers to SPEI_6 of September. It represents the SPEI of the growing season (April–September) and indicates water conditions. A positive SPEI value represents wet conditions, while a negative SPEI value indicates dry conditions.

Table 2. Sequential Mann–Kendall trend analysis of yield and enhanced vegetation index (EVI) at three different wild blueberry study zones: Airport (irrigated field, Deblois, ME), Baxter (non-irrigated field, Deblois, ME), and Maine wild blueberry fields (Washington County and Hancock County, ME).

Mann–Kendall Test	Airport, Deblois, ME (Irrigated Field)		Baxter, Deblois, ME (Non-Irrigated Field)		Maine WB Fields	
	Yield	EVI	Yield	EVI	Yield	EVI
Kendall's Tau	0.099	0.257	0.667	0.333	0.057	0.476
Mann–Kendall Stat (S)	9.000	54.000	52.000	70.000	12.000	100.000
Var (S)	333.667	1096.667	268.667	1096.667	1096.667	1096.667
p-value (two-tailed)	0.667	0.110	0.002	0.037	0.740	0.003
Alpha	0.050	0.050	0.050	0.050	0.050	0.050
Trend	Increasing (non-significant)	Increasing (non-significant)	Increasing (significant)	Increasing (significant)	Increasing (non-significant)	Increasing (significant)
Sen's Slope Q	54.10	0.003	89.91	0.003	0.301	0.003

3.2. Relationships between SPEI and Vegetation Indices in Wild Blueberry Fields of Maine

Based on the relationships of both the short-term and long-term average SPEI with the EVI and NDVI (Figures 6 and 7, and Table S1), the long-term SPEI showed a stronger influence on the vegetation indices (EVI in Figure 6 and NDVI in Figure 7) of wild blueberries compared to the short-term SPEI. While analyzing the impact of the short-term SPEI (SPEI_1_Year in Figures 6 and 7, and SPEI_1 to SPEI_11 in Table S1) on the EVI (Figure 6a–c) and NDVI (Figure 7a–c) during the growing season (May–September), no significant relationship was observed for the studied wild blueberry fields in Maine.

On the contrary, while observing the impact of the long-term SPEI (2 to 4 consecutive years) on both the EVI and NDVI of the wild blueberry fields during the growing season, we found both significant linear and quadratic relationships between the SPEI and the EVI (Figure 6d–l) as well as between the SPEI and the NDVI (Figure 7d–l). Among the significant linear and quadratic relationships between an average SPEI of 2 consecutive years (SPEI_2_Year) and vegetation indices for the Airport (Figures 6d and 7d), Baxter (Figures 6e and 7e), and studied wild blueberry fields in Maine (Figures 6f and 7f), the coefficient of determination (R^2) was higher for the quadratic relationships. Moreover, the strength (R^2 values) of both the linear and quadratic relationships was higher when considering more consecutive years, such as SPEI_3_Year (Figures 6g–i and 7g–i) and SPEI_4_Year (Figures 6j–l and 7j–l) compared to the SPEI_2_Year (Figures 6d–f and 7d–f). Although both the relationships between the SPEI and EVI (Figure 6) as well as the SPEI and NDVI (Figure 7) were significant when considering the long-term SPEI, the coefficient of determination (R^2) was higher for the relationships between the SPEI and EVI compared to the relationships between the SPEI and NDVI. Because of the stronger relationship between the SPEI and EVI, we further analyzed the impact of the short- and long-term water conditions (SPEI) on wild blueberry yield, as well as the influence of monthly water conditions (SPEI) during the growing season on EVI and yield. Interestingly, when considering the impact of the monthly SPEI (different temporal SPEI in Table S1) during the growing season, the SPEI of the early season (April–June) showed more impacts on the EVI of the wild blueberry fields compared to the SPEI later in the season (July–August).

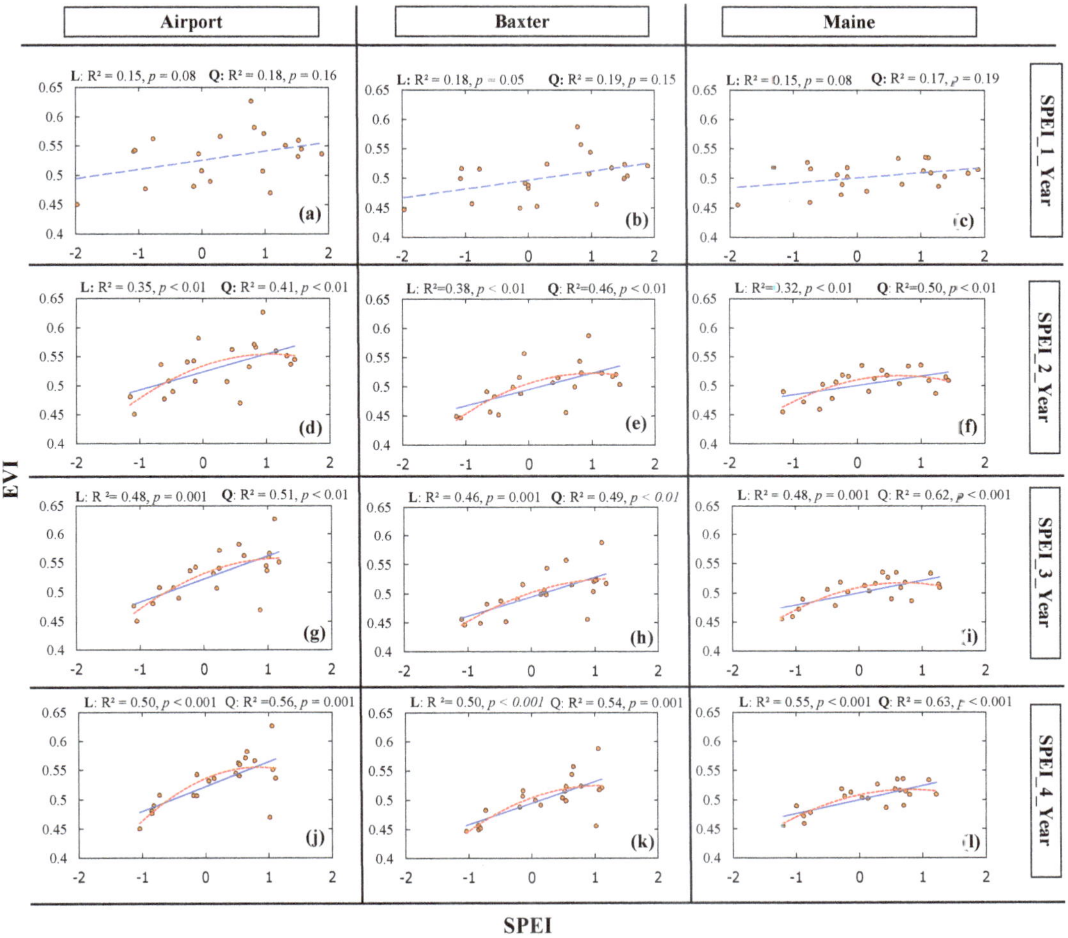

Figure 6. Average enhanced vegetation index (EVI) of wild blueberry fields during the growing season (May to September) for three different study zones: Airport (irrigated), Baxter (non-irrigated), and the major wild blueberry production region in Maine in relation to (**a–c**) SPEI_1_Year; (**d–f**) SPEI_2_Years (average SPEI of two consecutive years); (**g–i**) SPEI_3_Years (average SPEI of three consecutive years); (**j–l**) SPEI_4_Years (average of SPEI of four consecutive years). Here, SPEI refers to SPEI_6 of September and it represents the SPEI (water conditions) of the growing season (April–September). A positive SPEI value represents wet conditions, while a negative SPEI value indicates dry conditions. The blue solid lines indicate significant ($p < 0.05$) and blue dashed lines indicate marginally significant ($p < 0.10$) linear relationships. The dashed red lines indicate significant ($p < 0.05$) or marginally significant ($p < 0.10$) quadratic relationships. The time period of the EVI and SPEI data was from 2000 to 2020.

3.3. Relationships between SPEI and Yield of Wild Blueberry Fields in Maine

The impacts of the short-term and long-term SPEI on the wild blueberry yield (Figure 8 and Table S2) were different from the relationships between the SPEI and EVI during the growing season (April–September) (Figure 6 and Table S1). A significant and positive linear relationship was found between the short-term SPEI (SPEI_1_Year) and yield for the non-irrigated Baxter field (Figure 8b), whereas the relationship between the short-term SPEI (SPEI_1_Year) and yield was non-significant at the 95% confidence level (marginally significant at the 90% confidence level, $p = 0.058$) for the irrigated Airport field (Figure 8a). For the wild blueberry fields in Maine as a whole, we found a marginally significant ($p < 0.1$)

and positive linear relationship between the short-term drought index (SPEI_1_Year) and the wild blueberry yield (Figure 8c). We found a significant quadratic relationship between the short-term SPEI (SPEI_1_Year) and yield for the non-irrigated Baxter field (Figure 8b), but not for the irrigated Airport field or the studied wild blueberry fields of Maine as a whole. When considering the impact of monthly water conditions (different temporal SPEI in Table S2) during the growing season, the correlation between the SPEI and yield was significant for the non-irrigated Baxter field, whereas it was not significant for the irrigated Airport field.

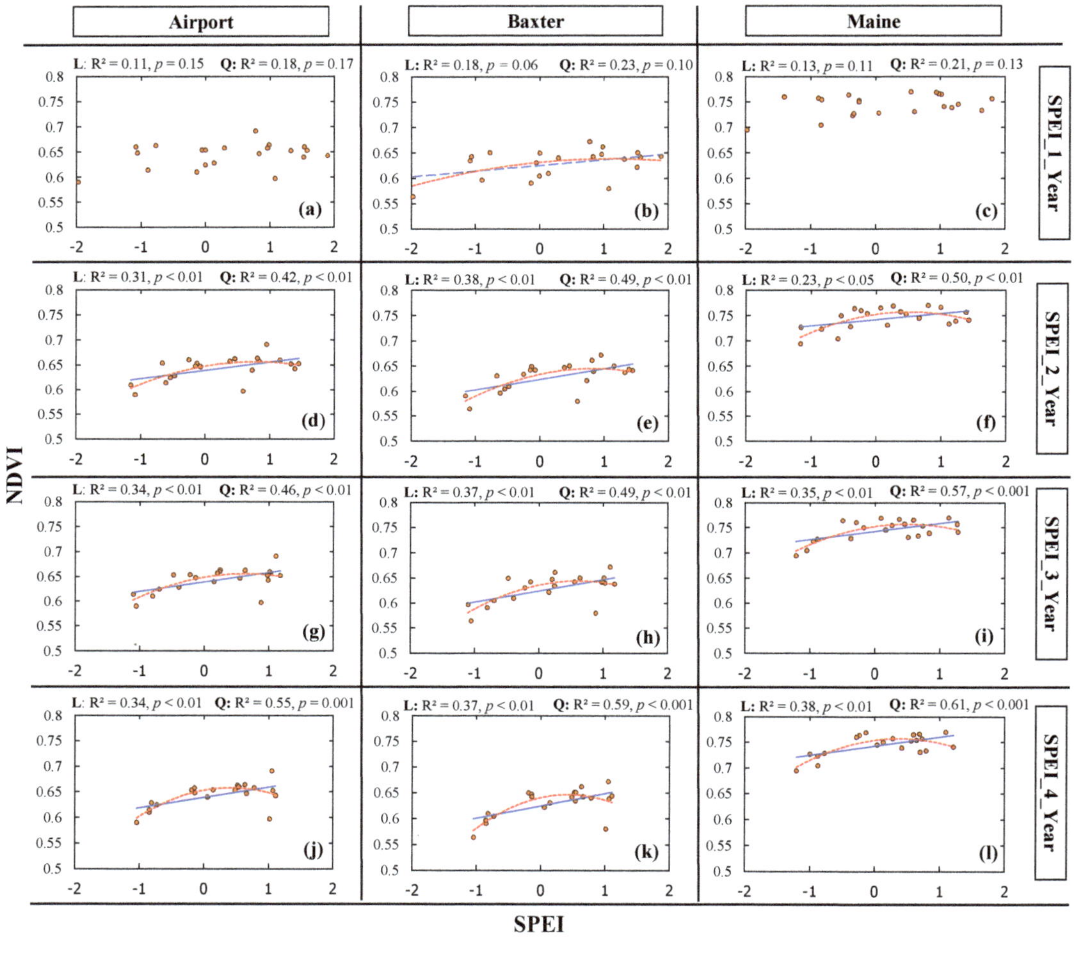

Figure 7. Average normalized difference vegetation index (NDVI) of wild blueberry fields during the growing season (May to September) for three different study zones: Airport (irrigated), Baxter (non-irrigated), and the major wild blueberry production region in Maine in relation to (**a–c**) SPEI_1_Year; (**d–f**) SPEI_2_Years (average SPEI of two consecutive years); (**g–i**) SPEI_3_Years (average SPEI of three consecutive years); (**j–l**) SPEI_4_Years (average SPEI of four consecutive years). Here, SPEI refers to SPEI_6 of September and it represents the SPEI (water conditions) of the growing season (April–September). A positive SPEI value represents wet conditions, whereas a negative SPEI value indicates dry conditions. The blue solid lines indicate significant ($p < 0.05$) and blue dashed lines indicate marginally significant ($p < 0.10$) linear relationships. The dashed red lines indicate significant ($p < 0.05$) or marginally significant ($p < 0.10$) quadratic relationships. The time period of the EVI and SPEI data was from 2000 to 2020.

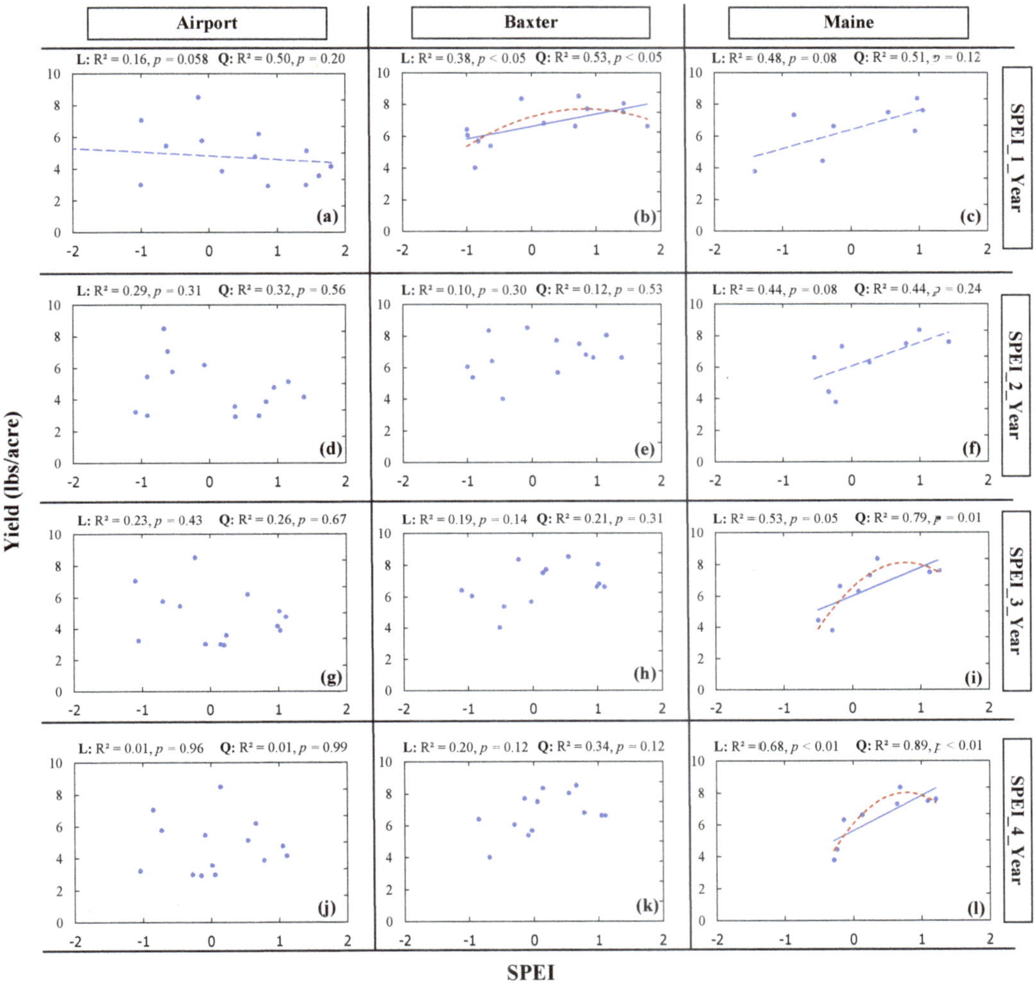

Figure 8. Average yield (lbs./acre) in a year for three different study zones: Airport, Baxter, and the major wild blueberry production region in Maine in relation to (**a–c**) SPEI_1_Year; (**d–f**) SPEI_2_Years (average SPEI of two consecutive years); (**g–i**) SPEI_3_Years (average SPEI of three consecutive years); (**j–l**) SPEI_4_Years (average of SPEI of four consecutive years). The SPEI data represent SPEI_6 of September. Here, SPEI refers to SPEI_6 of September and it represents the SPEI (water conditions) of the growing season (April–September). A positive SPEI value represents wet conditions, while a negative SPEI value indicates dry conditions. The numbers mentioned in the *Y*-axis were shortened by dividing the yield (lbs./acre) by 1000. The blue solid lines indicate significant ($p < 0.05$) and blue dashed lines indicate marginally significant ($p < 0.10$) linear relationships. The dashed red lines indicate significant ($p < 0.05$) or marginally significant ($p < 0.10$) quadratic relationships. The time period of the EVI and SPEI data was from 2000 to 2020.

While analyzing the impact of the long-term SPEI (2 to 4 consecutive years) on the yield of the irrigated Airport field and the non-irrigated Baxter field, no significant linear or quadratic relationships were found (SPEI_2_Year in Figure 8d,e; SPEI_3_Year in Figure 8g,h; SPEI_4_Year in Figure 8j,k). However, stronger relationships were observed between the long-term SPEI during the growing season and yield when considering the wild blueberry fields of Maine as a whole (Figure 8i–l). We found significant positive linear and quadratic relationships between the average yield of the wild blueberry fields in Maine and the long-term SPEI (Figure 8i–l), except for the SPEI_2_Year (Figure 8f), where the linear relationship was marginally significant ($R^2 = 0.36$, $p = 0.08$) and the quadratic relationship was not significant. In fact, the quadratic relationships were stronger between yield and the average SPEI of 3 and 4 consecutive years (Figure 8i–l). Moreover, while considering the cumulative impacts for more consecutive years, both the linear and quadratic relationships were observed to be stronger for the wild blueberry fields in Maine as a whole (Figure 8f,i,l).

3.4. Relationships between Vegetation Indices and Productivity

While comparing the influences of the vegetation indices (EVI and NDVI) on the yield of the irrigated Airport field and the non-irrigated Baxter field, no significant relationship was found between the yield and growing season EVI and NDVI for the Airport and Baxter wild blueberry fields during the prune and crop year (Figures 9 and 10). The only significant correlation was found between the mean EVI of the prune year and crop year for the Airport field and its yield when fitted with the quadratic relationship ($R^2 = 0.65$, $p = 0.03$), whereas the correlation between the mean NDVI of the prune year and crop year for the Airport field and its yield was non-significant.

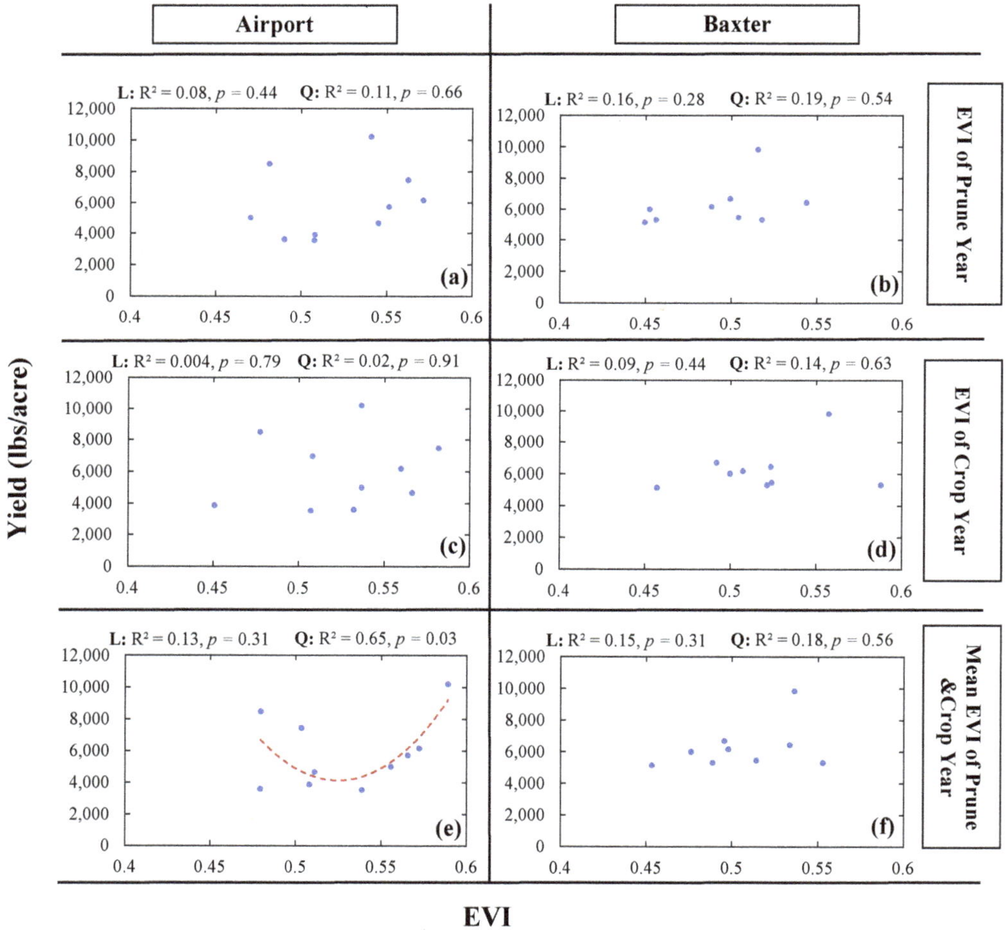

Figure 9. Relationship between wild blueberry yield (lbs./acre) and the average enhanced vegetation index (EVI) of (**a**,**b**) the prune year, (**c**,**d**) the crop year, and (**e**,**f**) the average of the prune and crop year from the Airport (irrigated) and Baxter (non-irrigated) fields in Deblois, Maine. The dashed red lines indicate significant ($p < 0.05$) or marginally significant ($p < 0.10$) quadratic relationships.

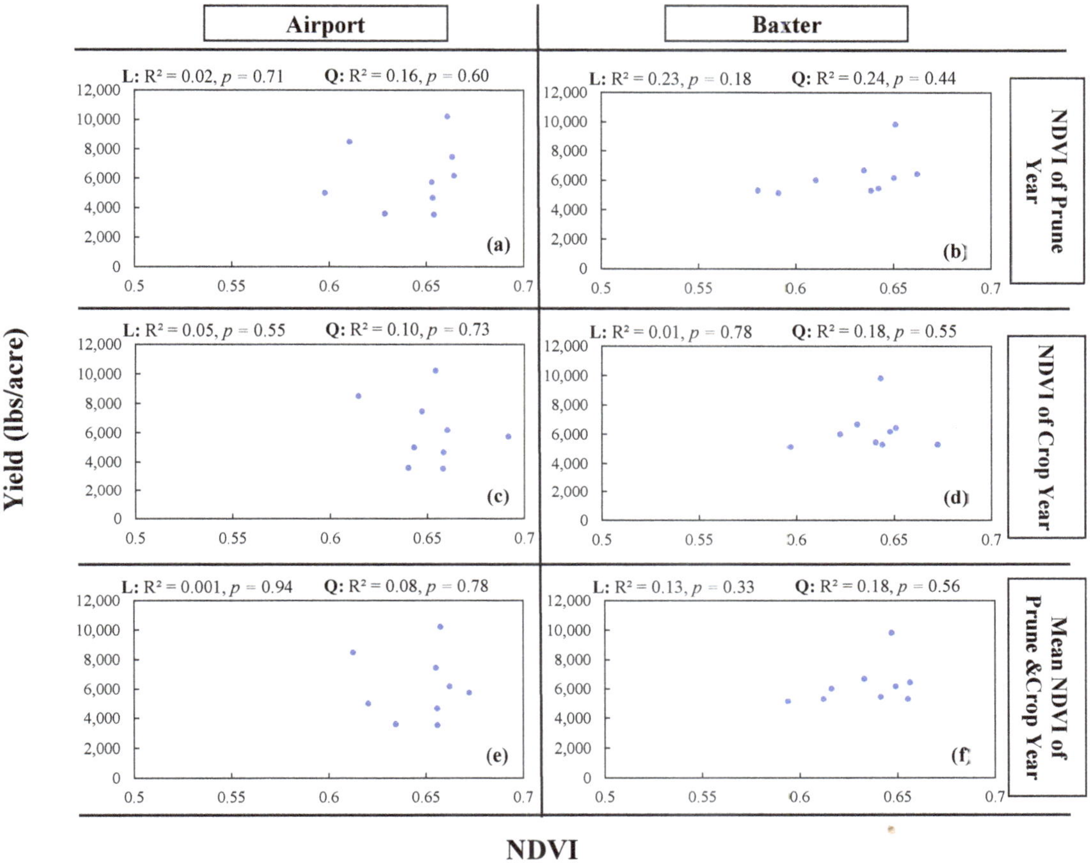

Figure 10. Relationship between wild blueberry yield (lbs./acre) and the average normalized difference vegetation index (NDVI) of (**a**,**b**) the prune year, (**c**,**d**) the crop year, and (**e**,**f**) the average of the prune and crop year from the Airport (irrigated) and Baxter (non-irrigated) fields in Deblois, Maine.

4. Discussion

Our study revealed that, despite significant warming in the past century, there were no significant changes in the drought patterns and drought impacts on the wild blueberry fields of Maine in the past 71 years. We also found that the water conditions (dry or wet, as indicated by the SPEI) in the growing season have significant impacts on wild blueberry vegetation vigor (as indicated by the NDVI and EVI) and production. The long-term water conditions (the long-term average SPEI) have substantial significant impacts on wild blueberry crop vegetation vigor (vegetation indices: NDVI and EVI) as well as their production (yield) in Maine, rather than the water conditions (SPEI) of the current growing season. The impact of the water conditions on the vegetation indices was more consistent and significant compared to the impact on yield. Interestingly, we also found that the water conditions of the early growing season (April–June) might decide the fate of crop vegetation vigor and production of wild blueberry later in the season (July–August). We further found that, in terms of vegetation status, water conditions had little impact on the irrigated field. The water conditions indicated by the SPEI had no impact on the yield of the irrigated field, suggesting irrigation effectively alleviated the impact of water deficits on the yield of wild blueberries. Based on our analyses, we also found that satellite-based vegetation indices (NDVI and EVI) cannot be used to predict wild blueberry crop production. However, several previous studies found significant correlations between high-resolution spectroradiometer-based vegetation indices and yield in different crops, i.e., maize, wheat, and soybean [32–34]. This could be because the yield of the wild blueberry crop is more affected by other factors such as pollination rather than the vegetation vigor. Moreover, further research could be carried out to test using drone-based high-resolution data to predict the yield of wild blueberries.

The absence of an increasing trend of the drought index SPEI in the wild blueberry fields could be associated with a lack of significant change in precipitation patterns during the growing season [6,15,35]. Although the atmospheric temperatures increased significantly in this region and in the wild blueberry fields in the past century [4–6,10,11], the warming pattern and subsequently increased evapotranspiration [6] have not resulted in a significant increase in drought impact. The studied fields are in a temperate climate region and they experience relatively low temperatures. The increase in evapotranspiration due to warming in this region has possibly not pushed the ecosystems here into the range of severe water deficits.

The water conditions of a relatively long period (SPEI of more than 11–12 months) showed significant and substantial impacts on vegetation vigor and the yield of wild blueberry crops. This could be because wild blueberries are a crop with large perennial underground stem systems called rhizomes, which can store sugar and nutrients [12–14,25], and their health and yield could mostly be determined by the sugar accumulation of previous years and not only that of the current growing season. Although the aboveground parts of the wild blueberries are pruned to the ground every two years, the belowground rhizomes and roots remain for a long time. As a result, the sugar stored underground could govern the effect of precipitation on the crop over the long term [36,37]. The wild blueberry crop requires only an inch of water per week [38] and is regarded as a drought-resistant crop [13,14]. This could be because of the large water and sugar storage in their underground tissues. The underground storage may weaken the effect of current year water conditions on crop health and yield.

The water conditions certainly affect the vegetation status and vigor of wild blueberries. The vegetation greenness or vigor of wild blueberries during both prune and crop years is affected by atmospheric temperature and precipitation during the growing period [6]. Also, precipitation directly affect the soil moisture availability to crops [39]. Furthermore, soil moisture availability affect the nitrogen uptake and accumulation in plants, which consequently determines leaf photosynthetic capacity [40], growth and yield of crops. However, a previous study on the wild blueberries in Eastern Canada found no correlations between the climate variables of that region and wild blueberry yield [41]. Further studies and analyses are needed to establish high-resolution relationships among climate variables, vegetation vigor, and yield.

Here, the vegetation indices (EVI and NDVI) are not good predictors of the yield of wild blueberries. This could be because, besides vegetation status, wild blueberry yield during the crop year is affected by many other important factors [42], such as pollination, insect pests, weeds, and pathogens. Though it has been found that vegetation indices are strongly correlated with yield in some crops [34], it might not be the case for wild blueberries. Vegetation indices are correlated with leaf chlorophyll content and photosynthesis capacity [43], and might be related to the number of developed flower buds [44]. However, there are a lot of other factors such as pollinator activity, weed coverage, and fruit set ratio, which are important in determining yield but can not be predicted by vegetation indices.

Water conditions (indicated by the SPEI) during the early growing season (April–June) have a larger impact on the vegetation status and yield of wild blueberry crops compared to that of the later growing season (July–September). This could be related to pollination. Precipitation intensity and frequencies, along with temperature and wind velocity, during the pollination period (April–May) in crop year would affect the bee pollination, which significantly affects the wild blueberry yield in July and August [42,45–47]. In addition, the availability of resources such as soil moisture and nutrients [42], determined by the precipitation and temperature [38], during fruit set and maturation (May–June) right after the pollination period ends, decides the fate of the final fruit production (July–August) [45–48].

Irrigation decoupled the relationship between the climatic water condition (SPEI) and yield. The positive relationship between SPEI and yield found in the non-irrigated Baxter field was not found in the irrigated Airport field, despite both fields being in the same location with same management practices (except irrigation). The positive correlation between the SPEI and the yield of the non-irrigated Baxter field suggests the importance of water conditions in determining yield and the need for effective irrigation practices to alleviate the impact of drought. The fields of the major wild blueberry region (which are mostly non-irrigated) showed similar patterns to the non-irrigated Baxter field. Thus, it suggests that the introduction of effective irrigation management practices might be useful to enhance the production of wild blueberries by mitigating drought. Additionally, wild blueberries respond more positively to precipitation frequency rather than total precipitation volume over the growing season [49]. Irrigation also decoupled the relationship between the SPEI and vegetation indices (EVI and NDVI), suggesting the positive effect of irrigation in mitigating the drought effects on vegetation vigor for wild blueberries. Meanwhile, the quadratic relationships between the SPEI and vegetation indices, as well as the SPEI and yield, suggest that when the optimum precipitation or water supply is reached, further increases in the water supply may have a negative effect on crop vigor and yield. Similar results were also reported between the EVI and precipitation from the wild blueberry fields in Downeast, Maine [6]. Hence, no overall significant differences in the vegetation indices or yield were observed between the irrigated Airport field and the non-irrigated Baxter field, but in drought years (e.g., 2003), the yield and EVI of the irrigated field were higher than that of the non-irrigated field.

5. Conclusions

Overall, our study suggests that though the temperature has been increasing significantly in the major wild blueberry production region of Maine, drought has not been increasing significantly over the last 71 years. However, accelerated warming and a projected decrease in soil water content [7] may result in an increase in drought impact on agricultural systems in the future [48]. The water conditions and drought severity quantified by the drought index SPEI had a stronger impact on the vegetation status of the non-irrigated field compared to the irrigated field. The short-term (one year) SPEI was positively related to the yield of the non-irrigated field, whereas no significant correlation was found for the irrigated field, suggesting the sensitivity of wild blueberry yield to water conditions and the effectiveness of irrigation. However, maintaining optimum soil moisture is a challenge due to the high spatial variability in soil water retention capacity in wild blueberry fields. Therefore, developing a precision irrigation system could be an efficient way to mitigate the effects of water deficits. Interestingly, we found that long-term water conditions determine the vegetation vigor and yield more than the short-term water conditions for wild blueberries. Thus, although the wild blueberry is regarded as a drought-tolerant species, maintaining good water conditions in the field during the growing season is important for securing a high yield for wild blueberries.

Supplementary Materials: The following are available online at https://www.mdpi.com/article/10.3390/cli9120178/s1: Table S1. Pearson correlation analysis between average enhanced vegetation index (EVI) of the growing season (May–September) and different scales of the SPEI from May to September at three different wild blueberry study zones: Airport (irrigated field, Deblois, ME), Baxter (non-irrigated field, Deblois, ME), and Downeast, Maine (all wild blueberry fields). Table S2. Pearson correlation analysis between average yield per year and different scales of the SPEI from May to September at three different wild blueberry study zones: Airport (irrigated field, Deblois, ME), Baxter (non-irrigated field, Deblois, ME), and Maine (all wild blueberry fields).

Author Contributions: Conceptualization, Y.-J.Z. and K.B.; methodology, K.B., R.T. and P.R.-B.; software, K.B.; validation, Y.-J.Z., K.B. and R.T.; formal analysis, K.B.; investigation, K.B.; resources, Y.-J.Z. and B.H.; data curation, K.B.; writing—original draft preparation, R.T. and K.B.; writing—review and editing, R.T., Y.-J.Z., P.R.-B., B.H. and K.B.; visualization, R.T. and Y.-J.Z.; supervision, Y.-J.Z.; project administration, Y.-J.Z.; funding acquisition, Y.-J.Z. All authors have read and agreed to the published version of the manuscript.

Funding: This project was supported by the USDA National Institute of Food and Agriculture, Hatch Project Number ME0-21832 and ME0-22021, through the Maine Agricultural and Forest Experiment Station. This research was also supported by the Wild Blueberry Commission of Maine, the Maine Department of Agriculture, Conservation and Forestry (SCBGP), and the UMaine Faculty Summer Research Award.

Data Availability Statement: Please refer to "Section 2.2" in this article for the sources of the publicly archived data products used in this study. Any data and codes used in this study are available upon request from the corresponding author (yongjiang.zhang@maine.edu).

Acknowledgments: We would like to acknowledge David Yarborough, Horticulture and Wild Blueberry Specialist, the University of Maine, for providing us with the KMZ file containing the wild blueberry field polygons in Maine to conduct this study.

Conflicts of Interest: The authors declare no conflict of interest.

References

1. Lesk, C.; Rowhani, P.; Ramankutty, N. Influence of extreme weather disasters on Global Crop Production. *Nature* **2016**, *529*, 84–87. [CrossRef]
2. Wilhelmi, O.V.; Wilhite, D.A. Assessing vulnerability to agricultural drought: A Nebraska case study. *Nat. Hazards* **2002**, *25*, 37–58. [CrossRef]
3. Solomon, S.; Manning, M.; Marquis, M.; Qin, D. *Climate Change 2007—The Physical Science Basis: Working Group I Contribution to the Fourth Assessment Report of the IPCC*; Cambridge University Press: Cambridge, UK, 2007; Volume 4.
4. Fernandez, I.; Birkel, S.; Schmitt, C.; Simonson, J.; Lyon, B.; Pershing, A.; Stancioff, E.; Jacobson, G.; Mayewski, P. Maine's Climate Future 2020 Update. Available online: Climatechange.umaine.edu/climate-matters/maines-climate-future/ (accessed on 17 May 2020).
5. NOAA–CAAG (National Oceanic and Atmospheric Administration–National Centers for Environmental Information). Climate at a Glance, US Climate Divisional Database. 2019. Available online: Ncdc.noaa.gov/cag/statewide/time-series (accessed on 17 May 2020).
6. Tasnim, R.; Drummond, F.; Zhang, Y.-J. Climate change patterns of wild blueberry fields in Downeast, Maine over the past 40 Years. *Water* **2021**, *13*, 594. [CrossRef]
7. Dai, A. Increasing drought under global warming in observations and Models. *Nat. Clim. Chang.* **2012**, *3*, 52–58. [CrossRef]
8. Carrão, H.; Naumann, G.; Barbosa, P. Global projections of drought hazard in a warming climate: A prime for Disaster Risk Management. *Clim. Dyn.* **2017**, *50*, 2137–2155. [CrossRef]
9. Sweet, S.K.; Wolfe, D.W.; DeGaetano, A.; Benner, R. Anatomy of the 2016 drought in the Northeastern United States: Implications for agriculture and water resources in humid climates. *Agric. For. Meteorol.* **2017**, *247*, 571–581. [CrossRef]
10. Wolfe, D.W.; DeGaetano, A.T.; Peck, G.M.; Carey, M.; Ziska, L.H.; Lea-Cox, J.; Kemanian, A.R.; Hoffmann, M.P.; Hollinger, D.Y. Unique challenges and opportunities for northeastern US crop production in a changing climate. *Clim. Chang.* **2017**, *146*, 231–245. [CrossRef]
11. Schattman, R.E.; Goossen, C.; Calderwood, L. *The 2020 Maine Drought and Agriculture Report*; University of Maine: Orono, ME, USA, 2021; pp. 1–30. [CrossRef]
12. About Maine Wild Blueberry. Available online: https://extension.umaine.edu/blueberries/about-2/ (accessed on 10 October 2021).
13. Glass, V.M. Drought Stress Effects in Wild Blueberries. Master's Thesis, Dalhousie University, Halifax, NS, Canada, 2000; 164p.
14. Percival, D.; Murray, A.; Stevens, D. Drought stress dynamics of wild blueberry (*Vaccinium angustifolium aiton*). *Acta Hortic.* **2003**, *618*, 353–362. [CrossRef]
15. Potop, V.; Možný, M.; Soukup, J. Drought evolution at various time scales in the lowland regions and their impact on vegetable crops in the Czech Republic. *Agric. For. Meteorol.* **2012**, *156*, 121–133. [CrossRef]
16. Illston, B.G.; Basara, J.B. Analysis of short-term droughts in Oklahoma. *Eos Trans. Am. Geophys. Union* **2003**, *84*, 157–161. [CrossRef]
17. Palmer, W.C. Meteorological droughts. U.S. Department of Commerce. *Weather Bur. Res. Paper* **1965**, *45*, 58.
18. McKee, T.B.; Doesken, N.J.; Kleist, J. The relationship of drought frequency and duration to time scales. In Proceedings of the 8th Conference on Applied Climatology, Anaheim, CA, USA, 17–22 January 1993; pp. 179–184.

19. Vicente-Serrano, S.M.; Beguería, S.; López-Moreno, J.I. A multiscalar drought index sensitive to global warming: The standardized precipitation Evapotranspiration index. *J. Clim.* **2010**, *23*, 1696–1718. [CrossRef]
20. Zarei, A.R.; Moghimi, M.M. Modified version for Spei to evaluate and modeling the agricultural drought severity. *Int. J. Biometeorol.* **2019**, *63*, 911–925. [CrossRef]
21. De Keersmaecker, W.; Lhermitte, S.; Hill, M.; Tits, L.; Coppin, P.; Somers, B. Assessment of regional vegetation response to climate anomalies: A case study for australia using Gimms Ndvi Time Series between 1982 and 2006. *Remote Sens.* **2017**, *9*, 34. [CrossRef]
22. Pettorelli, N.; Vik, J.O.; Mysterud, A.; Gaillard, J.-M.; Tucker, C.J.; Stenseth, N.C. Using the satellite-derived NDVI to assess ecological responses to environmental change. *Trends Ecol. Evol.* **2005**, *20*, 503–510. [CrossRef]
23. Son, N.T.; Chen, C.F.; Chen, C.R.; Minh, V.Q.; Trung, N.H. A comparative analysis of multitemporal MODIS EVI and NDVI data for large-scale rice yield estimation. *Agric. For. Meteorol.* **2014**, *197*, 52–64. [CrossRef]
24. Glenn, E.; Huete, A.; Nagler, P.; Nelson, S. Relationship between remotely-sensed vegetation indices, canopy attributes and plant physiological processes: What vegetation indices can and cannot tell us about the landscape. *Sensors* **2008**, *8*, 2136–2160. [CrossRef]
25. Drummond, F.; Smagula, J.; Annis, S.; Yarborough, D. Organic wild blueberry production. *MAFES Bull* **2009**, *852*, 43.
26. Arguez, A.; Durre, I.; Applequist, S.; Squires, M.; Vose, R.; Yin, X.; Bilotta, R. NOAA's US Climate Normals (1981–2010). *Bull. Am. Meteorol. Soc.* **2012**, *93*, 1687–1697. [CrossRef]
27. Maine Wild Blueberry Production Statistics. Available online: https://extension.umaine.edu/blueberries/factsheets/statistics-2/crop-production-statistics-2019/ (accessed on 10 October 2021).
28. Guttman, N.B. Comparing the palmer drought index and the standardized precipitation index. *J. Am. Water Resour. Assoc.* **1998**, *34*, 113–121. [CrossRef]
29. Didan, K. MOD13Q1: MODIS/Terra vegetation indices 16-day L3 global 250m grid SIN V006. NASA EOSDIS Land Processes DAAC. Available online: https://lpdaac.usgs.gov/products/mod13q1v006/ (accessed on 10 October 2021).
30. Lu, S.; Lu, F.; You, W.; Wang, Z.; Liu, Y.; Omasa, K. A robust vegetation index for remotely assessing chlorophyll content of dorsiventral leaves across several species in different seasons. *Plant Methods* **2018**, *14*, 1–15. [CrossRef] [PubMed]
31. Zhengxing, W.; Chuang, L.; Alfredo, H. From AVHRR-NDVI to MODIS-EVI: Advances in vegetation index research. *Acta Ecol. Sin.* **2003**, *23*, 979–987.
32. Teal, R.K.; Tubana, B.; Girma, K.; Freeman, K.W.; Arnall, D.B.; Walsh, O.; Raun, W.R. In-season prediction of corn grain yield potential using normalized difference vegetation index. *Agron. J.* **2006**, *98*, 1488–1494. [CrossRef]
33. Marti, J.; Bort, J.; Slafer, G.A.; Araus, J.L. Can wheat yield be assessed by early measurements of normalized difference vegetation index? *Ann. Appl. Biol.* **2007**, *150*, 253–257. [CrossRef]
34. Bolton, D.K.; Friedl, M.A. Forecasting crop yield using remotely sensed vegetation indices and crop phenology metrics. *Agric. For. Meteorol.* **2013**, *173*, 74–84. [CrossRef]
35. Belayneh, A.; Adamowski, J.; Khalil, B.; Ozga-Zielinski, B. Long-term SPI drought forecasting in the Awash River basin in Ethiopia using wavelet neural network and wavelet support vector regression models. *J. Hydrol.* **2014**, *508*, 418–429. [CrossRef]
36. Keyantash, J.; Dracup, J.A. The quantification of drought: An evaluation of drought indices. *Bull. Am. Meteorol. Soc.* **2002**, *83*, 1167–1180. [CrossRef]
37. Easterling, D.R.; Wallis, T.W.; Lawrimore, J.H.; Heim, R.R. Effects of temperature and precipitation trends on U.S. drought. *Geophys. Res. Lett.* **2007**, *34*, 20. [CrossRef]
38. Hunt, J.; Honeycutt, W.; Yarborough, D.; Starr, G. Guide to efficient irrigation of the wild blueberry. *Wild Blueberry Fact Sheet* **2009**, *631*, 11.
39. Tasnim, R.; Calderwood, L.; Annis, S.; Drummond, F.A.; Zhang, Y.J. *The Future of Wild Blueberries: Testing Warming Impacts Using Open-Top Chambers*; Spire: Orono, ME, USA, 2020.
40. Wright, I.J.; Reich, P.B.; Westoby, M.; Ackerly, D.D.; Baruch, Z.; Bongers, F.; Cavender-Bares, J.; Chapin, T.; Cornelissen, J.H.C.; Diemer, M.; et al. The worldwide leaf economics spectrum. *Nature* **2004**, *428*, 821–827. [CrossRef]
41. Hall, I.V.; Aalders, L.E.; Mcrae, K.B. Lowbush blueberry production in eastern Canada as related to certain weather data. *Can. J. Plant Sci.* **1982**, *62*, 809–812. [CrossRef]
42. Drummond, F.A. Factors That Affect Yield in Wild Blueberry, Vaccinium Angustifolium Aiton. *Agric. Res. Tech. Open Access J.* **2019**, *22*, 556212. [CrossRef]
43. Chen, J.-J.; Zhen, S.; Sun, Y. Estimating leaf chlorophyll content of buffaloberry using normalized difference vegetation index sensors. *HortTechnology* **2021**, *31*, 297–303. [CrossRef]
44. Sousa-Souto, L.; Bocchiglieri, A.; de Dias, D.; Ferreira, A.S.; Filho, J.P. Changes in leaf chlorophyll content associated with flowering and its role in the diversity of phytophagous insects in a tree species from a semiarid Caatinga. *PeerJ* **2018**, *6*, e5059. [CrossRef]
45. Qu, H.; Drummond, F. Simulation-based modeling of wild blueberry pollination. *Comput. Electron. Agric.* **2018**, *144*, 94–101. [CrossRef]
46. Obsie, E.Y.; Qu, H.; Drummond, F. Wild blueberry yield prediction using a combination of computer simulation and machine learning algorithms. *Comput. Electron. Agric.* **2020**, *178*, 105778. [CrossRef]
47. Qu, H.; Xiang, R.; Obsie, E.Y.; Wei, D.; Drummond, F. Parameterization and calibration of wild blueberry machine learning models to predict fruit-set in the northeast china bog blueberry agroecosystem. *Agronomy* **2021**, *11*, 1736. [CrossRef]

48. Samaniego, L.; Thober, S.; Kumar, R.; Wanders, N.; Rakovec, O.; Pan, M.; Zink, M.; Sheffield, J.; Wood, E.F.; Marx, A. Anthropogenic warming exacerbates European soil moisture droughts. *Nat. Clim. Chang.* **2018**, *8*, 421–426. [CrossRef]
49. 631-Guide to Efficient Irrigation of the Wild Blueberry—Cooperative Extension: Maine Wild Blueberries—University of Maine Cooperative Extension, n.d.. Cooperative Extension: Maine Wild Blueberries. Available online: https://extension.umaine.edu/blueberries/factsheets/irrigation/guide-to-efficient-irrigation-of-the-wild-blueberry/ (accessed on 28 November 2021).

climate MDPI

Article

Community-Level Impacts of Climate-Smart Agriculture Interventions on Food Security and Dietary Diversity in Climate-Smart Villages in Myanmar

Andrew Hanley [1], Galina Brychkova [1,2], Wilson John Barbon [3,*], Su Myat Noe [3], Chan Myae [3], Phyu Sin Thant [3], Peter C. McKeown [1,2], Julian Gonsalves [4] and Charles Spillane [1,2]

1 Masters Program in Climate Change, Agriculture & Food Security (MScCCAFS), Ryan Institute and College of Science & Engineering, National University of Ireland Galway, University Road, H91 REW4 Galway, Ireland; andrewhanley96@gmail.com (A.H.); galina.brychkova@nuigalway.ie (G.B.); peter.mckeown@nuigalway.ie (P.C.M.); charles.spillane@nuigalway.ie (C.S.)

2 Plant & AgriBiosciences Research Centre (PABC), Ryan Institute, National University of Ireland Galway, University Road, H91 REW4 Galway, Ireland

3 International Institute of Rural Reconstruction-Myanmar, Yangon 11052, Myanmar; suno0001@stud.slu.se (S.M.N.); chan.myae@iirr.org (C.M.); phyu.thant@iirr.org (P.S.T.)

4 International Institute of Rural Reconstruction Regional Center for Asia, Silang 4118, Philippines; juliangonsalves@yahoo.com

* Correspondence: wilsonjohn.barbon@iirr.org

Citation: Hanley, A.; Brychkova, G.; Barbon, W.J.; Noe, S.M.; Myae, C.; Thant, P.S.; McKeown, P.C.; Gonsalves, J.; Spillane, C. Community-Level Impacts of Climate-Smart Agriculture Interventions on Food Security and Dietary Diversity in Climate-Smart Villages in Myanmar. *Climate* **2021**, *9*, 166. https://doi.org/10.3390/cli9110166

Academic Editors: Christopher Robin Bryant, Andrea Vitali, Azzeddine Madani and Steven McNulty

Received: 25 September 2021
Accepted: 16 November 2021
Published: 21 November 2021

Abstract: Diversification of production to strengthen resilience is a key tenet of climate-smart agriculture (CSA), which can help to address the complex vulnerabilities of agriculture-dependent rural communities. In this study, we investigated the relationship between the promotion of different CSA practices across four climate-smart villages (CSVs) in Myanmar. To determine the impact of the CSA practices on livelihoods and health, survey data were collected from agricultural households (n = 527) over three years. Within the time period studied, the results indicate that some the CSA practices and technologies adopted were significantly associated with changes in household dietary diversity scores (HDDS), but, in the short-term, these were not associated with improvements in the households' food insecurity scores (HFIAS). Based on the survey responses, we examined how pathways of CSA practice adoption tailored to different contexts of Myanmar's four agroecologies could contribute to the observed changes, including possible resulting trade-offs. We highlight that understanding the impacts of CSA adoption on household food security in CSVs will require longer-term monitoring, as most CSA options are medium- to long-cycle interventions. Our further analysis of knowledge, attitudes and practices (KAPs) amongst the households indicated a poor understanding of the household knowledge, attitudes and practices in relation to nutrition, food choices, food preparation, sanitation and hygiene. Our KAP findings indicate that current nutrition education interventions in the Myanmar CSVs are inadequate and will need further improvement for health and nutrition outcomes from the portfolio of CSA interventions.

Keywords: climate-smart agriculture; food security; dietary diversity; climate-smart villages; HFIAS; HDDS

1. Introduction

Climate change is now recognized as a major threat to food security and adequate nutrition in the twenty-first century [1–3]. Extreme weather events that threaten food security, such as droughts, heat waves, floods, wildfires and storms, will also become more frequent and severe [4] Adverse climate change is already having direct effects on agricultural production, impacting food supply and food security [5]. The quantity and nutritional quality of products generated by agricultural systems is influenced by a range of factors, including, inter alia, soil quality, nutrient availability, temperature, water availability, CO_2 concentrations and the prevalence of pollinators [2,6,7]), many of which are undergoing changes due to climate change.

Changes in temperature and water availability are factors influenced by changing climates, particularly in vulnerable regions. The yields of most crop species are sensitive to alterations in temperature [8,9]. Indeed, when air temperatures exceed 30 °C, even for short periods, reductions in yields are expected in rainfed crops, regardless of the crop species [10,11]. Higher temperatures are also coupled with decreases in water availability due to increased evaporation and evapotranspiration, leading to crop yield reductions [9,12].

From a broader perspective, climate change can have a negative impact on the four pillars of food security, namely availability, access, utilization and stability (FAO et al. 2018). Food security is related to nutrition, and, consequently, malnutrition is an indicator of food insecurity. Dietary diversity is typically measured by the number of food groups eaten in the diet over a given time period. Overall, dietary diversity is often (although not always) a good indicator of micronutrient intake and associated malnutrition [13,14].

Dietary diversity outcomes are rarely considered when relating agricultural outputs to food security [15]. However, more ill health and mortality can be attributed to poor diet than to any other risk factor [16]. There are direct links between climate change, reduced access to food and diverse diets and increases in childhood stunting, wasting and low birth weights [14] as well as through direct temperature impacts on fetal health [17,18]. Stunting (height-for-age z-score < -2) occurs in children 5 years of age and below and can lead to shorter adult height, limited cognitive function and reduced adult income [19]. Childhood wasting (weight-for-height z-score < -2) is estimated to affect 10% of children globally and is associated with reduced lean mass and weaker immune systems, leaving children more susceptible to infections, which can result in death [20]. Low birth weights (<2500 g) are also associated with mothers and households who are food-insecure.

Food insecurity and micronutrient deficiencies associated with poor dietary diversity are major issues across Myanmar. Such challenges are attributed to diverse factors, such as conflict, poverty and vulnerability to natural disasters, which are becoming more frequent due to climate change [21]. According to the Myanmar Micronutrient and Food Consumption Survey 2017–2018, significant progress is needed to achieve the goals set by the World Health Organization for reducing wasting and stunting by 2025 [22]. The MMFC survey highlighted that nearly one in three children (26.7%) under the age of five are stunted in Myanmar, while 6.7% of children under the age of five are wasted and 19.1% of children in the same age bracket are underweight. Only 16% of babies aged 6–23 months receive the minimum acceptable diet for development at their age, while nearly 20% of adult men and 15% of adult women are underweight [23].

Over 23% of total anthropogenic greenhouse gas emissions are derived from agriculture, forestry and other land uses (AFOLU sector) [24–26]. Excluding land use change, agriculture contributes to approximately 11% of total anthropogenic GHG emissions, and requires up to 70% of our global fresh water supply [27]. Climate-smart agriculture (CSA) is a term used to describe a portfolio of practices that can reduce emissions and strengthen the adaptation of agricultural systems to climate change, while improving food security and livelihood outcomes [28]. The CSA approach anchors itself on three pillars that aim to jointly address food security and climate challenges, leading to systems that sustainably increase productivity and incomes while building resilience to climate variability, and seeking mitigation of GHG where possible [29,30].

Climate variability is experienced across most regions of Myanmar, with some regions receiving excessive rainfall, while other regions have insufficient rainfall, leading to drought periods during cropping cycles [31]. Access to safe and reliable water supplies, whether for irrigation, livestock or domestic use, is a key constraint to livelihoods and food production, with significant knock-on consequences for income [32]. Myanmar is also at increasing risk from a wide range of natural climate-influenced hazards, including cyclones, floods and droughts, that can have severe negative impacts on the livelihoods of the poor, contributing to seasonal food shortages. CSA programs in Myanmar to strengthen livelihood resilience will increasingly include diversification, including the increasing adoption of trees, livestock and off-farm incomes as risk aversion strategies for the rural poor.

Hence, the development, application and impact monitoring of climate-smart agriculture (CSA) strategies and programs is central to ensuring food system productivity to deliver key outcomes, including achieving food security, reducing malnutrition, reducing inequities and empowering the most vulnerable, while delivering resilience to climate change [33]. The impacts of climate change differ significantly across rural communities and agroecosystems. Hence, understanding, strategies and actions will need to take into account location-specific and community-based considerations [33,34].

The CGIAR Research Program on Climate Change, Agriculture and Food Security (CCAFS) developed and piloted the climate-smart village (CSV) approach in 2012 in Africa and South Asia, and later expanded CSV pilots to Latin America and Southeast Asia in 2014 [35]. The CSV approach was developed and promoted to address research gaps in climate-smart agriculture at the level of rural communities. This need arose as much of the knowledge on climate-smart agriculture technologies and practices has been initially developed in controlled environments of research farms and modeling. The CSV approach enables researchers to work in a participatory manner with local communities to test, demonstrate and generate evidence of which CSA practices can work for rural communities at the level of the CSV. The implementation of CSA in the CSVs includes testing and learning with farmers on a range of CSA interventions, including crop varieties, small livestock, small-scale aquaculture and improved farm management practices that consider climate change realities as experienced by the communities. CSA approaches place emphasis on the importance of soil, water and agro-biodiversity conservation within farms, as well as across larger landscape areas that determine the regional agroecology. The promotion of CSA practices in CSVs also includes a range of indirect agriculture interventions, including capacity development, and strengthening extension services (e.g., including agriculture finance and climate information services) that can enable farmers to transition towards climate-smart agriculture [36].

In Myanmar, the International Institute of Rural Reconstruction (IIRR), with support from CGIAR-CCAFS and the International Development Research Center in Canada, has taken a participatory action research (PAR) approach to establish four climate-smart villages in unique agroecologies around the country [33]. This PAR supports a process to establish CSVs in Myanmar, particularly to demonstrate the viability and impact of location-specific CSA in the four distinct agroecologies. The research further aimed to identify scaling pathways for CSA via CSVs, to enable the more widespread adoption CSA portfolio-based approaches by NGOs and government agencies in Myanmar.

This study investigates the relationship between the promotion of CSA practices implemented in four climate-smart villages (CSVs) across Myanmar and the changes in household food security and diet diversification during the time period of the CSA intervention. The key objectives of the study are to (1) monitor impacts on household food security and dietary diversity in CSVs, (2) identify routes to households becoming more food-secure with improved dietary diversity and (3) inform food security and nutrition programs on impacts and outcomes from the adoption of climate-smart agriculture practices and technologies in rural communities.

2. Methodology

2.1. Study Site: Myanmar Climate-Smart Villages

The implementation of the CSV approach was enhanced and adapted by IIRR by presenting it as not only a research for development approach that focused on CSA, but as a broader community development intervention package. The tailored CSV approach of IIRR followed the principles of participatory action research (PAR) and community-based adaptation, where community members are active participants in the process of understanding the challenges, finding and testing solutions and learning from doing.

The IIRR CSV approach in Myanmar follows a 3-step process that includes (1) understanding vulnerabilities and their drivers, (2) identifying and testing adaptation options and (3) social learning within the village and with other villages. For this process, IIRR developed a menu of "socio-technical" methods and tools to facilitate community processes along the 3-step process, consistent with the principles of PAR (Barbon et al. 2021). These socio-technical tools and methods include participatory climate vulnerability and risk assessments, community workshops to identify "no-regrets" options for climate change adaptation (vis a vis the experienced climate risks and vulnerabilities) as well as farmer field days and roving workshops to facilitate the cross-learning and cross-incubation of new ideas and new experiences of farmers working to adapt to climate change.

This study was undertaken across four climate-smart villages (CSVs) in Myanmar, each adopting a portfolio of climate-smart agriculture practices in the four agroecologies of the country. Table 1 provides an overview of the profile of the four Myanmar CSVs.

Table 1. Profile of the four climate-smart villages (CSVs) in Myanmar.

Village Name	Saktha	Htee Pu	Ma Sein	Taung Kamauk (TKM)
Agroecology	Highlands	Dry Zone	Delta	Upland
Major crops	Rice, corn, vegetables	Groundnut, pigeon pea, green gram	Rice, betel leaves/nuts	Rice, millet, corn
Township	Hakha	Nyaung-Oo	Bogale	Nyaung-Shwe
State/region	Chin	Mandalay	Ayeyarwaddy	Shan
Total households	200	275	103	94
Total population	865	11,180	453	405
No. of females	445	603	249	215
No. of males	420	577	214	190
Distance from nearest township	32 km	35 km	11 km	20 km
Ethnic group	Chin	Burmese	Burmese	Pa-o

(Source: Barbon et al., 2021).

Table 1 highlights that the four CSVs span the major diversity of agroecologies and agriculture systems across Myanmar. For instance, the farming system in Chin State, a highland region of Myanmar, is significantly driven by household consumption, as expected considering their isolation. This differs from the farming systems of the delta and dry zones, where production is primarily driven by markets. Agricultural production in the CSV in Shan is intermediate, driven by both household use and market sale, as this village is close to trading centers. Each of these four CSVs also experiences climate change differently, which is a key driver of IIRR's approach based on the importance of localized climate change adaptation in agriculture that is systems-oriented, rather than crop- or commodity-oriented. In systems-oriented approaches, broader consideration is made of the impact of soil, water, climate variability and extension services, all of which interact to determine the outcome, quality and livelihood impact of agriculture production.

As local communities experience climate change risks and vulnerabilities differently, adaptation approaches will also differ between communities. This is where the value of community-based approaches is significant, particularly by ensuring that CSA practices are tailored to the unique contexts of the participating communities. Consistent with this principle, IIRR has promoted a "portfolio" or "basket of options" approach" to CSA adoption by rural communities. The portfolio approach involves communities in considering a list of CSA adaptation options tailored to each of their specific vulnerabilities and risks. This menu of options can include, e.g., technological options, such as promoting stress-tolerant varieties of primary crops, or new platforms for agriculture production, such as integrating and improving small livestock production and vegetable production in homesteads (the patch of land around the household dwelling, which, in Southeast Asia, can sometimes comprise up to 200–400 square meters of land).

The portfolio of CSA practice options can also include practices such as the use of green manure to reduce the footprint of fertilizer use, integrating trees into the existing farming system to generate new sources of income, improving soil health and creating micro-climates around the farm to protect farms against strong winds during storms. The CSA practice portfolio approach also helps to ensure social inclusiveness (with the aim that no one member of the community is excluded) based on the identification of CSA options irrespective of the household context, e.g., for households with large land areas, households without farmland but with a homestead, women-headed households, households that are wealthier and households that are very poor.

In the process of developing the menu of CSA options, IIRR facilitators conducted consultations with farmers and other rural community researchers to produce portfolios of possible options as a response to their understanding of climate risks and vulnerabilities. The list of possible CSA options was further prioritized using the following criteria [33].

- Criteria 1: Is it climate-smart (i.e., reduces GHGs, enhances soil, agro-biodiversity, conserves and reduces risk of losses of the farms)?
- Criteria 2: Is it ecosystem friendly (environmentally friendly)?
- Criteria 3: Is it nutrition-sensitive?
- Criteria 4: Does it address food insecurity?

After each of the CSVs finalized their portfolio of options, IIRR provided a small grant facility (termed the CSV Adaptation Fund) to support the implementation and trials of the identified options. The implementation and trials were conducted for two annual production seasons during 2019 and 2020. Alongside the implementation of these CSA options in each of the CSVs, IIRR also supported capacity development and awareness building activities to maximize the potential of CSA to generate development outcomes. In relation to this, IIRR implemented community-based nutrition education activities.

2.2. Conceptual Framework

The conceptual framework of this study sought to understand the linkages by which climate-smart agriculture coupled with nutrition education can be better leveraged to achieve well-being outcomes for agriculture-based communities, such as food security and nutrition (Figure 1). For nutrition, we used dietary diversity as a proxy indicator for improved nutritional outcomes.

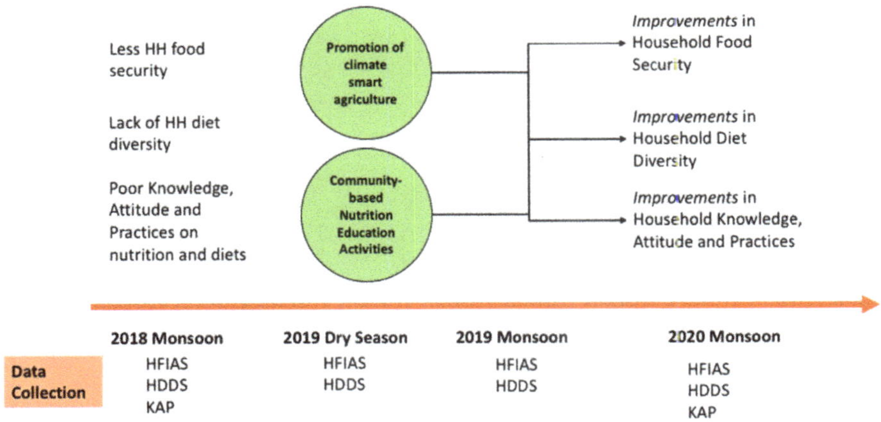

Figure 1. Conceptual framework for study. HH: household; HFIAS: Household Food Insecurity Access Scale; HDDS: Household Dietary Diversity Score; KAP: knowledge, attitudes and practices.

2.3. Household Food Insecurity Access Scale (HFIAS) and Household Dietary Diversity Score (HDDS)

To measure food security and diet diversity, the Household Food Insecurity Access Scale (HFIAS) and Household Dietary Diversity Score (HDDS) were used. Data were collected from households in the four climate-smart villages (CSVs) in Myanmar. IIRR and its local NGO partners facilitated and provided support for households to implement climate-smart agriculture options in the villages from 2018. The CSA options deployed relied heavily on fruit tree crops and small livestock as core components of diversification, along with intercrops of annual crops such as corn, sorghum, upland rice and vegetables (depending on location). The CSA interventions were tracked annually to determine the number of CSA options adopted by HHs in a given season. The data sets from 2018 (monsoon), 2019 (dry season) and 2019 (monsoon) were analyzed for HFIAS and HDDS.

The Household Food Insecurity Access Scale (HFIAS) is an approach to measure food insecurity at the household level. This approach is founded on the idea that when households experience food insecurity, it results in reactions and responses that can be collected and quantified in a structured community survey. Household food insecurity access was measured using a methodology designed and developed by a partnership of USAID and the Food and Nutrition Technical Assistance Project (FANTA) [37].

The Household Dietary Diversity Score HDDS is a metric used to measure the diversity of a household's diet. The HDDS is measured by the method developed by FAO Nutrition and Consumer Protection Division with support from EC/FAO and FANTA. Similar to the HFIAS questionnaire, HDDS uses a points-based system to calculate the diversity of a given diet. The recall period for HDDS surveys is 24 h, where respondents are asked to describe the foods (meals and snacks) that the household ate on the previous day, starting with the foods first eaten in the morning up until they went to sleep that night. A set of 12 food groups are used to guide the scoring as per the food items consumed (Table 2). Each food group is assigned a score of 1 if consumed or 0 if not consumed. The maximum score possible is hence 12, and the lowest is 1, meaning that the household only consumed one food type in that period. Food consumed outside of the home is not included [38].

Table 2. Food groups used in this study.

No.	Food Groups	No.	Food Groups
1	Cereals	7	Fish and seafood
2	White roots and tubers	8	Legumes, nuts and seeds
3	Vitamin A-rich vegetables, dark green leafy vegetables, other vegetables	9	Milk and milk products
4	Vitamin A-rich fruits, other fruits	10	Oils and fats
5	Organ meats, flesh meats	11	Sweets
6	Eggs	12	Spices, condiments and beverages

In addition to the HFIAS and HDDS surveys, the knowledge, attitudes and practices (KAP) of households were also assessed in the four climate-smart villages on nutrition, the importance of nutrition, food choices, food preparation and hygiene by inclusion of KAP questions included in the HFIAS and HDDS questionnaire. The data for KAP were collected and analyzed for the years 2018 and 2020.

2.4. Knowledge, Attitudes and Practices (KAP)

To assess the respondent's KAP, the respondents were asked whether they agreed or disagreed with each of the statements in the questionnaire. To assess KAP, there are a total of 45 statements, where 15 statements are each assigned as knowledge, attitudes and practices. The statements are also presented as either a positive or negative statement. This ensures that respondents will avoid giving responses that all agree to the statements. A positive statement ideally should be responded with an agreement and a negative statement a disagreement. The KAP results are presented as percentages (%) of the HHs agreeing to the statement. Data from both 2018 and 2020 were used. McNemar's test was used to determine whether any KAP increase or decrease between 2018 and 2020 was statistically significant.

2.5. Household Surveys

In this study, we used household survey data collected by IIRR for the years 2018, 2019 and 2020. The household surveys were conducted in full enumeration, where all households in the CSVs were included in the surveys. The survey questionnaire was prepared in English, translated into the Myanmar language and then pre-tested with other non-CSV farmers on-site to check the translation of the questionnaire. The questionnaire included information on household demographics, livelihoods, poverty and on HFIAS, HDDS and KAP. A total of 527 household respondents were included in the overall sample.

The survey data were encoded in Microsoft Excel and data analysis conducted using the Statistical Package for Social Sciences (SPSS). The following statistical analyses were performed.

1. *Analysis of Variance (ANOVA)* to determine statistically significant differences in HDDS and HFIAS across the 4 CSVs.
2. *Post-Hoc Tukey–Kramer test* to determine statistical differences in HDDS and HFIAS in the pairwise combination among CSVs.
3. *Likelihood Ratio Test* to determine which factors influenced the HDDS and HFIAS. The factors used in this analysis are based on the other data collected from secondary sources, such as temperature, rainfall and, from the survey data, the level of adoption of the household of CSA options.
4. *McNemar's Test* to determine statistical differences between 2018 and 2020 data is presented in percentages in the KAP. This test is used to analyze pre-test vs. post-test study designs, as well as being commonly employed in analyzing matched pairs and case–control studies.

3. Results and Discussion

3.1. Significant Differences in Household Food Insecurity (HFIAS) between CSVs

The ANOVA results showed that significant differences were found between the CSVs Htee Pu ($M = 1.29 \pm 0.11$), Ma Sein ($M = 3.89 \pm 0.22$), Saktha ($M = 7.01 \pm 0.22$) and TKM ($M = 4.48 \pm 0.32$) (Figure 2). On average, individuals in Saktha had the highest HFIAS scores, indicating that they tended to be the most food-insecure. Conversely, Htee Pu CSV had the lowest HFIAS scores, indicating that this community is the most food-secure out of the four CSVs. There was no significant difference among the HFIAS scores of the villages of Taung Kamau and Ma Sein. These results indicate that, in Myanmar, food security varies between CSV locations, where, within this study, the Saktha CSV in Chin State is the most food-insecure compared to the other CSVs.

3.2. Significant Differences in Household Dietary Diversity (HDDS) among CSVs

To identify any differences in the dietary diversity of households (HH) in the four CSVs, mean HDDS scores were calculated for each village, where a HDDS of 7 or higher indicates that a HH has an adequately diverse diet (Figure 3). ANOVA results indicated that there was no significant difference between the villages of Htee Pu ($M = 6.6 \pm 0.07$) and Ma Sein ($M = 6.7 \pm 0.12$). However, TKM ($M = 6.22 \pm 0.13$) and Saktha ($M = 5.4 \pm 0.09$) were instead both statistically different from each other and the other two villages. Our results indicate that the Htee Pu and Ma Sein have the best mean dietary diversity scores, while Saktha has the worst average HDDS.

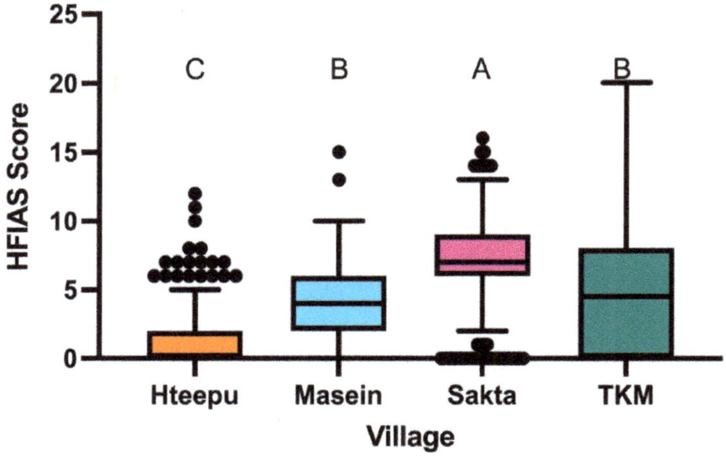

Figure 2. HFIAS Scores recorded from four Myanmar CSVs. The central line of each column represents the mean HFIAS Score for each CSV ± the standard error, with the outermost lines representing standard deviation. One-way ANOVA was used to determine statistical differences; villages with different letters are significantly different at a 95% confidence interval.

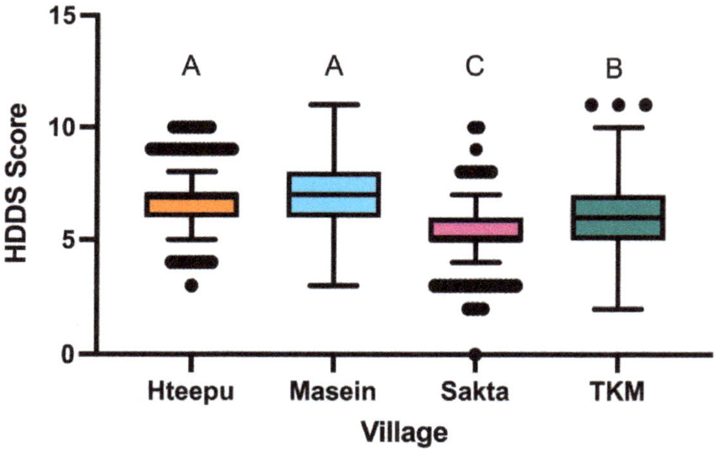

Figure 3. HDDS Scores recorded from four Myanmar CSVs. The central line of each column represents the mean HDDS Score for each CSV ± the standard error, with the outermost lines representing standard deviation. One-way ANOVA was used to determine statistical differences; villages with different letters are significantly different at a 95% confidence interval.

3.3. Number of CSA Options Adopted by the Households Correlates with HFIAS and HDDS

To investigate the impact of CSA introductions, the numbers of households that were considered to have diverse diets both before and after CSA introduction were considered. Across all four CSVs, 37% of households with no access to CSA obtained a score of 7 or higher, while, for households with access to at least one CSA intervention, this increased to 47%.

An effect likelihood ratio test (Table 3) confirmed that the location of each CSV had the most significant influence on the HFIAS scores, while the "numbers of CSA" were not significantly different. This suggests that the numbers of CSA interventions, carried out under these circumstances, had no influence on the HFIAS score that a household could achieve over the timescale of the intervention that was measured.

Table 3. Effect likelihood ratio test (ELRT) carried out to determine which factors influence the HFIAS scores of households across all four CSVs.

Source	Nparm	DF	L-R ChiSquare	Prob > ChiSq
Location 2	3	2	11.1549622	0.0038 *
Min. TEMP	1	0	0	-
Max. TEMP	1	0	0	-
Ave. Temp	1	0	0	-
Rainfall in inches	1	0	0	-
Rain days	1	0	0	-
Number of CSA	5	4	3.27635049	0.5127
CSA (all) YN	1	0	0	-

The impact of different variables on HDDS was also determined (Table 4) and the results indicated that both "location" and "number of CSA" options implemented had highly significant differences ($p = 0.0002$).

Table 4. Effect likelihood ratio test carried out to determine which factors influence the HDDS of households across all four CSVs.

Source	Nparm	DF	L-R ChiSquare	Prob > ChiSq
Location 2	3	2	16.6549429	0.0002 *
Min. TEMP	1	0	0	-
Max. TEMP	1	0	9.0949×10^{-13}	-
Ave. Temp	1	0	0	-
Rainfall in inches	1	0	0	-
Rain days	1	0	0	-
Number of CSA	5	5	23.8026591	0.0002 *

3.4. Contrasting Values of HFIAS and HDDS

Our study found no correlation between the number of CSA options adopted and food security, despite a strong correlation with dietary diversity. From the earlier 2010 Myanmar Census of Agriculture, rice is an important component of the Myanmar diet. Access to rice is often viewed as an indicator of food security. A reduction in access to rice will lead to an HFIAS response that food is inadequate for the household. Access to rice across much of Myanmar is achieved by purchasing this staple in markets, hence the importance of cash.

Many of the CSA options that have been promoted in the Myanmar CSVs are directed at diversifying accessible food at home and in the farm, relying on fruit trees, small livestock and vegetables, with relatively less reliance on rice as a CSA option (except in TKM, where upland rice is widely grown). The choice of commodities in the CSA project was focused on nutrient-dense products. Some CSA options with promised commercial returns (e.g., dryland horticulture in the dry zone Htee Pu CSV) will likely require more time (possibly years) for economic or nutritional benefits to be realized by the households. It should also be noted that there are other externalities beyond climate change and variabilities that affect the realization of economic benefits from the CSA options. For instance, there was a significant change in the markets for pulses in this period, which dry zone farmers (such as those in Htee Pu CSV) are heavily dependent on.

With regard to why the number of CSA options adopted contributes to changes in the HDDS, Table 5 highlights potential contributions to the dietary diversity of the household per CSA option.

Table 5. Contributions of climate-smart agriculture options to diet diversification.

No.	CSA Options Identified by the CSVs	Why Climate-Smart?	Potential Contributions to HHDS
1	Participatory Varietal Selection (PVS) of primary crops, i.e., rice, maize, pigeon pea, peanut	Enable the farmers to identify which varieties work in a specific climate scenario	
2	Diversification of farm production with vegetables; legumes with crop trials for newly introduced crops	Minimizes the risk of losses in case climate variability reduces yields of main crop	Provides food materials that are not necessarily for selling but end up consumed by the HH. For example, legumes as cover crops to protect soil (main purpose) can provide green beans for HH consumption. For producing several crops in the field—in TKM CSV—farms are planted with maize, peanuts and sunflower for selling and, if price is low, will end up being consumed by HH.
3	Integration of fruit trees in farms (avocado, mango, banana, jackfruit, oranges)	Minimize the risk of losses; trees are more tolerant to variability of rainfall and temperature; sequester more GHGs	Can supply fruits for selling for HH consumption too but these results are expected only in another 3 to 5 years
4	Planting of legume trees in farms and along boundaries (Alnus spp, Casia spp, Gliricidia spp)	Manages the soil degradation and erosion; minimizes dependence on artificial inputs; sequester more GHGs	No contribution to diet diversity but aimed at improvement of soil health
5	Homestead production of vegetables, fruits and cash crops	Addresses household food security and under nutrition in times of climate change stresses	Homestead production provides vegetables to the HH aside from vegetables for selling
6	Small livestock production in homesteads	Served as emergency assets in case of climate change shocks, provide opportunities for women	In Ma Sein, HH keep ducks, which provide eggs for the HH. In the other CSVs, they raise chickens, goats and pigs, which, in times of need, all can provide income as well as food to the HH.
7	Aquaculture (homestead and farm ponds)	Diversify income sources, provide opportunities for women	Same as #6. This was undertaken in Ma Sein and Saktha CSVs only.
8	Community-based animal propagation centers (pig, chicken, duck and fish)	Provide sustainable sources of stocks for HH level livestock production	Same as #6
9	School gardens (vegetables, fodder, fruit trees)	Served as source of planting materials, education tool for students on CSA	No contribution to HDDS
10	Improving water storage facilities	Reduces the risk of water shortages in dry conditions	No contribution to HDDS

3.5. Major Changes in Household Knowledge

The statistical significance of knowledge of the households in the four CSVs was assessed by McNemar's test (Table 6). Statements 1 and 8 relate to the understanding of the basic idea of nutrition, and the role of nutritious food in achieving a healthy body and longer life. The analysis revealed that only TKM CSV exhibited a significant improvement in the respondents' understanding of nutrition and nutritious food, while the other CSVs showed a poor understanding of these topics in 2020.

Table 6. Proportion of respondents who agree on the knowledge statements related to household nutrition in four CSVs.

Statements [a]	Researcher's Note [b]	Htee Pu			TKM (Shan)			Ma Sein			Saktha		
		2018	2020	McNemar's (*p*-Value) [c]	2018	2020	McNemar's (*p*-Value) [c]	2018	2020	McNemar's (*p*-Value) [c]	2018	2020	McNemar's (*p*-Value) [c]
1	Negative	17	33	0	40	28	0.144	15	60	0	35	65	0
2	Negative	16	14	0.596	32	29	0.868	2	17	0.002	21	20	1
3	Negative	98	96	0.302	88	95	0.18	93	98	0.289	94	89	0.607
4	Positive	77	91	0	60	49	0.243	84	91	0.286	87	85	1
5	Positive	88	95	0.007	80	95	0.007	82	94	0.019	93	98	0.07
6	Positive	100	100	1	86	96	0.035	90	90	1	95	98	0.453
7	Positive	100	100	1	78	85	0.327	100	97	0.25	95	77	0
8	Positive	99	99	1	79	99	0	93	99	0.063	94	99	0.07
9	Positive	98	92	0.015	64	69	0.532	93	98	0.219	67	86	0.001
10	Positive	17	59	0	25	26	1	17	66	0	77	66	0.153
11	Positive	80	86	0.104	79	80	1	79	93	0.017	91	84	0.23
12	Negative	76	84	0.051	56	52	0.755	64	92	0	77	74	0.755
13	Positive	98	97	0.581	98	96	1	98	93	0.289	97	96	1
14	Positive	98	96	0.302	73	78	0.571	97	98	1	95	95	1
15	Positive	86	70	0	44	48	0.643	64	83	0.018	84	88	0.345

(a) The statements used for HH knowledge were as follows.

1. Nutrition is about food preparation and malnourished children.
2. Anemia or lack of iron makes the child intelligent.
3. Fish, meat and eggs give a person energy.
4. Green and leafy vegetables are rich in Vitamins A, C and iron.
5. Vegetables and fruits help the person prevent diseases and infections.
6. Personal hygiene and cleanliness helps prevent diseases and infections.
7. Flies and other insects that come into contact with food may cause diseases to humans and also spoil the food.
8. Nutritious food is important for humans to be healthy and achieve longer life.
9. Parasitic worms contribute to malnutrition of children
10. Iron is important to the body as it helps in delivering oxygen to all parts of the body.
11. Green and leafy vegetables as well as brightly colored vegetables such as squash are good sources of Vitamin A for good eye sight and for growth and development.
12. Carbohydrates and fats are considered foods for growth.
13. Rice, corn, potatoes and peanut oil are important sources of energy for people.
14. Beans, groundnuts and meats are sources of protein needed for the growth of humans.
15. A good meal must contain food from three groups—energy foods, growth foods and protective foods.

(b) A positive statement ideally shall have move agree responses and a negative statement shall have less agree responses
(c) McNemar's test was conducted to determine whether there was a significant difference in the proportion (increase or decrease) over time.
If p-value < 0.05, then the proportion was statistically significant at 5%. If p-value < 0.01, then the proportion was statistically significant at 1%. Note: "No responses" were excluded from the analysis.

This suggests a need for more careful messaging and awareness building on what nutrition is, and why it is important.

Statements 2, 4, 10, 12 and 15 relate to the basic understanding of topics such as food groups, vitamins, minerals and anemia. Overall, there remains a lack of understanding of what anemia is (statement 2) and why it is important for ensuring nutrition in the households. While there was a lack of understanding of anemia, the Htee Pu and Ma Sein CSVs indicated some improvements in their understanding of the role of iron for a healthy body. However, overall, it is indicative that the concept of anemia and the role of iron are not well-understood across the four CSVs.

For the food groups (statements 4, 12 and 15), only Htee Pu and Ma Sein showed a significant improvement in their understanding of the three basic food groups. However, in the case of understanding carbohydrates and fats, there was no overall improvement in the respondents' understanding of these food groups. Only the Htee Pu CSV showed a significant improvement in understanding the important role of green and leafy vegetables as sources of vitamins A and C and iron.

Statements 5 and 11 relate to the role of vegetables and fruits in preventing disease and infection, and their dietary importance. All four CSVs indicated significant improvements in statement 5 (that vegetables and fruits prevent disease and infection) but only Ma Sein CSV indicated a significant improvement in understanding that green and leafy vegetables are important parts of the diet.

Statements 6, 7 and 9 relate to the importance of hygiene and cleanliness in addressing malnutrition. TKM CSV showed a significant improvement in understanding the important role of personal hygiene and cleanliness. Saktha CSV showed significant improvements in understanding the link of parasitic worms to malnutrition.

Overall, the Htee Pu and Ma Sein CSVs demonstrated the greatest number of improvements in their understanding of the food groups, the important role of fruits and vegetables in the diet and knowledge about vitamins and minerals.

3.6. Major Changes in Household Attitudes

To determine how household attitudes towards nutrition, food choices, food preparation and hygiene had changed, we tabulated responses from across the CSVs and used McNemar's test to assess statistical differences (Table 7). While we found various patterns of change, many CSVs displayed no or little improvement in understanding key aspects.

Table 7. Proportion of respondents who agree on the *attitude statements* in household nutrition in four CSVs.

Statements [a]	Researcher's Note [b]	Htee Pu			TKM (Shan)			Ma Sein			Saktha		
		2018	2020	McNemar's (*p*-Value) [c]	2018	2020	McNemar's (*p*-Value) [c]	2018	2020	McNemar's (*p*-Value) [c]	2018	2020	McNemar's (*p*-Value) [c]
1	Negative	68	46	0	38	36	1	44	74	0	52	49	0.677
2	Positive	92	95	0.281	80	99	0	99	93	0.125	85	94	0.078
3	Positive	98	84	0	88	80	0.189	93	68	0	92	76	0.009
4	Negative	58	88	0	45	48	0.77	43	68	0.002	75	81	0.324
5	Negative	74	93	0	56	74	0.015	79	60	0.015	84	69	0.018
6	Positive	71	92	0	54	66	0.144	83	92	0.134	90	85	0.556
7	Negative	99	93	0	95	62	0	98	82	0.001	94	41	0
8	Positive	85	96	0	69	94	0	82	95	0.004	87	94	0.143
9	Positive	100	98	0.031	86	94	0.118	94	98	0.453	97	97	1
10	Positive	99	98	0.289	100	93	0.031	100	97	0.25	99	95	0.219
11	Positive	97	100	0.07	74	98	0	76	100	0	95	96	1
12	Positive	88	92	0.145	48	52	0.77	55	90	0	75	83	0.31
13	Positive	84	75	0.027	73	91	0.009	66	69	0.735	81	91	0.041
14	Negative	98	35	0	72	48	0.005	97	28	0	93	47	0
15	Negative	48	31	0	80	79	1	71	52	0.015	57	43	0.112

(a) The statements used for HH attitudes were as follows.

1. I believe that proteins from beans such as pigeon pea, butter beans and green gram are not substitutes for protein from meat.
2. Eating vegetables and fruits is very important for good health.
3. I believe that eating the same food everyday is not enough to get good nutrition.
4. I like to eat meat because it gives me Vitamin C.
5. Preparing nutritious food for the family is very hard to do.
6. I believe that Vitamin A is very important to have very good eyesight.
7. It is normal children to have parasitic worms.
8. It is important to learn the right way to cook food to get the best nutrients from food.
9. It is important to give the right food to my children for them to grow well.
10. Parents should be role models to their children in eating the right and nutritious food.
11. It is important that the kitchen where food is prepared should be clean.
12. It is important to eat fruits and vegetables of different colors to get vitamins and minerals.
13. I believe that the best source of nutrition for babies up to 2 years old is breast milk
14. I believe that growing vegetables in the home is only doable in homes with big land.
15. It is alright to drink collected rain water as it is pure and clean already.

(b) A positive statement ideally shall have move agree responses and a negative statement shall have less agree responses

(c) McNemar's test was conducted to determine whether there was a significant difference in the proportion (increase or decrease) over time. If p-value < 0.05, then the proportion was statistically significant at 5%. If p-value < 0.01, then the proportion was statistically significant at 1%. Note: "No responses" were excluded from the analysis.

For example, Htee Pu CSV showed a significant improvement in considering beans and legumes as good substitutes for meat proteins (statement 1), while CSV and Ma Sein CSV showed significant improvements in their attitude towards consuming fruits and vegetables (statements 2, 12). Ma Sein CSV and Saktha CSV showed significant improvements in relation to food preparation for the family not being difficult to do (statement 5). All CSVs (except Saktha) showed significant improvements in believing that the way in which food is cooked is important for obtaining the best nutrients from it.

No significant improvement could be determined across the four CSVs with respect to the importance of feeding children the best foods, and the role of parents in being good role models to children about "eating right" (statements 9, 10). However, TKM and Saktha

CSVs showed significant improvements in their attitude towards the importance of giving breast milk to babies and infants up to 2 years old.

Across all CSVs, there were significant improvements in the attitude towards having home gardens, and in appreciating that having smaller landholdings is not necessarily a hindrance to having a home garden (statement 14).

In terms of hygiene, all CSVs showed significant improvements in their attitude that it is not normal for children to have parasitic worms (statement 7). TKM CSV and Ma Sein CSV improved in their attitude that kitchens where food is prepared should be clean all the time. Htee Pu and Ma Sein CSVs showed improvements in their attitudes that unprocessed rainwater is not a good source of drinking water (statement 15).

3.7. Improvements in Household Practices

Having identified several areas of improvement in attitudes towards nutrition, food preparation and hygiene, we also investigated improvements in related household practices across the CSVs, again using McNemar's test to evaluate the significance of any changes (Table 8).

Table 8. Proportion of respondents who agree on the *practice statements* in household nutrition in four CSVs.

Statements [a]	Researcher's Note [b]	Htee Pu			TKM (Shan)			Ma Sein			Saktha		
		2018	2020	McNemar's (*p*-Value) [c]	2018	2020	McNemar's (*p*-Value) [c]	2018	2020	McNemar's (*p*-Value) [c]	2018	2020	McNemar's (*p*-Value) [c]
1	Positive	58	91	0	58	91	0	57	99	0	75	91	0.003
2	Positive	77	87	0.002	91	91	1	74	91	0.009	54	61	0.263
3	Negative	12	30	0	61	20	0	21	30	0.216	45	37	0.337
4	Positive	66	81	0	54	55	1	59	64	0.551	72	93	0
5	Negative	24	27	0.428	31	33	0.874	16	36	0.007	44	28	0.022
6	Negative	66	73	0.137	41	54	0.136	54	48	0.532	54	59	0.401
7	Negative	78	92	0	46	69	0.004	49	79	0.001	48	93	0
8	Positive	66	54	0.011	52	38	0.112	39	36	0.742	52	50	0.885
9	Negative	39	25	0.001	46	52	0.522	49	53	0.775	64	33	0
10	Positive	98	99	0.688	86	95	0.077	95	98	0.688	97	96	1
11	Positive	100	100	1	91	92	1	99	100	1	97	99	0.625
12	Positive	100	97	0.039	55	44	0.212	44	87	0	95	92	0.581
13	Positive	97	80	0	69	81	0.1	71	72	1	93	59	0
14	Positive	100	98	0.375	93	92	1	100	99	1	100	96	0.125
15	Positive	98	93	0.019	75	81	0.441	93	100	0.031	85	93	0.041

(a) The statements used for HH practices were as follows:

1. Every person should drink at least 8 glasses of water every day in order to maintain good health.
2. I gave my children fruits, root crops and banana as snacks.
3. It is ok to wash vegetables and meat with any kind of water.
4. We have a vegetable garden at home.
5. Eating rice alone is enough to provide humans the proper nutrition for good health.
6. I have difficulty convincing my children to eat vegetables.
7. I sliced my vegetables first before I wash them.
8. I put oil into the food when cooking.
9. We only serve vegetables 3 times a week.
10. We wash our hands after we use the toilet, before we prepare food and before we eat.
11. We make sure that flies do not come to our food.
12. We boil our drinking water we got from rain and from the pond before we drink it.
13. My children are breast-fed for 2 years.
14. Kitchen and eating utensils must be washed with clean water to prevent diseases.
15. Deworming is important to make children healthy.

(b) A positive statement ideally shall have move agree responses and a negative statement shall have less agree responses
(c) McNemar's test was conducted to determine whether there was a significant difference in the proportion (increase or decrease) over time.
If p-value < 0.05, then the proportion was statistically significant at 5%. If p-value < 0.01, then the proportion was statistically significant at 1%. Note: "No responses" were excluded from the analysis.

Statements 1, 2, 5 and 9 relate to dietary diversification and to the consumption of clean drinking water. The Htee Pu CSV and Ma Sein CSVs exhibited significant improvements in the practice of giving children fruits, root crops and bananas as snacks. Htee PU CSV together with Saktha CSV also showed improvements in the practice of including vegetables in the diet more than three times a week, while Saktha CSV also showed a significant improvement in the practice of not only eating rice to ensure proper nutrition. All four

CSVs showed significant improvements in the practice of consuming the recommended amount of drinking water per day.

In agreement with the improved awareness, all four CSVs showed improvements in the proportion of households having home vegetable gardens, which were statistically significant for Htee Pu and Saktha (statement 4).

In relation to hygiene practices, not all CSVs showed significant improvements. The TKM CSV showed improvements in the practice of using clean water to wash vegetables (statement 3). The Htee Pu and Ma Sein CSVs also showed significant improvements in the practice of boiling rain and pond water before drinking (statement 12). Rain and pond water are important sources of water in the dry zone and delta regions, where the Htee Pu and Ma Sein CSVs are located, while upland and hilly villages may have more access to spring water for drinking. The Ma Sein and Saktha CSVs showed significant improvements in the practice of deworming children.

4. Conclusions

In this study, we investigated the value of promoting climate-smart agriculture (CSA) practices, coupled with community-level nutrition education and awareness building, to address food insecurity and inadequate nutrition for the overall enhancement of rural livelihoods in Myanmar. Our findings indicated that (based on data collected for two years across four climate-smart villages in Myanmar), CSA can contribute to diversifying and improving the quality of food consumed by households. Both diversification and intensification are key strategies in CSA efforts to sustain small farms, ecologically and economically, while generating critically important nutrition and food security benefits.

Most of the introduced and implemented CSA options that produce nutrient-dense foods (e.g., fruits, vegetables and small livestock) have not generated immediate benefits to households. It is likely that rural communities in Myanmar equate food security with rice, a commodity that was not a focus of the CSA project. In future studies, further consideration of the local food system dimensions, particularly in terms of how households access food, is warranted. Our findings suggest that community education efforts could help communities to understand the benefits that farm diversification can confer in establishing resilience and for fostering local adaptation to climate change manifestation.

Our analysis of KAP indicated that while there is a mix of improvements, there is a poor understanding of households' knowledge, attitudes and practices in relation to nutrition, food choices, food preparation and sanitation and hygiene.

We also observed that the improvements from the CSA interventions were different across the four CSVs. This may suggest that community-level nutrition education can be further improved, possibly by customizing it according to the particular food system and agro-ecosystem features of each CSV. Such education will likely be necessary to more effectively communicate the potential of leveraging climate-smart agriculture for nutrition.

Author Contributions: The field research conceptualization, methodological design and data collection was overseen and conducted by W.J.B., J.G., S.M.N., C.M. and P.S.T. The data collected were analyzed by A.H., G.B., W.J.B., P.C.M., J.G. and C.S. The initial figures and tables were generated by G.B. and A.H., and subsequently finalized by W.J.B. The paper was drafted by A.H. under supervision of C.S., P.C.M., W.J.B. and J.G., with final drafts generated by W.J.B. and finalized and edited by W.J.B., P.C.M. and C.S. Funding acquisition was by W.J.B. and J.G. All authors have read and agreed to the published version of the manuscript.

Funding: This research was funded by the International Development Research Center (IDRC) Canada, grant number 108748-001. The APC was funded by the same grant.

Institutional Review Board Statement: The studies involving human participants were reviewed and approved by International Development Research Center Canada. The patients/participants provided their written informed consent to participate in this study.

Informed Consent Statement: All subjects gave their informed consent for inclusion before they participated in the study. The study protocol was approved by the International Development Research Center-Canada and the International Institute of Rural Reconstruction.

Data Availability Statement: The data supporting the conclusions of this study are available upon reasonable request from the authors, with the exception for data that identifies the personal information of the research participants.

Acknowledgments: This research was funded by the International Development Research Center (IDRC) Canada and the Consultative Group of International Agricultural Research Centers—Climate Change, Agriculture and Food Security Program (CGIAR-CCAFS). The study was part of a project, Climate and Nutrition Smart Villages as Platforms to Address Food Insecurity in Myanmar, led by the International Institute of Rural Reconstruction (IIRR). The authors would like to acknowledge the work of the research teams in Myanmar for their contribution and support during the study.

Conflicts of Interest: The authors declare no conflict of interest.

Abbreviations

CSV	climate smart villages
CSA	climate smart agriculture
HFIAS	householkd food insecurity and access score
HDDS	household diet diversity score
KAP	knowledge, attitudes and practices
TKM	Taungkhamauk

References

1. Seager, R.; Ting, M.; Held, I.; Kushnir, Y.; Lu, J.; Vecchi, G.; Huang, H.-P.; Harnik, N.; Leetmaa, A.; Lau, N.-C.; et al. Model projections of an imminent transition to a more arid climate in Southwestern North America. *Science* **2007**. *316*, 1181–1184. [CrossRef]
2. Myers, S.S.; Smith, M.R.; Guth, S.; Golden, C.D.; Vaitla, B.; Mueller, N.D.; Dangour, A.D.; Huybers, P. Climate change and global food systems: Potential impacts on food security and undernutrition. *Annu. Rev. Public Health* **2017**, *38*, 259–277. [CrossRef]
3. Fanzo, J.; Davis, C.; McLaren, R.; Choufani, J. The effect of climate change across food systems: Implications for nutrition outcomes. *Glob. Food Secur.* **2018**, *18*, 12–19. [CrossRef]
4. Davis, K.F.; Downs, S.; Gephart, J.A. Towards food supply chain resilience to environmental shocks. *Nat. Food* **2021**, *2*, 54–65. [CrossRef]
5. *The State of Food Security and Nutrition in the World 2020*; FAO; IFAD; UNICEF; WFP; WHO: Geneva, Switzerland, 2020. [CrossRef]
6. Smith, M.R.; Myers, S.S. Impact of anthropogenic CO_2 emissions on global human nutrition. *Nat. Clim. Chang.* **2018**, *8*, 834–839. [CrossRef]
7. Marshman, J.; Blay-Palmer, A.; Landman, K. Anthropocene crisis: Climate change, pollinators, and food security. *Environments* **2019**, *6*, 22. [CrossRef]
8. Sánchez, B.; Rasmussen, A.; Porter, J.R. Temperatures and the growth and development of maize and rice: A review. *Glob. Chang. Biol.* **2014**, *20*, 408–417. [CrossRef]
9. Zhao, C.; Liu, B.; Piao, S.; Wang, X.; Lobell, D.B.; Huang, Y.; Huang, M.; Yao, Y.; Bassu, S.; Ciais, P.; et al. Temperature increase reduces global yields of major crops in four independent estimates. *Proc. Natl. Acad. Sci. USA* **2017**, *114*, 9326–9331. [CrossRef]
10. Wang, J.; Vanga, S.; Saxena, R.; Orsat, V.; Raghavan, V. Effect of climate change on the yield of cereal crops: A review. *Climate* **2018**, *6*, 41. [CrossRef]
11. Lamaoui, M.; Jemo, M.; Datla, R.; Bekkaoui, F. Heat and drought stresses in crops and approaches for their mitigation. *Front. Chem.* **2018**, *6*, 26. [CrossRef]
12. Grossiord, C.; Buckley, T.N.; Cernusak, L.A.; Novick, K.A.; Poulter, B.; Siegwolf, R.T.W.; Sperry, J.S.; McDowell, N.G. Plant Responses to rising vapor pressure deficit. *New Phytol.* **2020**, *226*, 1550–1566. [CrossRef]
13. Steyn, N.; Nel, J.; Nantel, G.; Kennedy, G.; Labadarios, D. Food variety and dietary diversity scores in children: Are they good indicators of dietary adequacy? *Public Health Nutr.* **2006**, *9*, 644–650. [CrossRef]
14. Neufeld, L.M.; Beal, T.; Larson, L.M.; Cattaneo, F.D. Global landscape of malnutrition in infants and young children. In *Nestlé Nutrition Institute Workshop Series*; Michaelsen, K.F., Neufeld, L.M., Prentice, A.M., Eds.; Karger Publishers: Basel, Switzerland, 2020; Volume 93, pp. 1–14. [CrossRef]
15. Nicholson, C.F.; Stephens, E.C.; Kopainsky, B.; Jones, A.D.; Parsons, D.; Garrett, J. Food security outcomes in agricultural systems models: Current status and recommended improvements. *Agric. Syst.* **2021**, *188*, 103028. [CrossRef]
16. Hossain, M.S.; Ferdous, S.; Raheem, E.; Siddiqee, M.H. The Double burden of malnutrition—Further perspective. *Lancet* **2020**, *396*, 813–814. [CrossRef]

17. Kuehn, L.; McCormick, S. Heat exposure and maternal health in the face of climate change. *Int. J. Environ. Res. Public Health* **2017**, *14*, 853. [CrossRef]
18. Lappé, M.; Jeffries Hein, R.; Landecker, H. Environmental politics of reproduction. *Annu. Rev. Anthropol.* **2019**, *48*, 133–150. [CrossRef]
19. Victora, C.G.; Adair, L.; Fall, C.; Hallal, P.C.; Martorell, R.; Richter, L.; Sachdev, H.S. Maternal and child undernutrition: Consequences for adult health and human capital. *Lancet* **2008**, *371*, 340–357. [CrossRef]
20. Briend, A.; Khara, T.; Dolan, C. Wasting and stunting—Similarities and differences: Policy and programmatic implications. *Food Nutr. Bull.* **2015**, *36* (Suppl. 1), S15–S23. [CrossRef]
21. Ministry of Agriculture, Livestock and Irrigation (MOALI). Myanmar Agriculture Development and Investment Plan 2018/2019 –2022/2023. 2018. Available online: https://www.lift-fund.org/download/file/fid/3787 (accessed on 10 July 2020).
22. WHO. Global Nutrition Targets 2025 Policy Brief Series, Geneva. 2014. Available online: https://www.who.int/publications/i/item/WHO-NMH-NHD-14.2 (accessed on 20 October 2021).
23. National Nutrition Centre, Department of Public Health and Ministry of Health and Sport. Myanmar Micronutrient and Food Consumption Survey (MMFCS) (2017–2018) Interim Report, February 2019. Available online: https://www.mohs.gov.mm/page/7339 (accessed on 5 July 2020).
24. Smith, P.; Clark, H.; Dong, H.; Elsiddig, E.A.; Haberl, H.; Harper, R.; House, J.; Jafari, M.; Masera, O.; Mbow, C.; et al. Chapter 11—Agriculture, forestry and other land use (AFOLU). In *Climate Change 2014: Mitigation of Climate Change. IPCC Working Group III Contribution to AR5*; Cambridge University Press: Cambridge, UK, 2014.
25. Springmann, M.; Clark, M.; Mason-D'Croz, D.; Wiebe, K.; Bodirsky, B.L.; Lassaletta, L.; De Vries, W.; Vermeulen, S.J.; Herrero, M.; Carlson, K.M.; et al. Options for keeping the food system within environmental limits. *Nature* **2018**, *562*, 519–525. [CrossRef]
26. IPCC. *Global Warming of 1.5 °C. An IPCC Special Report on the Impacts of Global Warming of 1.5 °C above Pre-Industrial Levels and Related Global Greenhouse Gas Emission Pathways, in the Context of Strengthening the Global Response to the Threat of Climate Change, Sustainable Development, and Efforts to Eradicate Povearty*; Masson-Delmotte, V., Zhai, P., Pörtner, H.-O., Roberts, D., Skea, J., Shukla, P.R., Pirani, A., Moufouma-Okia, W., Péan, C., Eds.; Intergovernmental Panel on Climate Change: Geneva, Switzerland, 2018.
27. Lamb, W.F.; Wiedmann, T.; Pongratz, J.; Andrew, R.; Crippa, M.; Olivier, J.G.J.; Wiedenhofer, D.; Mattioli, G.; Khourdajie, A.A.; House, J.; et al. A review of trends and drivers of greenhouse gas emissions by sector from 1990 to 2018. *Environ. Res. Lett.* **2021**, *16*, 073005. [CrossRef]
28. Lipper, L.; Thornton, P.; Campbell, B.M.; Baedeker, T.; Braimoh, A.; Bwalya, M.; Caron, P.; Cattaneo, A.; Garrity, D.; Henry, K.; et al. Climate-smart agriculture for food security. *Nat. Clim. Chang.* **2014**, *4*, 1068–1072. [CrossRef]
29. FAO. *Climate-Smart Agriculture Policies, Practices and Financing for Food Security, Adaptation and Mitigation*; FAO: Rome, Italy, 2010. Available online: https://www.fao.org/3/i1881e/i1881e00.htm (accessed on 10 July 2020).
30. De Pinto, A.; Cenacchi, N.; Kwon, H.-Y.; Koo, J.; Dunston, S. Climate smart agriculture and global food-crop production. *PLoS ONE* **2020**, *15*, e0231764. [CrossRef]
31. Mie Sein, Z.M.; Ullah, I.; Syed, S.; Zhi, X.; Azam, K.; Rasool, G. Interannual variability of air temperature over Myanmar: The influence of ENSO and IOD. *Climate* **2021**, *9*, 35. [CrossRef]
32. Lar, N.M.; Pumijumnong, N.; Roachanakanan, R.; Arunrat, N.; Tint, S. An assessment of climate variability on farmers' livelihoods vulnerability in ayeyarwady delta of Myanmar. *App. Environ. Res.* **2018**, *40*, 1–12. [CrossRef]
33. Barbon, W.J.; Myae, C.; Vidallo, R.; Thant, P.S.; Monville-Oro, E.; Gonsalves, J. Applying Participatory Action Research Methods in Community-Based Adaptation With Smallholders in Myanmar. *Front. Clim.* **2021**, *3*, 734053. [CrossRef]
34. Htwe, N.M.; Htwe, N.M.; The, N.E.M.; Naing, N.N.Z.; Hein, Y. Documenting the application of the Myanmar Climate-Smart Agriculture Strategy. CCAFS Working Paper No. 292. Wageningen, the Netherlands: CGIAR Research Program on Climate Change, Agriculture and Food Security (CCAFS). 2019. Available online: www.ccafs.cgiar.org (accessed on 15 October 2021).
35. CGIAR. Brochure. Climate-Smart Villages: An AR4D Approach to Scale up Climate-Smart Agriculture. 2017. Available online: https://hdl.handle.net/10568/79353 (accessed on 15 July 2021).
36. CCAFS; UNFAO. *Questions & Answers: Knowledge on Climate-Smart Agriculture*; United Nations Food and Agriculture Organisation (UNFAO): Rome, Italy, 2014.
37. Coates, J.; Swindale, A.; Bilinsky, P. *Household Food Insecurity Access Scale (HFIAS) for Measurement of Household Food Access: Indicator Guide (v. 3)*; FANTA: Washington, DC, USA, 2007; FHI360/FANTA. Available online: https://www.fantaproject.org/sites/default/files/resources/HFIAS_ENG_v3_Aug07.pdf (accessed on 15 July 2020).
38. Swindale, A.; Paula, B. *Household Dietary Diversity Score (HDDS) for Measurement of Household Food Access: Indicator Guide (v.2)*; FANTA: Washington, DC, USA, 2006; FHI 360/FANTA.

climate

MDPI

Article

Geospatial Assessment of Flood-Tolerant Rice Varieties to Guide Climate Adaptation Strategies in India

Nisha Koppa [1] and Giriraj Amarnath [2,*]

[1] International Water Management Institute, New Delhi 110012, India; nishakoppaa@gmail.com
[2] International Water Management Institute, Battaramulla 10120, Sri Lanka
* Correspondence: a.giriraj@cgiar.org; Tel.: +94-11-2880000

Abstract: Rice is the most important food crop. With the largest rain-fed lowland area in the world, flooding is considered as the most important abiotic stress to rice production in India. With climate change, it is expected that the frequency and severity of the floods will increase over the years. These changes will have a severe impact on the rain-fed agriculture production and livelihoods of millions of farmers in the flood affected region. There are numerous flood risk adaptation and mitigation options available for rain-fed agriculture in India. Procuring, maintaining and distributing the newly developed submergence-tolerant rice variety called Swarna-Sub1 could play an important role in minimizing the effect of flood on rice production. This paper assesses the quantity and cost of a flood-tolerant rice seed variety- Swarna-Sub1, that would be required during the main cropping season of rice i.e., *kharif* at a district level for 17 major Indian states. The need for SS1 seeds for rice production was assessed by developing a geospatial framework using remote sensing to map the suitability of SS1, to help stakeholders prepare better in managing the flood risks. Results indicate that districts of Bihar, West Bengal and Uttar Pradesh will require the highest amount of SS1 seeds for flood adaptation strategies. The total estimated seed requirement for these 17 states would cost around 370 crores INR, less than 0.01 percent of Indian central government's budget allocation for agriculture sector.

Keywords: remote sensing; GIS; flood tolerant seeds; Swarna-Sub1 Rice; climate adaptation

Citation: Koppa, N.; Amarnath, G. Geospatial Assessment of Flood-Tolerant Rice Varieties to Guide Climate Adaptation Strategies in India. *Climate* **2021**, *9*, 151. https://doi.org/10.3390/cli9100151

Received: 12 August 2021
Accepted: 3 October 2021
Published: 13 October 2021

1. Introduction

Floods are amongst the most common natural disaster across the globe. They pose a threat not only to the environment but also to society as they endanger lives, properties and livelihoods of the people. The report and analysis collected by UNISDR and the Belgian-based Centre for Research on the Epidemiology of Disasters (CRED) highlighted that in the year 2021, the impacts of floods were felt heavily across the developing countries in Africa and Asia [1]. For populous continent such as Asia, around 42% of the global flood events occurred between 1950–2020, which affected around 3.65 billion people and economic losses accounted to USD 556 billion [1,2]. Within Asia, the south Asian region is highly vulnerable to flood impacts. Recent estimates for the South Asian region shows that, between 2000 and 2020, these countries have experienced 11% of the world's natural disasters and 12% of floods and droughts, making over 700 million people and 190 million ha of agricultural land vulnerable [1]. Considering the increasing global temperature, unplanned urban growth and environmental degradation, it is likely that the frequency and severity of flood risks will increase in the exposed countries such as Bangladesh, India and Nepal [3]. Additionally, for largely agrarian countries such as India and Bangladesh, these changes especially threaten the agriculture sector, as it increases the ambiguity for the small-scale and poor farmers whose livelihoods are dependent on the agricultural production in these regions.

For a large and populous country like India, the increasing weather variability and the subsequent impact of disasters such as floods is concerning. For instance, official statistics reveal that 15% of the total area in India (which amounts to approximately 49.82 million hectares) is extremely vulnerable to floods [4]. Moreover, the variable summer monsoon in India has often precipitated floods, especially in the basins of the Himalayan rivers. These large river basins, such as the Indus, Ganges and Brahmaputra, cause significant monsoon runoff, leading to immense flooding in the plains [5]. Considering that these rich and fertile plains are used for agriculture production, frequent floods in the region affects the people dependent on agriculture. One of the most commonly grown crops in the fertile plains of Indo-Gangetic River basin is 'Rice' (also referred to as paddy in this study). Currently, rice is grown across 43.86 million hectare of area and the production level is 104.80 million tonnes in India [6]. The rice crop which requires a lot of water, is commonly sown during the months of July-October i.e., during the monsoon season in India. The rice farmers in this region take a heavy toll as the recurring floods just after crop sowing leads to crop losses. While rice crop can thrive well in flooded soils, the crop is still vulnerable to complete submergence for longer days and around 16% of the world's rice production area is affected by recurring submergence due to flash floods [7–9]. These recurring impacts of floods in India necessitate improving the farmer's knowledge with regards to adapting and coping methods along with improving flood-resilient infrastructure to reduce the damaging impacts on farming communities.

1.1. Extent and Impact of Flooding on Rice Production

Rice production in rain-fed low lands is often severely affected as the crop at different growing stages suffers from various stresses, such as limited gas diffusion, effusion of soil nutrients, mechanical damage, increased susceptibility to pests and diseases and stresses due to low-light due to flooding (also called as submergence) or water-logging [10,11]. The frequent flooding during rice cultivation (which occurs during the monsoon months) in rain-fed lowland areas of South Asia leads to a complete submergence of the rice crop for approximately 10–15 days. While rice has some adaptive traits for tolerance to submergence, the low-land rice cultivars used in South Asian countries are still sensitive to complete submergence [12]. In India, the Indo-Gangetic River basin, which is a favourable belt for rice cultivation, is also the most flood-affected region in India. Moreover, around 30% of the total rice growing area, which amounts to 12–14 million ha is prone to flash flooding with an average productivity of only 0.5–0.8 tonnes per ha as compared to 2 tonnes per ha in favourable lowlands [13]. With a high incidence and severity of floods, small and poor farmers incur heavy economic losses. Figure 1 shows the average agriculture area (in hectare) affected by floods in 17 major Indian States. It can be seen from Figure 1 that Bihar is the worst flood hit state in India followed by Uttar Pradesh, West Bengal and Assam-all situated in the Indo-Gangetic plains, i.e., each has one of the two major rivers (or its tributaries) of Ganga and Brahmaputra rivers flowing from the Himalayas. In Bihar, nearly 73% of the total geographical area and 76% of its total population are constantly under the threat of flood [14]. Almost every year, there is severe flooding in the state of Bihar which causes loss to lives, properties and livelihood [15]. With the onset of the monsoon, the rivers originating from Himalayas flow down with massive force, causing rivers such as *Koshi* and *Ganges* to rise above the danger level, which leads to severe floods in northern parts of Bihar.

In Figure 2, the graph plots the state-wise five-year average rice production in India. It can be seen that West Bengal produces around 15,000 tonnes of rice on average per year, followed by Uttar Pradesh, Punjab, Andhra Pradesh and Odisha. Comparing this with the agricultural flooding, it can be seen that states which experience high levels of flooding are also amongst the top producers of rice, except Bihar. For instance, the eastern state of West Bengal ranks first in rice area and production in the country. However, around 30% of the rice growing area in this state comes under the rain-fed lowlands which suffer from frequent flash floods due to unpredictable rainfall during the major rice growing season (*kharif*), leading to a drastic reduction in yield [13,16].

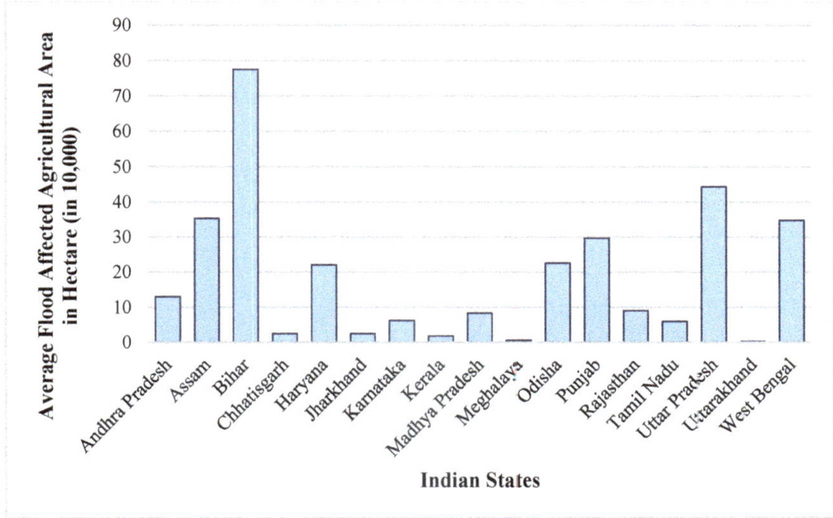

Figure 1. Average flood affected agricultural areas between 2000 to 2018 for the 17 Indian States in India. Source: IWMI.

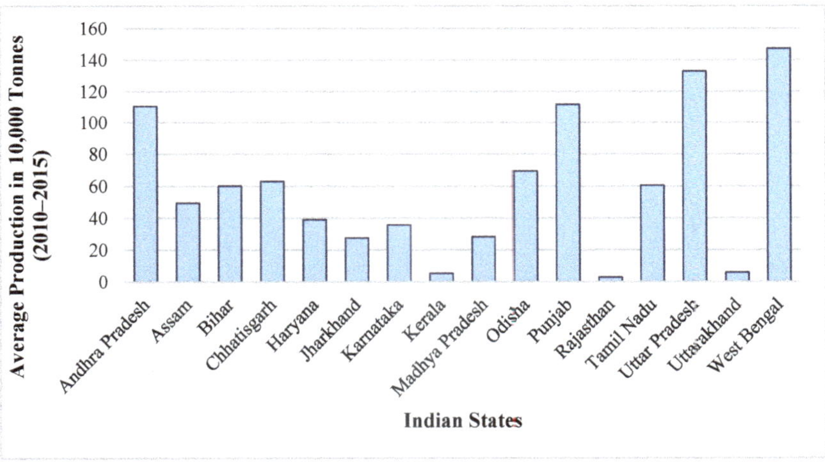

Figure 2. State-wise average rice production in India, 2010–2015. Source: Author's elaborations using data from the Directorate of Economics and Statistics, India.

1.2. Flood-Tolerant Seed Variety: Swarna-Sub1

Quality seeds are the most basic and crucial input that is required in agriculture production, as the response of other inputs of production is dependent on the quality of seeds being used. Several poor farmers in the flood prone regions are switching from high-yielding varieties of seeds to traditional and local varieties which can withstand submergence to cope with flooding. However, studies indicate that these local and traditional varieties often give very low yield, making it unprofitable for the farmers [17]. With floods affecting every year, in some places, farmers often abandon cultivation and leave their fields fallow during the monsoon season [18]. In areas where high-yielding but submergence-intolerant rice varieties are cultivated, farmers suffer from heavy crop losses caused by recurrent flooding. Amongst recent strategies adopted to overcome the problems of flooding in agricultural areas is the development and dissemination of high-yielding varieties that are flood-tolerant along with acceptable agronomic and quality traits [13].

The work on the development of flood-tolerant rice varieties was started in the year 1987 at International Rice Research Institute (IRRI), and the submergence-tolerant gene-SUB1A- was developed [18,19]. Since then, several new rice varieties have been developed by introgressing SUB1 gene into high yielding and high yielding rice varieties [13]. The Sub1 gene was fused into several other popular varieties of rice grown in South and South-East Asia, such as *Swarna, Sambha mahsuri, BR11*, etc., which can ensure rice production in flood-prone areas [13,17,20]. Amongst these new varieties, Swarna-Sub 1 (SS1)–a submergence-tolerant rice variety–is considered extremely viable in the flood affected regions in India and has been distributed to rice farmers in eastern India since 2008 [21]. SS1 survives full submergence for up to 14 days as it was developed by introgressing a single quantitative trait locus that causes submergence tolerance in Swarna, which is a popular rice variety in eastern India [19]. Even under normal conditions, SS1 is considered to show no significant differences in agronomic performance, grain yield or grain quality as compared to Swarna [7,21]. Moreover, it is already available in the markets for commercial cultivation. This new variety can ensure rice production in flood-prone areas owing to its tolerance to submergence.

Several studies have documented the performance of SS1 amongst farmers in both experimental and non-experimental settings to understand the seed's yield advantages under different submergence stages. In a non-experimental setting, a study on the farmers in eastern Uttar Pradesh and Odisha on rice production in 2011 *Kharif* season reveals a yield advantage of SS1 compared to Swarna under medium-duration submergence i.e., 8 to 14 days of submergence [19]. In a comparative study by [7] for all Sub1 varieties, including SS1, the study shows that all submergence-tolerant seed varieties were promising, with either similar or higher yield than their non-Sub1 counterparts. Conducting a randomized control trial in flood affected fields of Orissa, ref. [21] found that SS1 had a positive impact on the yield when fields were flooded compared to Swarna. By effectively replacing Swarna with SS1 seeds, a significant improvement in rice production is expected of approximately 9 to 12% [21]. Furthermore, a wide scale adoption of SS1, prior to floods would have resulted in an approximate increase in rice production by 26% [21]. Apart from flood tolerance, the SS1 seeds also reduce the risk of yield loss during the normal growing period i.e., when there is no flood impact [20,22,23]. Overall, the development of submergence-tolerant varieties allows farming communities to become more resilient to existing and growing flooding risks [24].

To cope with recurring flood impacts, flood-tolerant seeds are increasingly being adopted by farmers to bring productivity gains to flood prone areas. Potential benefits of adopting SS1 seed varieties have been explored across India and it has been estimated that a large-scale development and dissemination of SS1 seed varieties can be beneficial for 30–40% or 12–14 million ha of the 44 million ha of rice cultivated area which is exposed to recurrent flooding in the Gangetic basin in India alone [21]. In such conditions, seeds that can withstand flood submergence for a longer period can indeed be a game changer in making small holder farmers resilient to frequent floods. Technology such as Swarna-Sub1 are already available for use [22], however in regions where flooding is predictable, there is a requirement to do a needs assessment for such stress-tolerant seeds and arrange for seed procurement accordingly. Therefore, it is imperative to identify the quantity of certain important seed varieties such as SS1, which can be beneficial in those agriculture areas. Considering the potential benefits offered by the SS1 seed variety in terms of coping with floods, an assessment of the amount of SS1 seed requirement will allow the governments to ensure seed security and cater to the flood-affected farmers need. As per our knowledge there are no systematic in-depth within-country assessments of seed requirement for flood risk management.

Following from the above discussion, this paper presents an assessment of the required flood-tolerant rice seed variety called Swarna-Sub1 (SS1), as an adaptive flood risk management method during the main rice cropping season called Kharif (or monsoon) at the district level in India. Using a combination of flood area estimates derived from remote sensing data from the period 2000–2018 and land-use data from the government database, this paper provides an estimate of SS1 seeds that would be required as an adaptive flood strategy in seed banks for 17 major states in India. Specifically, the paper presents an estimate of the amount of SS1 seeds that will be required to be maintained in the seed banks and the cost implication on the exchequer for procuring the seeds. Furthermore, these estimates can also be useful for planning in-season flood risk management through the revival of crop production and in maintaining food supply. The quantity of seeds required across different districts show large differences based on the cultivated area and the severity of floods in that district. The estimates from this study provides valuable information to the policy makers, who can make informed investment decisions to establish new seed banks in locations where floods are recurring with a high probability or store additional seeds in existing seed banks. In addition, our analysis shows how remote sensing data can be used in complement with land use data to obtain reliable estimates of seed requirement needs that can improve the preparedness of government departments in procuring necessary stress-tolerant seeds in areas with the most urgent demand.

2. Data and Methodology

2.1. Mapping Flood Extent Using Satellite Data

To map the extent of long-term flood records the study employed NASA MODIS (MOD09A1) eight-day composite surface reflectance product with 500 m spatial resolution between 2001 and 2018. Furthermore, to determine the severity and the duration of the flooding, two major water indices, namely, Enhanced Vegetation Index (EVI) and Land Surface Water Index (LSWI), were applied to differentiate land and water pixels using the threshold approach [25,26]. In the case of EVI, a threshold value of less than or equal to 0.05 and an LSWI less than or equal to 0, as the first criteria, were adapted. The second criteria applies if the EVI value less than or equal to 0.3 and the difference in the value of the EVI and LSWI (DVEL) values is less than or equal to 0.05, to estimate the overall inundation extent on each 8 day MODIS product. Steps involved in image processing and computation of land and water indices and its thresholds can be referred here [27].

Figure 3 shows the comparison of MODIS Terra satellite data with flood inundation extent for the 2010 flood event clearly shows good agreement with the satellite data. Using ArcGIS Spatial Analyst toolbox, the time series inundation product was produced monthly and annually for the flood extent information, to identify the flood duration and its occurrences over 18 years (2001–2018).

2.2. Data for Estimating Seed Requirement

2.2.1. Land-Use Data

Focusing on the cultivation of rice during the monsoon season, the district-wise data on land use was collected from Directorate of Economics and Statistics, Ministry of Agriculture and Family Welfare, Government of India from the year 2000 to 2014. From this, data on the area under paddy sown during the *Kharif* season and the net area sown under all crops was collected. Net area sown, defined as the total area which is sowed at least once in the same year, was used for the analysis. The dataset provides details on the area sown with rice crop by seasons, which are broadly categorised as autumn, winter and summer. The main rice growing season in India is the *Kharif* season and roughly 84% of the rice is grown during this season. The sowing time of winter (*Kharif*) rice is in June–July which is when monsoon arrives and it is harvested between the months of November–January). Given that the focus of this study is on the *Kharif* season, when most of the flooding happens in India, the land-use data on area during *Kharif* season was used.

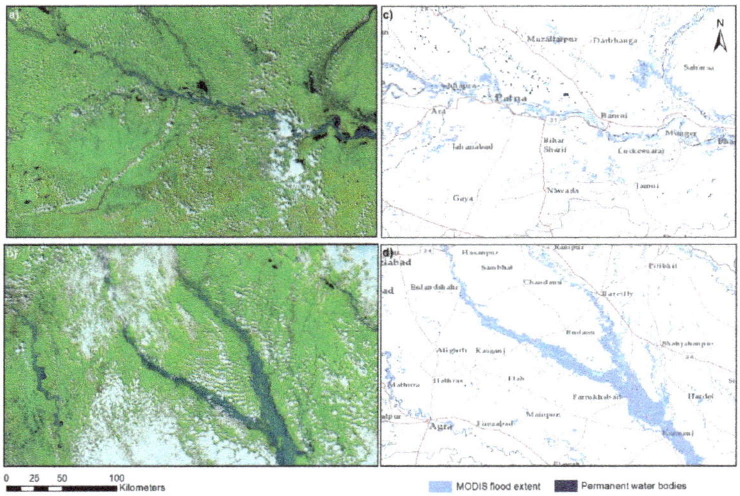

Figure 3. (**a**,**b**) MODIS TERRA satellite data taken on 22 September 2010; (**c**,**d**) are the flood inundation extent covering the states of Uttar Pradesh and Bihar, India. Source: IWMI.

Furthermore, the study considers only 17 states for analysis, which are Andhra Pradesh, Assam, Bihar, Chhattisgarh, Haryana, Jharkhand, Karnataka, Kerala, Madhya Pradesh, Meghalaya, Odisha, Punjab, Rajasthan, Tamil Nadu, Uttar Pradesh, Uttarakhand and West Bengal. These 17 states represent approximately 89% of the rice-growing area across the Indian states. The states of Maharashtra and Gujarat are considerably big, both in terms of their economic capacity and agricultural share; however, rice is not the primary grown crop in these regions owing to their agro-climatic conditions which are better suited for growing cotton, sugar cane, millet and so on. Overall, the rice growing area is 5% for both the states combined. The remaining states that were not included in the study were excluded either because district level land use data was not available, for states such as Gujarat, Maharashtra, Manipur and Telangana; or because average area used for growing rice was negligible, such as for the states of Arunachal Pradesh, Himachal Pradesh, Jammu and Kashmir.

2.2.2. Data on Seed Rates

Along with the area under paddy, our estimates will also require the seed rates for paddy cultivation. Generally, the seed rate is understood as the quantity of seed that is required to sow a unit area of land for optimum crop-production. The seed rates vary with the methods of crop establishment by farmers. In India, there are three primary ways of establishing rice for cultivation; *dry seeding*, *wet seeding* and *transplanting* [28,29]. In the dry seeding method, the dry seeds are sown by either broadcasting, i.e., scattered across the field; drilling, wherein the seeds are drilled manually or mechanically into the field; or dibbling, which is practiced along the mountain slopes using traditional methods and tools to sow the seeds [28]. In the wet-seeding method, pre-germinated seeds are sown in wet puddled soils using the same techniques as broadcasting or drilling. Both the dry- and wet-seeding methods are collectively known as the *direct seeding method*, as the seeds are sown directly using broadcasting, drilling and dibbling [28].

In the transplanting method, the seedlings are first grown in nurseries and later replanted to the main fields, which requires a lot of labour input. One of the reasons for employing such labour-intensive methods for rice cultivation is due to a higher labor supply resulting from population growth [30]. Alternatively, the incentives of adopting the direct seeding method increase when both labour and water availability is low, it and is especially adopted during the dry seasons or in dry regions. Given the higher proportion of rain-fed rice area in India, it has subsequently led to a lesser adoption of the direct seeding method in India [28,30,31]. Consequently, transplanting became a dominant method for rice establishment in India. The estimates of the percentage of total rice area established by direct seeding method reveals that only 28% of the total paddy area in India is established using this method [28]. Due to unavailability of data on the percentage of paddy area under different sowing methods, this paper assumes that 72% of total rice area in India is established using the transplanting method and 28% of the total rice area is established using the direct-seeding method, following from the above discussion. This is an important assumption in this study, as the seed rates for both the methods vary significantly.

Based on the above discussion, the data on seed rates for each state were collected from the reports published by the state department of agriculture, which includes the details of rice production and output. The data on seed rates based on the method of direct seeding/transplanting are presented for each state in Table 1 (below). For direct seeding, wherever data was available for broadcasting, dibbling or more, an average was calculated. It should be noted that, due to a lack of data on the seed rates for the states of Uttarakhand, Meghalaya and Rajasthan, the data for their neighbouring states were used. For instance, for Uttarakhand, the seed rates used for Uttar Pradesh were used; for Meghalaya, the seed rates of Assam were used; and for Rajasthan, only for direct sowing method, seed rates were unavailable and therefore the seed rates used in Gujarat were used. The final estimation for seed rates is done by taking weighted averaged of the two rice establishment methods, i.e., 28% is cultivated through direct seeding and 72% through transplanting. For example, (see Table 1): for the state of Madhya Pradesh the weighted average will be equivalent to: $(80 \times 0.28) + (50 \times 0.72) = 61.5$, where 0.28 is the percentage of direct seeded rice establishment area and 0.72 is the percentage of transplanted rice establishment area. This was done for all the states and the results are presented in Table 1 (column 3).

Table 1. Seed rates for different rice establishment method.

Sr. No.	States	Seed Rate Using Direct Sowing Method	Seed Rate Using Transplantation	Weighted Average
1	Madhya Pradesh	80	50	58.4
2	Andhra Pradesh	70	50	55.6
3	West Bengal	70	43.33	50.8
4	Odisha	80	30	44
5	Bihar	90	30	46.8
6	Punjab	17.5	25	22.9
7	Uttar Pradesh	75	21	36.12
8	Karnataka	80	62	67.04
9	Kerala	80	60	65.6
10	Tamil Nadu	80	43.33	53.59
11	Haryana	100	35	53.2
12	Chhattisgarh	80	30	44
13	Jharkhand	80	40	51.2
14	Assam	75	40	49.8
15	Uttarakhand	75	21	36.12
16	Himachal	80	30	44
17	Rajasthan	50	25	32

2.3. Estimation of Seed Requirement

Using the satellite-derived agriculture mask data from National Remote Sensing Centre (NSRC), India and data on net area sown for rice during the *Kharif* season collected from Directorate of Economics and Statistics, Ministry of Agriculture and Family Welfare, Government of India, the following estimation method was developed to derive the seed requirement for the districts of 17 Indian states.

The estimation of flood-affected *Kharif* paddy area was performed using the combined data on flood inundation and land-use data. Firstly, using the district-level land-use data from 2000–2014, an average percentage of the *Kharif*-paddy sown area as a percentage total net sown area was estimated. From the time series data on flood inundation, the average agricultural flooded area from the year 2001 to 2018 was estimated. Finally, assuming the proportion of flood affected area that is under paddy to be the same as the proportion of net sown area under paddy, we derive the extent of flood-affected *Kharif*-paddy area in each district. From these district-wise estimations of the flood-affected paddy area and using the average seed rates in each state, we estimate the requirement for flood-tolerant seeds that will be required if we want to provide entire flood-affected paddy areas with Swarna Sub1. Further the estimations were also extended to calculate the cost for procuring the required amount of SS1 seeds at the rate of 40 INR per kg (this is the average cost of procuring SS1 seeds in India).

3. Results

3.1. Flood Inundation Mapping

In this study, the spatio-temporal extent of flooding was assessed for all of India using MODIS TERRA satellite data, which is shown in Figure 4. The map shows the extent of flood severity and its likely impacts on people and agriculture given the large stretch of area is highly prone to flooding and several states are frequently affected due to trans-boundary floods. The flood recurrent map of India shows two main hotspots as the Ganges and the Brahmaputra. These rivers' major tributaries flowing across the states of Bihar, Uttar Pradesh, Assam and West Bengal cause heavy flooding during the monsoon season. As noted in earlier sections, these states are amongst the most flood-prone states in India and are annually affected by flood impacts. Furthermore, from the spatial assessment, weekly flood inundation maps were aggregated into monthly and annual flood inundation maps to derive the frequency of flooding in each pixel across India for a period of 18 years (2001–2018) to produce a flood recurrence map (Figure 4). These flood hotspots quantify the frequent occurrence of flood events in a 20-year mapping period.

3.2. Seed Requirement Estimations

Following the methodology described in the previous section, the seed requirement was estimated at district and state level. The state-wise estimates of total seed requirement, total cost of procuring the seeds (at the rate of 40 Rs/kg) and total cost in both Indian Rupee (INR) and US Dollars (USD (1 USD = 70 Rs)) is provided in Table 2.

Overall, our estimates show that 92,764 tonnes of Swarna Sub1 seeds is required for the flood affected paddy areas in 17 Indian states. The cost of procuring the seeds would be approximately USD 53.01 million. Given that Bihar is the most flood-affected state, the seed requirement is also highest for Bihar, i.e., 21,888 tonnes, and the cost of procuring this is estimated at USD 12.51 million. It is estimated that 18,684 tonnes of seeds, costing USD 10.68 million, would be required. Figure 5 presents an all-India map depicting state-wise seed requirements for better illustration.

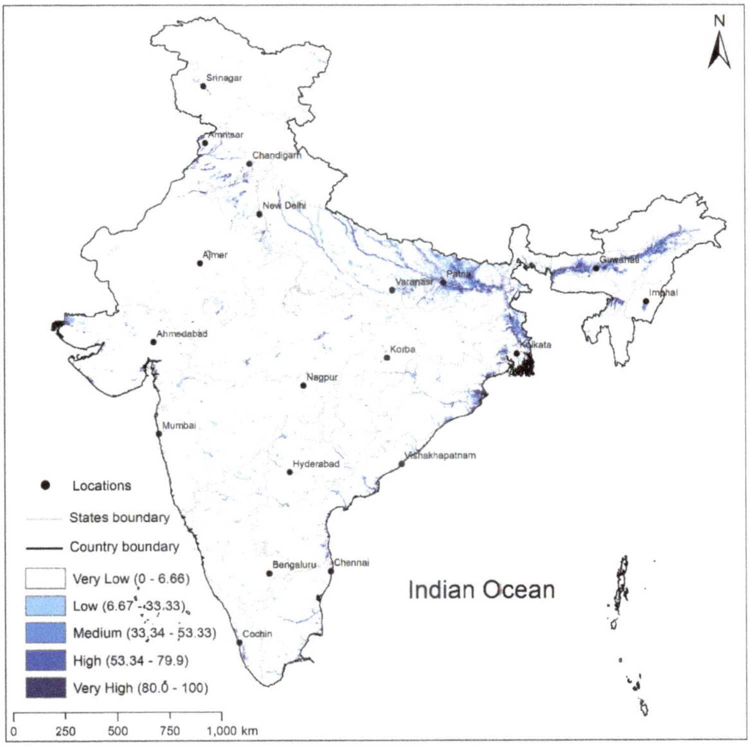

Figure 4. Recurrent flooded areas mapped using MODIS satellite data for India.

Table 2. State-wise total Swarna-Sub 1 seed requirement and cost of procuring in INR and USD.

State	Flood Affected Agricultural Area (ha)	Flood Affected Paddy Area (ha)	Seed Requirement (Tonnes)	Cost in Million INR	Cost in Millions USD	Ranking of States Based on Seed Requirement
Andhra Pradesh	180,152.63	58,571.68	3256	130.24	1.86	8
Assam	484,214.47	305,204.90	15,200	608	8.69	3
Bihar	1,063,189.47	467,756.26	21,888	875.52	12.51	1
Chhattisgarh	31,644.74	26,402.19	1162	46.48	0.66	11
Haryana	183,084.21	65,467.11	3478	139.12	1.99	7
Jharkhand	28,696.05	21,289.74	1092	43.68	0.62	12
Kerala	19,019.74	1796.07	116	4.64	0.07	14
Madhya Pradesh	91,780.26	20,983.37	1224	48.96	0.70	10
Meghalaya	6477.63	1449.99	70	2.80	0.04	15
Odisha	291,390.79	248,963.81	10,958	438.32	6.26	4
Punjab	298,363.16	211,419.15	4844	193.76	2.77	6
Rajasthan	68,869.74	885.49	28	1.12	0.02	16
Tamil Nadu	85,164.47	37,249.61	1996	79.84	1.14	9
Uttar Pradesh	486,614.30	227,700.22	8216	328.64	4.69	5
Uttarakhand	2217.11	605.42	22	0.88	0.01	17
West Bengal	542,385.73	367,816.48	18,684	747.36	10.68	2
Karnataka	76,267.11	7914.67	530	21.20	0.30	13
Total	**3,939,531.60**	**2,071,476.16**	**92,764**	**3,710.56**	**53.01**	

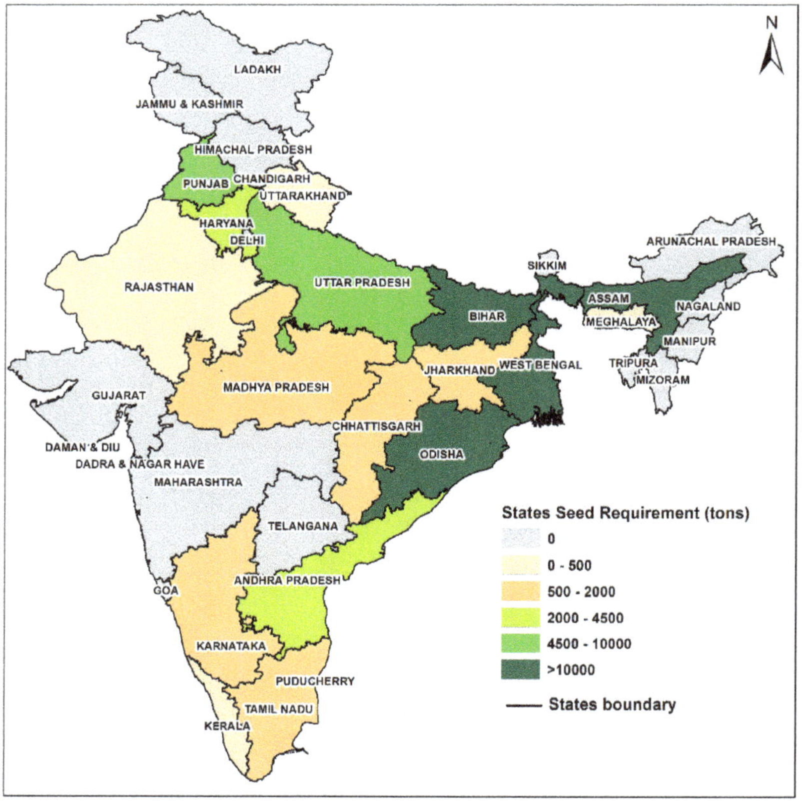

Figure 5. State-wise requirement of Swarna-Sub1 seeds across the study area.

The following figure (Figure 6) provides a district-wise map of India illustrating the seed requirements (in tonnes) for each district. At an aggregate level, the five most flood-affected states requiring highest number of seeds are *Bihar, West Bengal, Assam, Odisha* and *Uttar Pradesh.*

Interestingly, it can be seen that while West Bengal is the largest producer of rice (refer Figure 2), Bihar requires the highest amount of seed for adaption. This is because the area in the *Kharif* rice area affected by flooding is higher in the state of Bihar. Delving further into the state-wise requirement, the results appear consistent with the regional variation in flood impacts. For instance, in the state of Bihar, districts like Gaya and Jamui are drought-prone and the estimates reveal a lower seed requirement for these districts. On the other hand, the districts of Patna, Muzaffarpur and Madhubani in Bihar require the highest quantity of seeds owing to large paddy areas affected by flood. In West Bengal, most districts reveal a high seed requirement given that it is the largest rice producer in the country. Only for the district of Darjeeling, which is at a higher altitude (around 2000 m above sea level), is the seed requirement negligible. In Uttar Pradesh, the districts of Siddharth Nagar and Gorakhpur, situated on the eastern part of the state, show the highest seed requirement. In Odisha, the Kendrapara district requires the highest amount of seeds which is approximately 2574 tonnes. In Assam, Cachar and Lakhimpur top the list of districts requiring seeds to cope with floods. Amongst the Southern states, Andhra Pradesh, which is a top rice-producing state, requires an overall 3256 tonnes of SS1 for its flood affected farms. For the districts of Karnataka and Kerala, the requirement of SS1 is comparatively low. Moreover, in the northern and norther-eastern states such as Haryana,

Meghalaya and Uttarakhand, the seed requirement for some districts is zero, which is most likely because these districts are either not affected by floods or they do not cultivate rice.

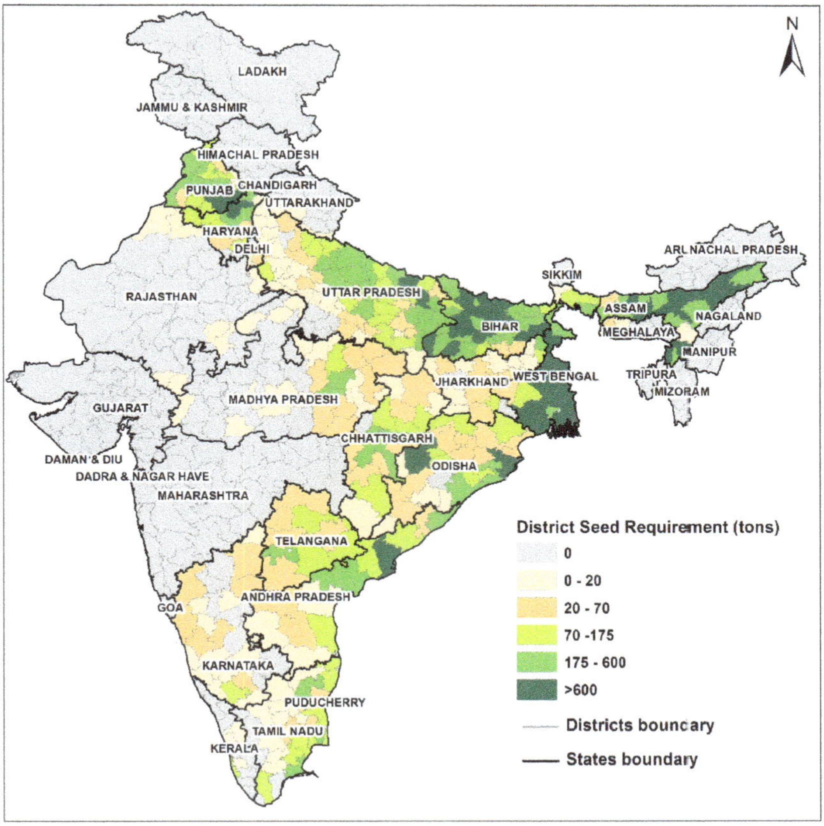

Figure 6. District-wise seed requirement estimation for Swarna-Sub 1 for the selected states of India.

4. Discussion

Despite growing economically, Indians are still hugely dependent on agriculture for their livelihoods. Floods have been a major natural disaster (along with droughts) which impacts Indian agriculture sector heavily. Between 1950–2020, India had 304 major flood events, which affected around 895 million people in the country and cost USD 84 billion [1]. With exacerbating changes in global climate, the severity and intensity of floods will likely increase further. Therefore, there is an urgent need to find ways to adapt to floods and secure poor and smallholder farmers' livelihood. With technological advancement, scientists have developed several viable and stress-tolerant seed varieties which can withstand disasters such as floods and droughts. Given the availability of such seed technologies, there is still a need to understand the amount of seeds required by farmers. With this regard, this study explored the amount of flood-tolerant seeds of the *Swarna-Sub 1* rice variety that would be required by rice cultivators (during the monsoon season) at the district level of India's 17 major states. The estimates reveal an overall requirement of 92,764 tonnes of SS1 rice variety for the flood-affected *Kharif*-paddy areas across all the states. In monetary terms, this would cost approximately 370 crore INR (or USD 53 million) to the exchequer. Considering the government's total spending on the agriculture sector, USD 53 million amounts to less than 0.01 percent of the Indian central government's budget allocation to the Ministry of Agriculture and Farmers' Welfare in 2020. These estimates can be benefited by potential private sector investors at the regional, national and international level, and stakeholders in the seed sector, while making decisions focusing on flood-risk management in agriculture. While great progress has been made in developing advanced flood-tolerant seed varieties, the question remains if these estimates are viable considering if the government or private sector can establish a seed distribution system to ensure all the flood-affected farmers benefit from this.

Focusing on climate change adaptation, an immediate focus should be given to the agriculture areas which are severely affected by floods annually. As noted earlier, around 12 to 14 million ha of rain-fed rice cultivation area could be benefited if the SS1 variety was adopted by the farmers, as it offers better yield even during flood events [21]. With regards to SS1′s adoption and awareness amongst the farmers in India, it can be noted that these seeds have been distributed to farmers, especially in eastern India, since 2008. Since then, the distribution of SS1 seeds has expanded significantly, in particular when the National Food Security Mission, which is a central government initiative to increase the annual production of rice, wheat and pulses, included these seeds in its eastern India programs in 2010 [19,24]. From 2010–2012, around 38,000 tons of paddy seeds were distributed, which reached 1.3 million farmers in Eastern India [21]. Another such initiative to promote and increase SS1′s adoption amongst farmers called the 'Stress Tolerant Rice for Africa and South Asia' (STRASA) was funded by the Bill and Melinda Gates Foundation [21]. These initiatives are slowly bringing in changes and an increasing demand and adoption for the same can be seen in the future.

However, as estimated in this study, the production and dissemination of approximately 90,000 tons of SS1 seeds is still a major concern. In India, the production of seeds has a well-established channel for production [32]. The production is done on the basis of indents from either the private or public sector organizations placed with state or central government institutions such as the Department of Agriculture Cooperation (DoAC), Government of India or State Agriculture University/ National Seed Producers, who consolidate the indents and forwards them to Indian Council of Agriculture Research (ICAR) [33]. These seeds are then supplied to indenting organizations on the basis of allocation by DoAC for multiplication of seeds which are later made available to the farmers. This entire process takes at least 3 years, which implies that the public sector seed companies/state governments should have a pre-determined requirement of seeds at least 3 years in advance [34]. Assessing the requirement of seeds in case of contingency can provide a way to pre-plan and procure large quantities of seeds. Amongst other ways of making these SS1seeds available at appropriate times and affordable prices is by deploying a suitable model, such as a participatory seed production method involving both farmers and local stakeholders in the process; enabling partnership with private sector; seed village scheme; and creating awareness through self-help groups and community-based organizations [33–35]. This can be achieved through a continuous interaction between various institutions, policymakers and concerned stakeholders, which can further strengthen the existing local seed systems. Moreover, these interactions will enhance seed productivity and availability, thereby enabling the distribution of flood-tolerant seeds to farmers in distress across the regions [33].

Apart from these structural problems in procurement, distribution and dissemination of the seeds, there is little understanding of how the socio-economic characteristics of the farmers play a role in the adoption of this technology. For instance, in the field experiment conducted in Orissa to understand the yield variability of SS1, the authors point that caste, a marker of social status, in India-played an important role in the adoption of stress-tolerant seeds [21]. Firstly, the study found that the plots cultivated by farmers belonging to marginalized caste groups were already exposed to more flooding. Despite being more vulnerable to flooding, the study found a lower adoption rate of the SS1 seeds amongst them, which was attributed to the high incidence of poverty amongst the lower-caste groups. Thus, there are substantial social barriers in effective adoption of new technology such as SS1 amongst farmers. This will require designing policies that can make stress-tolerant seeds affordable for poor farmers through subsidies or other benefits.

5. Conclusions

The incidences of flood have increased over the years owing to global climate change. Enhancing livelihood options for the people depending on agriculture in several flood affected regions in India is challenging. Despite the technological improvements in terms of development of new tolerant seed varieties, or techniques, poor farmers often adapt to these flood impacts by abandoning farming altogether. To enhance the capacity of the farmers to deal with extreme flood events, advanced flood-tolerant seed varieties can be adopted. To this end, this study provides estimation of one such flood-tolerant rice variety–Swarna-Sub 1 for rice–which can be adopted by farmers during the major rice growing season called *Kharif* in India, at the district level for 17 major Indian states. The total seed requirement for the 17 states is approximately 92,800 tonnes, and the cost for procuring this amount was estimated to be INR 3800 million.

However, our estimates are the first attempt (as far as our knowledge) to quantify the potential need for flood-tolerant seeds such as Swarna-Sub1 that would need to be procured if major flood prone paddy/rice areas in India had to be supplied with flood-tolerant seeds. Our methodology of combining remote sensing data for flood affected areas with land use pattern data can be used for need assessment by government departments to be better prepared in procuring seeds and allocating budget in enhancing agriculture resilience and flood proofing the vulnerable smallholder farmers. Future work would necessitate collecting more data on seed rate usage and sowing methods that can facilitate more accurate estimates. Moreover, future work can also explore the requirement for other major food grains along with considering the impact of droughts.

The procurement of stress-tolerant seeds is not enough by itself to ensure adoption by farmers, who will need these stress-tolerant seeds to become climate resilient. There is need to design integrated flood management policy to attract more farmers in these floods affected areas to adopt stress-tolerant seeds such as Swarna Sub1, and promote the use of climate information services and good agronomic practices in reducing crop losses. Although national and international agencies are beginning to recognize the extent to which extreme weather events such as flooding will affect agricultural production in India, their initiatives to adapt and cope with floods have been relief-oriented and rather short term [36]. Given the increasing flood occurrences, there is a need to develop a comprehensive seed production and dissemination strategy for rapid and targeted distribution of this flood-tolerant rice variety, Swarna-sub1, especially amongst the poor smallholder farmers residing in the flood-prone regions of India. Additionally, workshops and awareness drives on specific climate resilient technologies suitable for that region need to be undertaken to enable farmers to cope with extreme weather events and further enhance their adaptive capacity. In addition, training and capacity-building programs are necessary to enable farmers to adopt best practices and resilient technologies to increase yields.

Author Contributions: G.A. and N.K. developed the ideas for the project, analyzed data, interpreted results, and wrote most of the manuscript. N.K. also spent considerable time in data collection. All authors have read and agreed to the published version of the manuscript.

Funding: This research was funded by the CGIAR Research Program (CRP) on Climate Change, Agriculture and Food Security (CCAFS), and CGIAR Research Program (CRP) on Water, Land and Ecosystems (WLE), which is carried out with support from the CGIAR Trust Fund and through bilateral funding agreements. For details, please visit https://ccafs.cgiar.org/donors and https://wle.cgiar.org/donors.

Institutional Review Board Statement: Not applicable.

Informed Consent Statement: Not applicable.

Data Availability Statement: The data presented in this study are available on request from the corresponding author.

Acknowledgments: The authors would like to thank the Indian Council of Agricultural Research (ICAR) and Japan's Ministry of Agriculture, Forestry, and Fisheries (MAFF). Further, the authors would also like to thank Archisman Mitra and Alok Sikka for their support in conducting the research and providing feedback. We thank Niranga Alahacoon, IWMI in producing GIS maps. Authors would like to thank the reviewers who helped in improving the manuscript.

Conflicts of Interest: The authors declare no conflict of interest.

References

1. EM-DAT: The CRED/OFDA International Disaster Database. Available online: http://www.emdat.be/ (accessed on 12 June 2021).
2. Amarnath, G.; Alahacoon, N.; Smakhtin, V.; Aggarwal, P. *Mapping Multiple Climate-Related Hazards in South Asia. Colombo, Sri Lanka*; IWMI Research Report 170; International Water Management Institute (IWMI): Gujarat, India, 2017; p. 41.
3. IPCC. Summary for policymakers. In *Climate Change: Impacts, Adaptation, and Vulnerability*; Cambridge University Press: Cambridge, UK; New York, NY, USA, 2014; pp. 1–32.
4. Central Water Commission. *Hand Book for Flood Protection, Anti-Erosion and River Training Works*; Flood Management Organization, Government of India: New Delhi, India, 2012.
5. Amarasinghe, U.; Amarnath, G.; Alahacoon, N.; Ghosh, S. How Do Floods and Drought Impact Economic Growth and Human Development at the Sub-National Level in India? *Climate* **2020**, *8*, 123. [CrossRef]
6. National Food Security Mission, Government of India. Available online: https://www.nfsm.gov.in/ (accessed on 10 January 2021).
7. Dar, M.H.; Zaidi, N.W.; Waza, S.A.; Verulkar, S.B.; Ahmed, T.; Singh, P.K.; Iftekharuddaula, K.M. No yield penalty under favorable conditions paving the way for successful adoption of flood tolerant rice. *Sci. Rep.* **2018**, *8*, 9245. [CrossRef] [PubMed]
8. Adkins, S.W.; Shiraishi, T.; McComb, J.A. Submergence tolerance of rice—A new glasshouse method for the experimental submergence of plants. *Physiol. Plant.* **1990**, *80*, 642–646. [CrossRef]
9. Setter, T.L.; Ellis, M.; Laureles, E.V.; Ella, E.S.; Senadhira, D.; Mishra, S.B.; Sarkarung, S.; Datta, S. Physiology and genetics of submergence tolerance in rice. *Ann. Bot.* **1997**, *79*, 67–77. [CrossRef]
10. Greenway, H.; Setter, T.L. *Is There Anaerobic Metabolism in Submerged Rice Plants? A View Point*; University of Agriculture and Technology and International Rice Research Institute: Manila, Philippines, 1996; pp. 11–30.
11. Singh, R.K.; Reddy, J.N.; Singh, V.N.; Pani, D.R.; GHOSH, R.; Mackill, D.; Singh, U.S. Swarna-Sub 1-A boon for Submergence-Prone Rainfed Lowlands of Eastern India. In Proceedings of the 14th Australian Plant Breeding and 11th SABRAO Congress, Cairns, Australia, 10–14 August 2009; pp. 10–14.
12. Nishiuchi, S.; Yamauchi, T.; Takahashi, H.; Kotula, L.; Nakazono, M. Mechanisms for coping with submergence and waterlogging in rice. *Rice* **2012**, *5*, 2. [CrossRef] [PubMed]
13. Bhowmick, M.K.; Dhara, M.C.; Singh, S.; Dar, M.H.; Singh, U.S. Improved management options for submergence-tolerant (Sub1) rice genotype in flood-prone rainfed lowlands of West Bengal. *Am. J. Plant Sci.* **2014**, *5*, 14. [CrossRef]
14. Portal for Disaster Compensation Disbursement, Government of India. Available online: http://aapda.bih.nic.in/(S(uymf3i2rsqqij0yapjfcb2kh))/Disaster2020/DisasterTypeDesc.aspx (accessed on 30 June 2021).
15. Frenken, K. Irrigation in Southern and Eastern Asia in figures: AQUASTAT Survey-2011. *Water Rep.* **2012**. Available online: http://www.rkmp.co.in/ (accessed on 13 January 2021).
16. Dana, I.; Chatterjee, S. Swarna-Sub1: A Boon to the Farmers of West Bengal. *STRASA News* **2012**, *5*, 5.
17. Ismail, A.M.; Singh, U.S.; Singh, S.; Dar, M.H.; Mackill, D.J. The contribution of submergence-tolerant (Sub1) rice varieties to food security in flood-prone rainfed lowland areas in Asia. *Field Crops Res.* **2013**, *152*, 83–93. [CrossRef]
18. Dar, M.H.; Chakravorty, R.; Waza, S.A.; Sharma, M.; Zaidi, N.W.; Singh, A.N.; Singh, U.S.; Ismail, A.M. Transforming rice cultivation in flood prone coastal Odisha to ensure food and economic security. *Food Secur.* **2017**, *9*, 711–722. [CrossRef]
19. Yamanot, T.; Malabayabas, M.; Gumma, M.K. Adoption, yield, and ex ante impact analysis of Swarna-Sub 1 in Eastern India. *STRASA Econ. Briefs* **2013**, *2*, 3–10.
20. Mackill, D.J.; Ismail, A.M.; Singh, U.S.; Labios, R.V.; Paris, T.R. Development and rapid adoption of submergence-tolerant (Sub1) rice varieties. *Adv. Agron.* **2012**, *115*, 299–352.
21. Dar, M.H.; De Janvry, A.; Emerick, K.; Raitzer, D.; Sadoulet, E. Flood-tolerant rice reduces yield variability and raises expected yield, differentially benefitting socially disadvantaged groups. *Sci. Rep.* **2013**, *3*, 3315. [CrossRef]
22. Bailey-Serres, J.; Fukao, T.; Ronald, P.C.; Ismail, A.M.; Heuer, S.; Mackill, D. Submergence Tolerant Rice: SUB1's Journey from Landrace to Modern Cultivar. *Rice* **2010**, *3*, 138–147. [CrossRef]
23. Dash, S.R.; Routray, B.; Mohanty, S.K.; Behera, N. Evaluation of excess water tolerant rice varieties Swarna sub-1 and CR-1009 sub-1 under Head to Head Project in East and South- Eastern Coastal Plain zone of Odisha. *Curr. Agric. Res. J.* **2020**, *8*, 39–45. [CrossRef]
24. Singh, U.S.; Dar, M.H.; Singh, S.; Zaidi, N.W.; Bari, M.A.; Mackill, D.J.; Collard, B.C.Y.; Singh, V.N.; Singh, J.P.; Reddy, J.N.; et al. Field performance, dissemination, impact and tracking of submergence tolerant (Sub1) rice varieties in South Asia. *SABRAO J. Breed. Genet.* **2013**, *45*, 112–131.

25. Amarnath, G. An algorithm for rapid flood inundation mapping from optical data using a reflectance differencing technique. *J. Flood Risk Manag.* **2014**, *7*, 239–250. [CrossRef]
26. Amarnath, G.; Rajah, A. An evaluation of flood inundation mapping from MODIS and ALOS satellites for Pakistan. *Geomat. Nat. Hazards Risk* **2016**, *7*, 1526–1537. [CrossRef]
27. Amarnath, G.; Ameer, M.; Aggarwal, P.; Smakhtin, V. Detecting spatio-temporal changes in the extent of seasonal and annual flooding in South Asia using multi-resolution satellite data. In Proceedings of the International Society for Optics and Photonics (SPIE), Amsterdam, The Netherlands, 1–6 July 2012.
28. Pandey, S.; Velasco, L. Economics of direct seeding in Asia: Patterns of adoption and research priorities. *Int. Rice Res. Notes* **1999**, *24*, 6–11.
29. Tabbal, D.; Bouman, B.; Bhuiyan, S.; Sibayan, E.; Sattar, M. On-farm strategies for reducing water input in irrigated rice; case studies in the Philippines. *Agric. Water Manag.* **2002**, *56*, 93–112. [CrossRef]
30. Pandey, S.; Mortimer, M.; Wade, L.; Tuong, T.P.; Lopez, K.; Hardy, B. *Direct Seeding: Research Strategies and Opportunities*; IRRI: Los Baños, Philippines, 2002.
31. Singh, H.N.; Pandey, S.; Villano, R.A. Rainfed rice, risk, and technology adoption: Microeconomic evidence from eastern India. In *Characterizing and Understanding Rainfed Environments*; IRRI: Los Baños, Philippines, 2000; pp. 323–338.
32. Trivedi, R.K.; Gunasekaran, M. Compendium on seed legislations. In *Seeds Division, Department of Agriculture and Co-Operation*; Ministry of Agriculture, Government of India: New Delhi, India, 2014; p. 148.
33. Chauhan, J.S.; Prasad, S.R.; Pal, S.; Choudhury, P.R.; Bhaskar, K.U. Seed production of field crops in India: Quality assurance, status, impact and way forward. *India J. Agric. Sci.* **2016**, *86*, 563–579.
34. Chauhan, J.S.; Satinder, P.; Prasad, S.R. Recent trends in breeder seed production of selected food and oilseed crops. In Proceedings of the 7th National Seed Congress, 'Quality Seeds for Successful Agriculture', Bhopal, Madhya Pradesh, India, 25–27 September 2014; pp. 103–105.
35. Chauhan, J.S.; Prasad, S.R.; Pal, S.; Choudhury, P.R. Seed Systems and Supply Chain of Rice in India. *J. Rice Res.* **2017**, *10*, 9–15.
36. IPCC. Climate Change 2014: Synthesis Report. In *Contribution of Working Groups I, II and III to the Fifth Assessment Report of the Intergovernmental Panel on Climate Change*; Core Writing Team, Pachauri, R.K., Meyer, L.A., Eds.; IPCC: Geneva, Switzerland, 2014; p. 151.

climate

MDPI

Article

Impacts of Climate Change on the Water Resources of the Kunduz River Basin, Afghanistan

Noor Ahmad Akhundzadah [1,*], Salim Soltani [2] and Valentin Aich [3]

1 Faculty of Environment, University of Kabul, Kart-e-Sakhi, Kabul 1001, Afghanistan
2 Institute for Geography and Geology, University of Würzburg, Am Hubland, 97074 Würzburg, Germany; mohammad_salim.soltani@stud-mail.uni-wuerzburg.de
3 Potsdam Institute for Climate Impact Research (PIK), Am Telegraphenberg, 14473 Potsdam, Germany; vaich@wmo.int
* Correspondence: noorahmad.akhundzadah@ku.edu.af; Tel.: +93-(0)-707083359

Received: 30 August 2020; Accepted: 16 September 2020; Published: 23 September 2020

Abstract: The Kunduz River is one of the main tributaries of the Amu Darya Basin in North Afghanistan. Many communities live in the Kunduz River Basin (KRB), and its water resources have been the basis of their livelihoods for many generations. This study investigates climate change impacts on the KRB catchment. Rare station data are, for the first time, used to analyze systematic trends in temperature, precipitation, and river discharge over the past few decades, while using Mann–Kendall and Theil–Sen trend statistics. The trends show that the hydrology of the basin changed significantly over the last decades. A comparison of landcover data of the river basin from 1992 and 2019 shows significant changes that have additional impact on the basin hydrology, which are used to interpret the trend analysis. There is considerable uncertainty due to the data scarcity and gaps in the data, but all results indicate a strong tendency towards drier conditions. An extreme warming trend, partly above 2 °C since the 1960s in combination with a dramatic precipitation decrease by more than −30% lead to a strong decrease in river discharge. The increasing glacier melt compensates the decreases and leads to an increase in runoff only in the highland parts of the upper catchment. The reduction of water availability and the additional stress on the land leads to a strong increase of barren land and a reduction of vegetation cover. The detected trends and changes in the basin hydrology demand an active management of the already scarce water resources in order to sustain water supply for agriculture and ecosystems in the KRB.

Keywords: climate change; Kunduz River Basin; trend analysis; river discharge; landcover changes

1. Introduction

Afghanistan is a semi-arid country with high variability and irregularity in precipitation. Based on the morphological and hydrological systems of Afghanistan, its surface water is divided into five major river basins: Kabul, Helmand, Harirud-Murghab, Northern, and Amu-Darya River Basins [1] (Figure 1). The Kunduz river is one of the main tributaries of the Amu Darya in North Afghanistan. It is mainly nourished by snow and glaciers melting during spring and summer (Figure 1). Similar to other tributaries of the Amu-Darya, it is the main water resource for drinking, irrigation, and hydropower usages in the basin and the river plays an important role for all ecosystems in the basin [2–4]. However, riverine floods and flash floods are common disasters in the Kunduz River Basin (KRB), because of the extreme climate regime in the Hindu Kush Mountains. Severe riverine flooding in the lowlands and upper parts of the catchments occur regularly during spring due to glacier and snow melt and spring rainfall. In the year 2019, early rainfall in upper parts of the catchments, combined with increased snowmelt due to high temperatures, caused strong flooding in most river basins of the Amu

Darya tributaries in Afghanistan, with approximately 124,500 people affected and many killed [5]. Little literature is available on climate change impacts in Afghanistan; some recently conducted studies indicate a distinct warming trend and a decrease of rainfall in some parts of the country [6,7]. The first detailed and systematic analysis of climate data for Afghanistan that was conducted by Aich et al. (2017) showed a warming by 1.8 °C for Afghanistan between 1951 and 2010; the temperature in Afghanistan increased by 1.8 °C, which is higher than the global mean. These changes severely affected the key sectors, including water resources, agriculture, energy, and it imposed flash flood, drought, soil erosion, and environmental degradation [8–11].

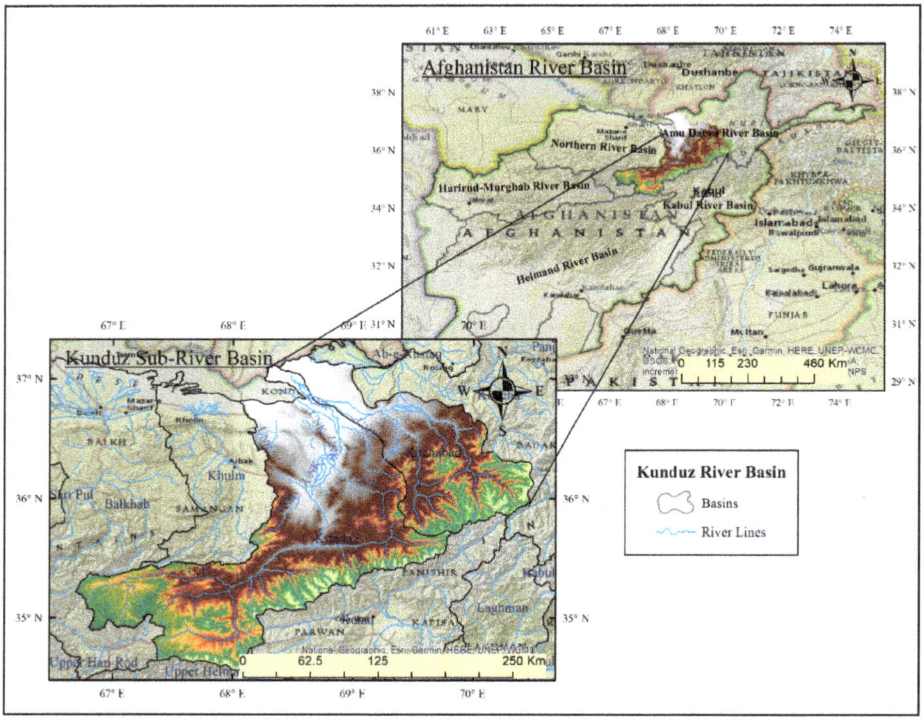

Figure 1. Afghanistan main river basins and Kunduz river watershed location.

Because about 80% of Afghanistan's population depends on agriculture for their livelihoods, and agriculture contributes to almost half of the GDP [12], these changes directly affected livelihoods, food security, and the socio-economy of the country [10,13]. The changing climate has changed the hydrological condition and land cover of the Amu-Darya River Basin [11,14]. The increase in temperature has been melting glaciers and permafrost in Himalayan and Hindukush mountains [1,15,16]. The decrease in precipitation and glaciers melting has reduced the volume of water in the Amu-Darya and KRB [2,4,17].

The climate change impacts, compounded by the past four decades of war and conflict, have destroyed the country's infrastructure and institutions, and it has led to underdevelopment that collectively contributes to Afghanistan's vulnerability to climate change impacts. Now, any climate change study in Afghanistan is faced with the challenge of lack of reliable historical meteorological data, with more than two decades gaps in the historical data records during the war and conflict in the country [18]. The related uncertainties are also reflected in global reanalysis products [9].

However, so far, no study has addressed the impacts of these different factors on the water resources of the catchment with all available observed data. Therefore, the main focus of this study is to investigate climate change impacts in the hydrology of the KRB with a focus on temperature, precipitation, river discharge, and land use and land cover (LULC) change, while taking into account the lack of data and the resulting uncertainty. Therefore, the trends of these variables are analyzed and the results integrated in a discussion. The observed data from the KRB are mainly available for the period 1960s–1980s and then again from the 2000s until now with a large gap in between due to the political conflicts in Afghanistan, which hinders the trends analysis. The limitation in data availability and its implications on the study and its results are discussed when interpreting the results. Finally, conclusions for water resource management in the basin are drawn while taking the data constraints and the related uncertainty into account.

2. Study Site

2.1. Kunduz River Basin

The Kunduz River is one of the main tributaries of the Amu Darya. It originates from the North side of the Hindukush Mountain and flows through the wide lowlands of Baghlan to finally join the main Amu Darya stream in Qala-i-Zal area (Figure 2). The Kunduz watershed has an area of 28,024 km^2, which is 4.5% of the country [19] and about 1.9% of the population of the country live in the River Basin [20]. The KRB covers the mountainous area of the Hindukush, with elevation ranging from up to 4000 m a.s.l. in the upper, Southern parts of the Basin. Lowland areas are about 600 m a.s.l. in Baghlan and 400–350 m a.s.l. in Kunduz provinces. The soils of the KRB are characterized by Palaeogene and Neogene sediments and covered by Loess deposits about 30 m to more than 100 m thickness in the center. Alluvial deposits consist of gravel, sands, and silt spread around floodplain in the basin. The area adjacent to the mountains are covered by coarse deposits of gravel, pebble, cobble, and other detritus deposits [21,22]. The higher altitude areas in the basin are partly used for rain-fed agriculture, but they mostly consist of deforested areas [23]. The flood plains consist of highly fertile medium drained soils with good agricultural land, which comprises the main economic center of the basin [24].

Arable land covers 38% (10,344 km^2) of the total area of the KRB (28,024 km^2). The Takhar province has more arable land as compared to the Kunduz and Baghlan provinces. Bamyan province is located in the high mountain area of the KRB and it has the least arable land [23]. The main crops cultivated in the arable area of the KRB are wheat, maize, barely, and rice. The crops are mostly planted during March to May and harvesting during July to September [25]. Watermelon, melon, potatoes, and onions are the main vegetables crops. Apples, grapes, berries, and peaches are the major fruits, and Cotton is the major industrial crop. There is an increasing number of pistachio- and almond plantations grown in the KRB.

Recently, the Ministry of Agriculture, Irrigation and Livestock (MAIL) conducted qualitative and quantitative investigations on climate change impacts on the agriculture sector in Afghanistan. The results presented a significant reduction in crops production, which was likely due to a decrease of precipitation and rising temperatures within the North-East agro-climatic zone that covered the KRB (e.g., 10–20% reduce in wheat) [25].

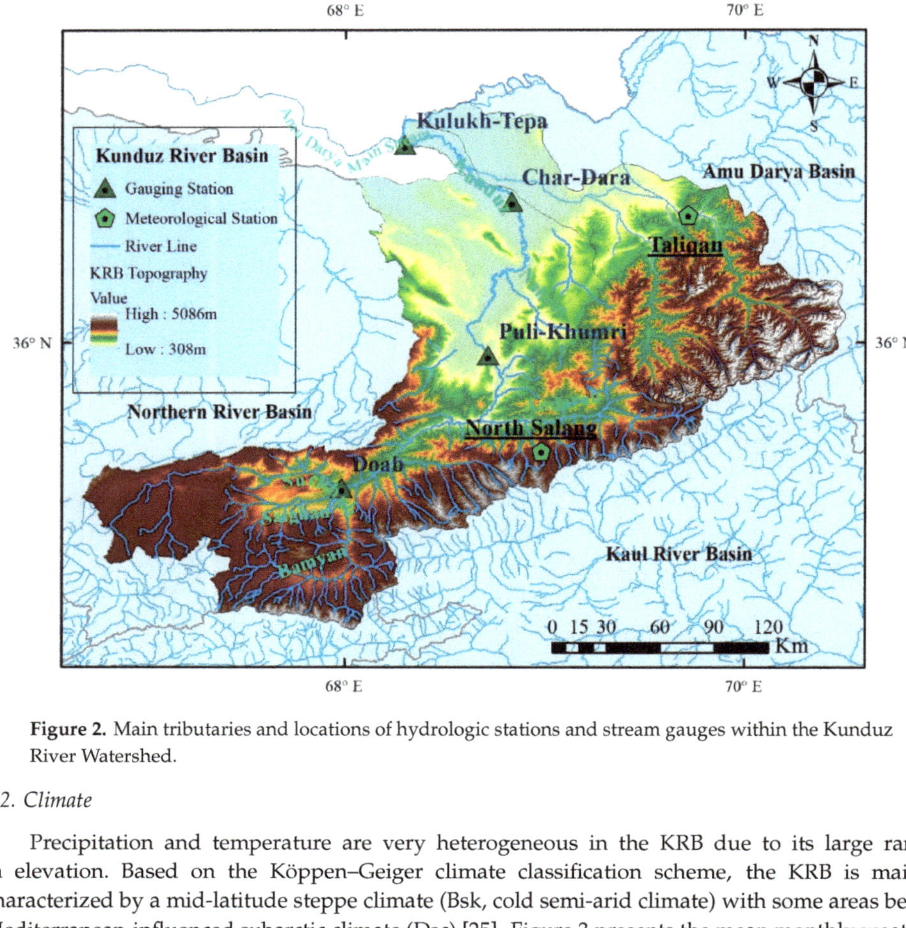

Figure 2. Main tributaries and locations of hydrologic stations and stream gauges within the Kunduz River Watershed.

2.2. Climate

Precipitation and temperature are very heterogeneous in the KRB due to its large range in elevation. Based on the Köppen–Geiger climate classification scheme, the KRB is mainly characterized by a mid-latitude steppe climate (Bsk, cold semi-arid climate) with some areas being Mediterranean-influenced subarctic climate (Dsc) [25]. Figure 3 presents the mean monthly weather average of the recent decade (2009–2019) mean monthly weather average, recorded in North Salang and Kunduz stations. The data were provided by the Afghanistan Meteorological Department [26]. The mean annual temperature in North Salang (3400 m a.s.l.) is around 1 °C and it is 19 °C in Kunduz (991 m a.s.l.). The mean annual rainfall is recorded 71 mm in North Salang and 32 mm in Kunduz. From June to September are mainly dry months with very little precipitation and most of the annual precipitation falls from January to April. At North Salang, the annual average precipitation is around 200 mm and 100 mm at the Kunduz station. July is the warmest month of the year, in North Salang the average temperature in July is 11 °C and, in Kunduz, it is 33 °C. January is the coldest month of the year, with −10 °C and 5 °C in North Salang and Kunduz, respectively. In Kunduz, the temperature extremes can rise to over 40 °C during the warmest months and fall to −20 °C during the cold season. There are occasions of heavy precipitation events, for example, over 400 mm/d in North Salang (e.g., March of 2019) and 350 mm/d in Kunduz (e.g., February 2008). High precipitation during spring 2019 caused severe flash floods in the main river basins, including the Kunduz sub-river basin [27].

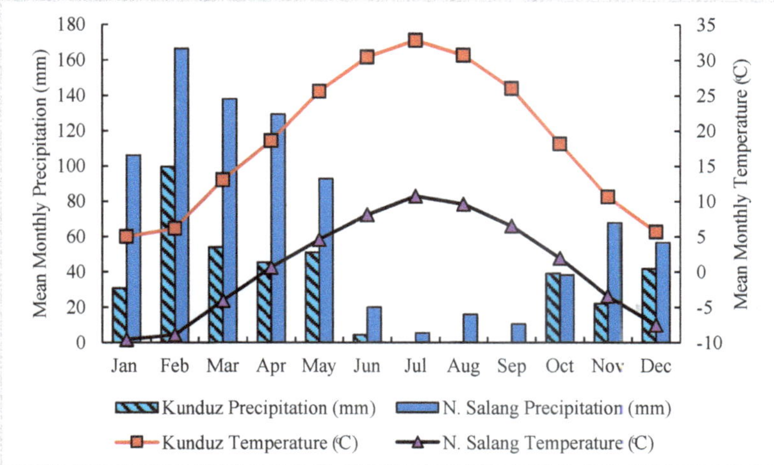

Figure 3. Average monthly precipitation and temperature recorded in North Salang and Kunduz stations during the period 2009–2019.

2.3. Hydrology

The Kunduz River is a tributary of the Amu Darya River in North Afghanistan. The upper part of the KRB is characterized by high mountains and steep valleys. In the upper part, the KRB is fed by the rainfall, snow, and small glaciers of the Koh-e-Baba range and the Hindu-Kush mountains [24] (Figure 2). The KRB has a number of tributaries, including the Khinjan, Andarab, and Bamyan rivers [28]. Upstream of the Kunduz province, the Kunduz river is called the Pul-I Khumri River. Another small tributary, the Nahrin River, has its sources in Nahrin district and it joins the Kunduz river near the town Baghlan-i Kohna. Finally, the Kunduz River reaches the Amu Darya main stream at Qala-i Zal (Figure 3). The KRB covers all of Baghlan province, the western part of Bamiyan province, and parts of Kunduz and Takhar Provinces [23]. Two hydropower dams have been built on the Pul-i Khumri in 1943 [29].

The hydrology of the KRB is mainly controlled by the high mountains of the Hindukush. Upstream, channels are generally narrow and deep and flowing throughout the whole year [12]. The runoff regimes are largely controlled by snow-melt, with high discharge from April to June and only close to glaciers in the upstream parts of the catchment, the small glaciered area has significant influence on the flow regime (e.g., Doab station). Precipitation in the KRB mainly occurs in the form of rain, drizzle, snowfall, and hail, and it is high during the winter months [24]. The water carried by the river supports an intensive irrigated agriculture, which is the main economic basis of the region. There are a number of river gauging stations within the watershed, as shown in Figure 2.

Figure 4 presents the mean monthly discharge of the recent five years from 2014 to 2018 recorded in the four main gauging stations, Doab, Puli-Khumri, Char-Dara, and Kulukh-Tepa (for locations, see Figure 2). Historically, the monthly peak flows generally occurred during April through July, which resulted in very high discharge at the downstream drainage outlet (Figure 4). The Doab gauge is located in the most upper part at 1468 m a.s.l. It covers a small watershed and has low discharge, being mainly fed by small glaciers. The peak monthly discharge at that gauge from 2014 to 2018 was 36 m^3/s during June. The gauge at Puli-Khamri is downstream at 634 m a.s.l. and its peak monthly discharge during this period was 199 m^3/s. Char-Dara gauge, further downstream at 401 m a.s.l., the peak discharge is 138 m^3/s and 177 m^3/s at Kulukh-Tepa gauge (320 m a.s.l.). The Kulukh-Tepa gauging station is located at the confluence of the Kunduz River and the Amu Darya mean stream (Figure 3). The peak average monthly discharge at Puli-Khamri gauge is higher than the Kulukh-Tepa at the outlet of the KRB. This can be explained by the high temperature and related high evaporation during June, July, and August in the lowland downstream area and diverging small portion of the stream to irrigation as well.

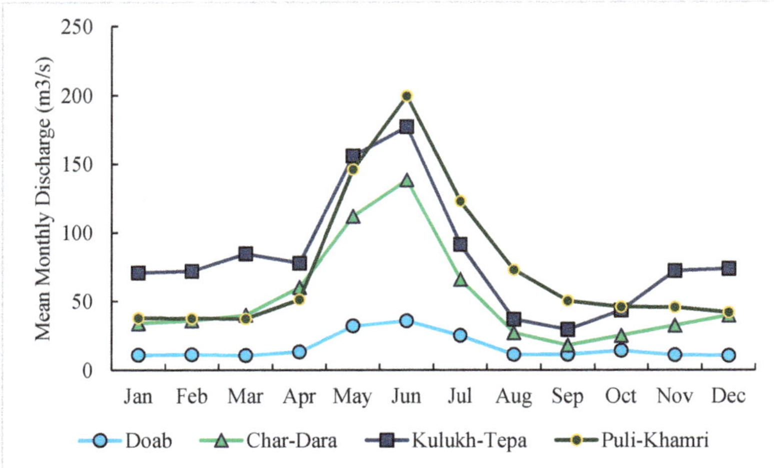

Figure 4. Comparison of flow discharge at Doab, Char-Dara, Puli-Khamri, and Kulokh-Tepa gauges.

3. Data and Methods

3.1. Data

For the analysis, historical temperature, precipitation, and river discharge data recorded from about 1960 to 1979 at gauging stations installed within the Kunduz River Basins are analyzed. Between 1979 and 2009, or even later, there are no data records due to political turmoil in the country available. The recent data from 2009 to 2019 are, with some exceptions, available. The data for available years are listed in Table 1. Only meteorological and river discharge stations with over 20 years of data records have been used in the study and stations with less data available neglected in order to have a minimum of confidence in the time series (discussed in more detail in Section 5.1). The hydrological and meteorological data were provided by the Ministry of Energy and Water and Afghanistan Meteorological Department. The river discharge data are also provided by the same ministry [30].

Table 1. Overview of gauges and meteorological stations in the Kunduz River basin, including drainage area, elevation, and the period for which data are available.

River Gauge Stations							
Station Name	**Lat.**	**Long.**	**Elevation (m)**	**Drainage Area (km^2)**	**Record Period**	**Record Period**	**N° Years**
Doab	35.2666667	67.9833333	1468	5005	1968–1979	2009–2018	22
Puli-Khumri	35.9333333	68.7166667	639	17405	1950–1968	2009–2018	29
Char-Dara	36.7000000	68.8333333	401	24820	1964–1980	2007–2018	29
Kulokh-Tepa	36.9833333	68.3000000	320	37100	1966–1980	2014–2018	20
Meteorological Stations							
North Salang	35.4528396	68.9852142	3400	Met-Station	1960–1978	2010–2019	29
Taliqan	36.6333333	69.7166667	991	4110	1969–1978	2010–2019	20

For topographical and hydrological mapping, remote-sensing data and satellite images from sources, including National Geographic and Esri, were accessed and processed using ArcGIS software (https://www.arcgis.com/home/item.html?id=b9b1b422198944fbbd5250b3241691b6). For the LULC classification, Landsat 5 Thematic Mapper (TM) scenes and Landsat 8 Operational Land Imager (OLI) have been used [31].

3.2. *Trend Analysis for Temperature, Precipitation and River Discharge*

Linear trends in the time series were analyzed using the Mann–Kendall test [32]. It was chosen, because it is a robust nonparametric test and it can handle missing data as well as it has higher power for non-normally distributed data, which are common in hydrological and meteorological data [33]. Each element is compared with its successors and ranked as larger, equal, or smaller. Based on this analysis, the statistical significance of rejecting the null hypothesis that there is no monotonic trend is tested (for all tests $\alpha = 0.05$). The R package "Kendall" was used for the calculation [34].

The Theil–Sen approach was used in order to quantify the linear trend [35,36]. It computes the slope for all pairs of the ordinal time points of a time series and then used the median of these slopes as an estimate of the complete slope. This approach is commonly combined with the Mann–Kendall test and estimates the trend slope of a time series in its original unit. The R package "zyp" was used in order to calculate the Theil–Sen trend and includes a pre-whitening according to Ye et al. (2002) [37] if autocorrelation occurs [38].

3.3. *Land Cover Classification*

The supervised land cover classification has been carried out in two time steps, 1992 and 2019, while using Landsat 5 TM for the earlier date and for the latter Landsat 8 OLI. To account for annual variation in the snow and glacier coverage, data from August and September, when snow and glacier coverage have their annual minimum, have been used. A cloud mask was applied to remove cloud contamination.

The Random Forest Classifier (RFC) method [39] was applied using Google Earth Engine (GEE) for the classification [40]. For constructing the study wide cloud free mosaic, the median function of GEE has been used, which takes the median value of each pixel in available image temporal stack. In order to achieve higher classification accuracy, we followed the method of [41]: Gray-level co-occurrence matrix (GLCM) texture features [42,43] and spectral indices were produced in order to serve as collective variable predictors for the classification algorithm. The texture characterizes the variance of the pixel DN value over space, so it needs to be measured in a multiple pixel neighborhood. Within this neighborhood of pixels can be found the following three elements: tonal (DN) difference between pixels, the distance over which this difference is measured, and directionality [42]. These neighboring pixels are considered a window or kernel, and usually have a square area with an odd number of pixels for practical reasons. It is important to define the window size, because the larger window size can include edges or patches with different textures; this is particularly applicable for larger window sizes. For the first time in 1973, Haralick et al. [42] proposed GLCM textures, which are co-occurring or second-order texture measures. The calculation is based on tonal (DN) differences in a spatially defined relationship between pairs of pixels, taking into account all pixel pairs within the neighborhood. Hall-Beyer explains that second-order measurements can distinguish two pixels wide vertical stripes from one pixel wide stripes, given uniform DN values in each stripe; first-order texture measurements are not able to perform this [42]. The GLCM can account for all three elements of texture and that is one of its advantages. GLCM can be calculated while using single input layer and defined window size (i.e., 7×7), selected by the user, and can deliver to one or more output layers based on the selected measurements (i.e., variance, homogeneity, entropy, etc.). Based on the empirical result, a window size of 7×7 yielded a better result for generation of GLCM textures and the following textures features were generated: Variance, Inverse Difference Moment, which measures the homogeneity, Contrast, Dissimilarity, Entropy, Correlation, and Angular Second Moment. which measures the number of repeated pairs [42]. The GLCM of band 3 and Band 4 of Landsat 8 were used for 2019 land cover classification, and GLCM band 4 of Landsat 5 TM was used for 1992 to generate the texture features. The spectral indices used include: Modified Normalized Difference Water Index (MNDWI) [44], Enhanced Vegetation Index (EVI) [45], Normalized Different Moisture Index (NDMI) [46], Green Optimized Soil Adjusted Vegetation Index (GOSAVI) [47], Built-up Area Extraction Index (BAEI) [48], and the Normalized Difference Bareness Index (NDBai) [49].

The Smile RFC method was applied using 200 decision trees and eight variables per split, which accounts for two-third of all variables. The number of input variables for both years have been filtered according to the variable importance function of the RFC and only the variables that contributed most have been selected in order to produce the final study area land cover for the list of input variables.

Training data were collected from annual land cover data by ESA [50]. Stratified random sampling techniques were used to collect 500 points per class with a total 10,000 points. The overall classification accuracy reached over 80% for all time steps.

4. Results

4.1. Change in Temperature and Precipitation

In the KRB, two weather stations with more than 20 years of data are available, North Salang and Taliqan. Historical data are not available for the Kunduz meteorological station (see Figure 2). North Salang is located in the upstream, in a very high altitude with high precipitation and low temperature; Taliqan lies in the lowland area near of Kunduz.

Figure 5 shows a strong and statistically significant increase in the mean annual temperature within the KRB since the 1960s by 1.45 °C (see Table 2). All temperatures increase; however, the increase of the winter temperature is less and not statistically significant. Precipitation shows a very strong and significant trend by −35.02% (−412.56 mm) (see Figures 6 and 7).

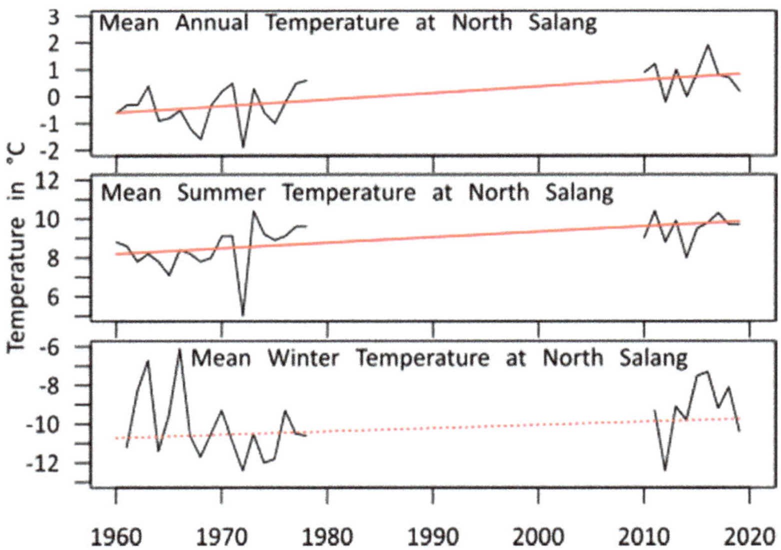

Figure 5. Mean annual, summer (J,J,A) and winter (D,J,F) temperature at station North Salang. Significant trends ($\alpha = 0.05$) are depicted as solid red line.

Table 2. Trends in temperature and precipitation for the stations North Salang and Taliqan. Statistically significant trends are bold (all but winter of North Salang).

	Trend Mean Annual Temperature	Trend Mean Annual Spring Temperature (MAM)	Trend Mean Annual Summer Temperature (JJA) 1969–2019	Trend Mean Annual Autumn Temperature (SON) 1969–2019	Trend Mean Annual Winter Temperature (DJF)	Trend Precipitation 1960–2019
North Salang	**1960–2019: +1.45 °C**	1960–2019: **+1.66 °C**	1960–2019: **+1.69 °C**	1960–2019: **+1.8 °C**	1961–2019: +1 °C	**−412.56 mm (−35.02%)**
Talaqin	1969–2019: **+2.73 °C**	1969–2019: **+2.56 °C**	1969–2019: **+2.87 °C**	1969–2019: **+2.0 °C**	1970–2019: **+3.68 °C**	**−26.03 mm (−57.73%)**

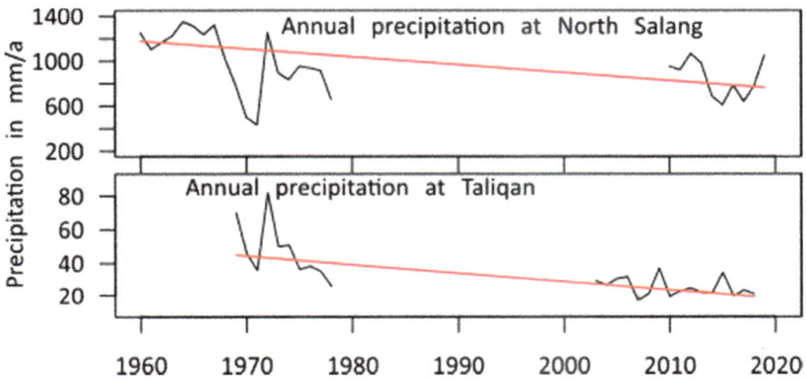

Figure 6. Mean annual precipitation at stations North Salang and Taliqan. All trends are significant t ($\alpha = 0.05$) and depicted as solid red line. Please note that only 18 years of data are available for Taliqan.

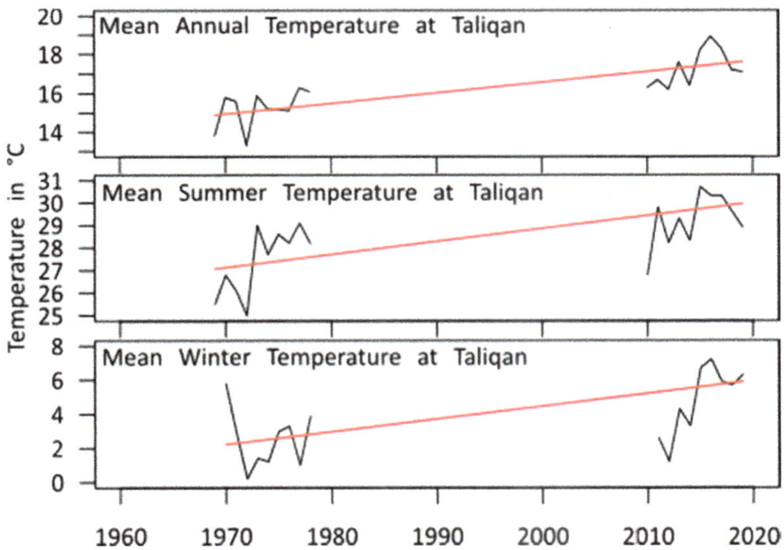

Figure 7. Mean annual, summer (J,J,A) and winter (D,J,F) temperature at station Taliqan. All of the trends are significant ($\alpha = 0.05$) and depicted as solid red line. Please note that only 18 years of data are available for Taliqan.

For the Taliqan station in the lowland, where only 18 years are available, all of the trends are significant and extreme. Mean annual temperature increased according to the data for the period from 1969 to 2019 by 2.73 °C and summer (+2.87 °C) and winter (+3.68 °C) temperature even more.

Precipitation decreased in the same period by 57.72% (−26.03 mm). These trends have to be interpreted with caution due to the limited number of years available (see Section 5.1).

4.2. Changes in Discharge

For this study, data from four gauging stations with at least 20 years of data in the KRB have been analyzed. The highland Doab station (see Figure 8, Table 3) shows a strong and significant increase in the mean and minimum annual streamflow, with over 100%, whereas the maximum flow is still strong, but due to the limited data not significant.

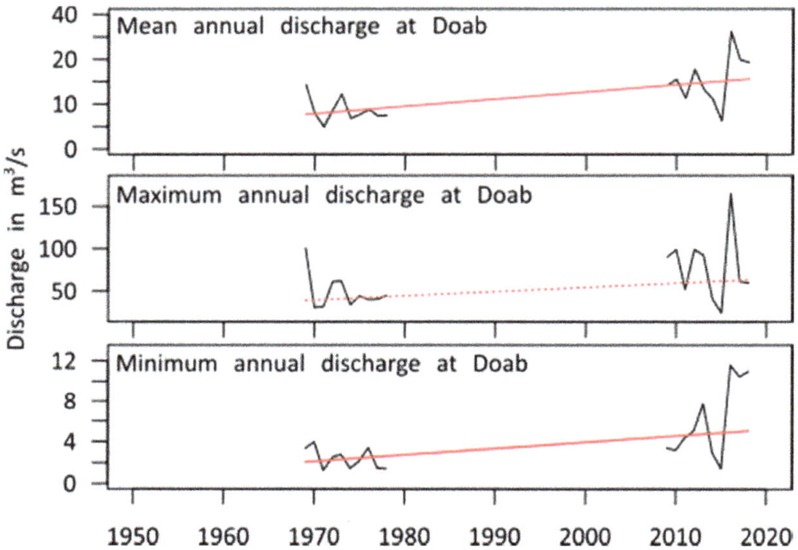

Figure 8. Mean, maximum, and minimum annual discharge for Doab gauging station. Significant trends ($\alpha = 0.05$) are depicted as solid red line.

Table 3. Trends in mean, maximum, and minimum annual discharge for the gauging stations Doab, Pul-i-Khumri, Chahar Dara, and Kulokh Tepa. Statistically significant trends are bold ($\alpha = 0.05$).

Gauging Station	Trend Mean Annual Discharge	Trend Maximum Annual Discharge	Trend Minimum Annual Discharge
Doab	+7.95 m^3/s (+103.12%)	+24.5 (+62.74%)	+2.98 m^3/s (+143.27%)
Puli-Khumri	+5.38 m^3/s (+7.86%)	−82.46 m^3/s (−23.45%)	+11.39 m^3/s (+53.34%)
Chahar Dara	−9.57 m^3/s (−18.40%)	−125.53 m^3/s (−43.05%)	−5.98 m^3/s (−46.36%)
Kulokh Tepa	−27.46 (−25.30%) (significant at $\alpha = 0.1$)	−334.61 (−58.51%)	−15.47 (−66.20%)

The Puli-Khumri station, (Figure 9, Table 3) further downstream in the lowland of the KRB, shows inhomogeneous trends with an again strong and significant increase in minimum flow with over 50% decrease, whereas the maximum annual discharge is significantly decreasing by over 20% and the mean annual flow is consequently levelled out without a significant trend.

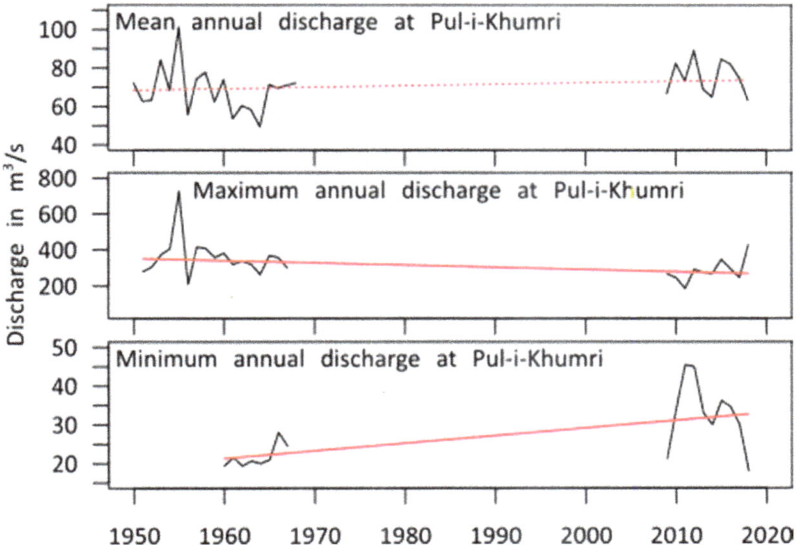

Figure 9. Mean, maximum, and minimum annual discharge for Doab gauging station. Significant trends ($\alpha = 0.05$) are depicted as solid red line.

The Chahar Dara gauging station (Figure 10, Table 3) further downstream shows strong decreasing trends throughout the year; however, only for the maximum flow this decrease is significant with over 40% reduction.

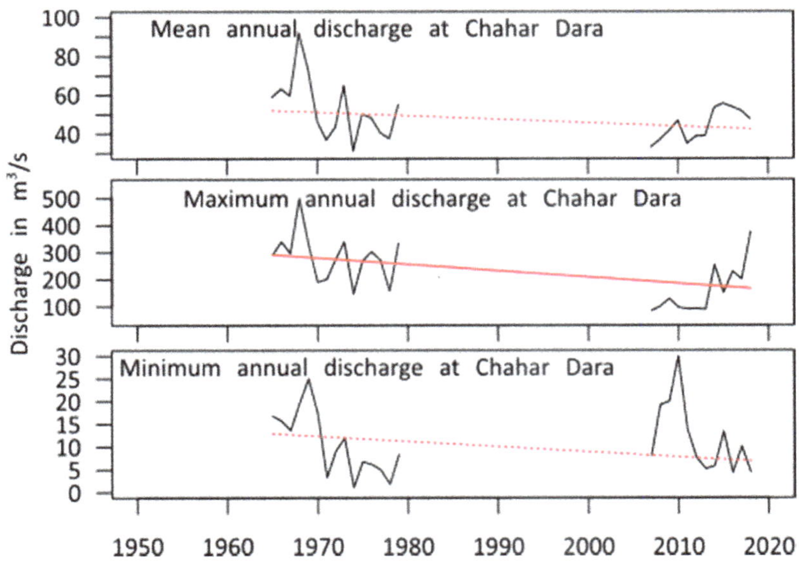

Figure 10. Mean, maximum, and minimum annual discharge for Chahar Dara gauging station. Significant trends ($\alpha = 0.05$) are depicted as solid red line.

The Kulokh Tepa station (Figure 11, Table 3) at the confluence of Kunduz the Amu Darya River shows similar decreasing patterns with a significant decrease in the maximum flow by almost 60% and a slightly less significant decrease ($\alpha = 0.1$) for the mean annual discharge by around 25%.

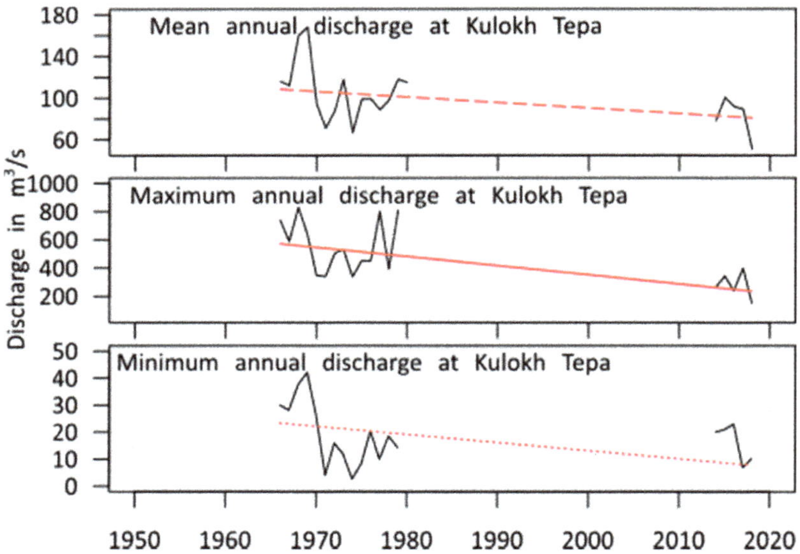

Figure 11. Mean, maximum, and minimum annual discharge for the Kulokh-Tepa gauging station. The significant trend with $\alpha = 0.1$ is depicted as dashed, the significant trend with $\alpha = 0.05$ is depicted as solid red line.

4.3. Change in Landcover

LULC trends in the KRB are assessed by comparing changes between the years 1992 and 2019 (Figure 12). Figure 13 shows the areal changes of the ten defined LULC types. Since 1992, irrigated agriculture, forest/trees, shrubland, urban coverage, as well as barren land and water surfaces, have increased substantially. At the same time, rainfed agriculture, grasslands, and snow/glacier coverage drastically decreased. Table 4 shows landcover classification area in Km^2 and the change in landcover percentage between 1992 and 2019.

Table 4. Comparison of 1992 and 2016 Landcover areas [23,28].

Class Name	Landcover Area km^2(1992)	Landcover Area km^2 (2019)	Change in %
Rainfed agriculture	6382	4461	−30.1
Irrigated agriculture	2064	2377	+15.2
Mosaic Vegetation	12,847	12,488	−2.8
Forest, tree	464	973	+109.7
Shrubland	1859	3602	+93.8
Grassland/Rangeland	9942	7361	−26
Urban	266	548	+106
Bare land	4964	5877	+18.4
Water	174	249	+43.1
Snow/Glacier	994	668	−32.8

Forest/tree area also includes fruit trees and the doubling of this coverage can be explained by a massive expansion of fruit tree plantations, such as almond and pistachio trees. Grassland was mainly degraded to barren land or shrub land. There is also a shift from rainfed to irrigated agriculture, even though the decrease in rainfed agriculture cannot fully be explained by this shift. Large areas of rainfed agriculture seemed to shift into shrublands and barren land.

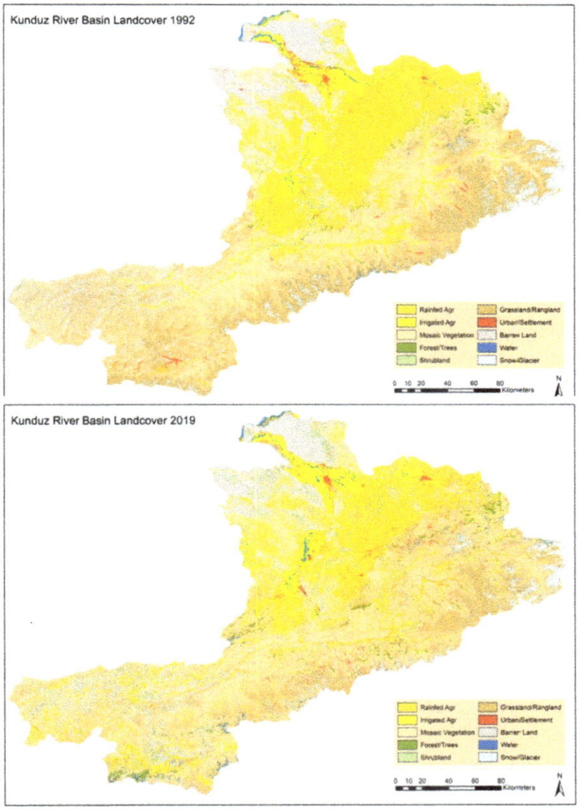

Figure 12. Land cover maps of 1992 and 2019 of the Kunduz River Basin derived from Landsat data.

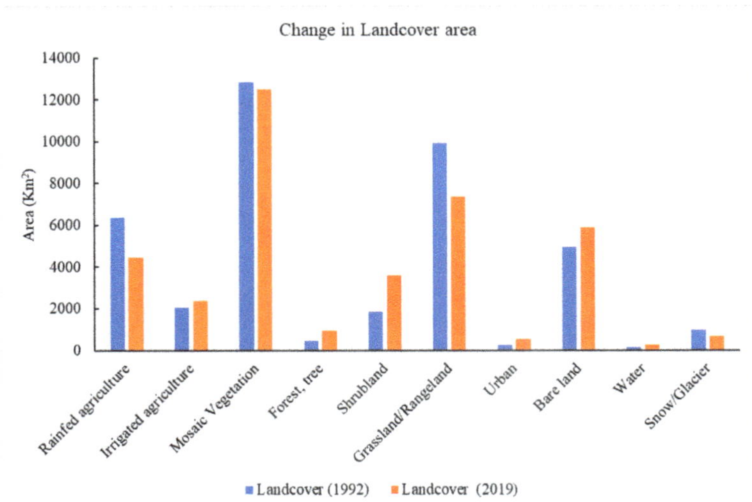

Figure 13. Changes in land use and land cover between 1992 and 2019 in the Kunduz River Basin.

5. Discussion

5.1. Constraints Due to Limited Data Availability

The availability of data is very limited in the study region as well as for the whole of Afghanistan for many reasons. The station density of meteorological as well as river gauging stations has always been low, due to the low population density, underdevelopment, and the relatively low influence of central government in many regions of the country. Characteristic for Afghanistan are, in addition the long periods of conflict and foreign rule, which hindered sustained observations or fragmented them. For example, weather observations from during the Soviet occupation, which still have taken place according to local knowledge, are not currently available. The lack of data also substantially reduces the quality of climate reanalysis in the region. Comparisons of observations with reanalysis for the available stations in the KRB showed the same results as for Aich et al. 2017 [9], which found that, for central Afghanistan, monthly precipitation in reanalysis deviated by up to 30% from the observations. For this reason, only observed station data are used for this study. We selected all of meteorological and river stations with at least 20 years of observations, since the IPCC AR5 used the period from 1986 to 2005 as modern baseline and deemed 20 years to be long enough to average over natural variations [51]. This filtering limited the time series for analysis to only two meteorological and four river gauging stations in KRB. Another constraint is the long gaps within the time series, which fragment the time series in two parts and make a continuous trend analysis impossible. The authors decided to use the data, despite these strong constraints, since it is still the currently best available data, which, in summary, still allow careful interpretation. The uncertainty of the temperature trend at the station North Salang is acceptable, since almost 30 years of data (29) are available, the value that the World Meteorological Organization (WMO) recommends for climate studies [52]. For precipitation, the uncertainty is slightly higher due to the strong interannual variability and long period of missing data. For Taliqan, the uncertainties are markedly higher, since only 18 years of data are available. Still, the temperature measurements give plausible results, even though the absolute numbers should be interpreted with caution. This holds even more for the extreme precipitation, which might be only natural variability.

However, both meteorological stations show consistent trends, which also confirm the findings from other studies with strongly increasing temperature and a reduction of precipitation [9]. This gives some confidence when interpreting the results and this holds also for the river stations. However, the climatic trends have the expected impacts on the river discharge in the KRB, even though the absolute numbers can be doubted. Finally, the individual time series can be questioned due to the mentioned constraints, but, all together, they show a coherent picture of a strong warming trend and drier conditions, which are also reflected by the changes of LULC.

In order to improve the situation and make more data available, we urge data rescue initiatives, like idare (https://www.idare-portal.org), to include Afghanistan in their efforts and particularly the integration of existing data in archives of the former Soviet Union might be promising.

5.2. Climate Change Impacts

The results of the temperature and precipitation trend analysis are, in general, in line with former studies, like Aich et al. 2017 [9]. The extreme increase in temperature by significantly over 1 °C in the central highland and even over 2 °C in the lowland of the KRB. The temperature increase is more pronounced in summer, accompanied by a not less extreme decrease in precipitation by over 35%, respectively, 50% during the second half of the last century until now (see Table 2). As discussed in Section 5.1, uncertainties with regard to the magnitude of trends is large, particularly for precipitation; however, the direction of trend seems to be plausible and in line with observations from other countries in the region. Still, the general decrease is significant and has, similar to the strong temperature increase, a strong impact on the water resources.

River discharge results are more heterogeneous for different parts of the catchment. In the headwaters of the catchment (Doab station), the discharge is significantly increasing, which can be explained by the increase of glacier melt due to the higher temperatures. The LULC analysis shows an extreme reduction of 359 km^2 (−35%) of glaciered area between 1992 and 2019. With the accelerated warming trend, the melting of the glaciers is also expected to accelerate and, at a tipping point, the increase in discharge in these upstream catchments will stop and discharge abruptly be reduced. Studies in other catchments in the Hindukush area show exactly this behavior, with a current increase in discharge in the headwaters, but project a strong decrease on the long run [2,53]. The warming is, in general, altering the flow regime in the whole catchment, since the period of snowfall is reduced and precipitation, which is usually stored until spring as snow cover, feeds as direct runoff into the river systems.

In the Puli-Khumri station, which is already in the lowland of the catchment, the decrease of precipitation already leads to a decrease in maximum annual discharge, even though this is leveled out overall by the additional discharge through the glacier melt. For the other stations further downstream, the increase in evapotranspiration that is caused by the increased temperature and the strong reduction of precipitation leads to strong decrease in streamflow. This holds for both maximum and minimum discharge, but it is most pronounced during the summer discharge peak.

This interpretation of the results is also supported by the trends in change of landcover, which show a general tendency to drier conditions and a significant increase in human activities. The reduced rainfall and increased evaporation caused a reduction of grassland and an increase of barren land. Parts of rainfed agriculture have been turned into irrigated agriculture, but large parts have also been abandoned and turned into shrubland and barren land. A plausible explanation for this observation might be the drier conditions, which do not allow rainfed agriculture in many parts of the basin anymore. On the other side increased forest and tree cover, which can be explained by the substantial increase of fruit tree cultivation, which are more resilient to the drier conditions in the catchment. In addition, urban settlements increased strongly, which likely puts even more pressure on the land and available water resources.

6. Conclusions and Recommendations

The study results indicate that, since the 1960s, the annual average temperature in the KRB has been increasing, while precipitation and river discharge have been decreasing, with the exception of glacier-fed headwaters. The increase in the discharge in the upper catchment will continue until the small glaciers that still exist are melted and then a dramatic decrease in summer discharge where it is most needed for irrigation can be expected for the whole catchment, similar to other catchments [54]. In addition, there has been a drastic and significant change in landcover since 1992, most likely due to climate change impacts as well as environmental degradation and human impact. This leads to more direct run-off of precipitation which increases the risk of floods. In combination, these processes negatively impact the livelihoods and wellbeing of its communities.

About 1.9 million people live in the KRB and their livelihoods mostly rely on agriculture. Climate change impacts therefore affect food security, particularly of those depending on the household farming. Decreasing precipitation results in a depletion of water resources, in some cases leading to water scarcity. In addition, the combination of climate change impacts and strong pressure on the land use during the long period of war and conflict in the country has led to a degradation of vegetation cover in the KRB. Afghanistan is traditionally an agrarian country, with 22% of the national GDP produced in this sector. Approximately 79% of the population is engaged in farming. Agriculture is an important source of livelihood and local economy rely on that [55,56]. Agriculture and farmers are more affected by the impacts of climate change in Afghanistan [57]. The main obstacles are war and conflict in the country and a lack of effective investment and management in agriculture and irrigation sectors. Additionally, land use and land cover change due to socio-economic changes through political and economic transformation and climate change impacts is a critical issue in Afghanistan and a number of studies have been conducted on LULC in Afghanistan [58–61]. A LULC study undertaken by FAO as compared LULC for Afghanistan between1993 and 2016 showed changes in the KRB land cover that are in line with the results of this study [19,23].

In turn, this affects the capacity of people and the environment to adapt to climate change. The strong warming trend in winter and spring lead to an earlier snow melt, which again increases the risk of flash flooding. However, there are also positive signals visible. There is a strong increase of fruit trees, which are more resilient to harsher climatic conditions and may even locally have the positive effect the microclimate. In addition, irrigated agriculture has also increased. Both of the signals show that farmers adapt automatically autonomously to the changing conditions.

In addition, the study shows that the annual discharge of the KRB is sufficient for developing the watershed if the water resources are managed in an integrated and sustainable way. The downstream part of the KRB covers a wide area with large agriculture potential, for example by multiple cropping through irrigation. At the same time, the downstream part of the KRB is very vulnerable to flash floods and droughts, which affect the livelihood and socio-economy of the community living within the watershed deeply. Therefore, integrated water resources management is key for the agricultural development, livelihoods, and local economy. Measures, like reforestation, could reduce the risk of flash floods and droughts. Other measures, which have proven their effectiveness for many catchments in a developing context, could include guidelines on best practices, the establishment of a river basin council, and adapted community-based participation approaches. Using approaches that directly involve the communities in management and decision-making processes, these collectively can improve the socio-economy and livelihoods of the people within the KRB. However, a comprehensive IWRM strategy is still missing for Afghanistan and particularly the KRB. Therefore, we hope that the results of this study contribute to informing sustainable water resources development and watershed management. In conclusion, this study argues for establishing an Integrated Water Resources Management Plan for the KRB to trigger sustainable development [62].

Author Contributions: Conceptualization, N.A.A. and V.A.; Methodology, N.A.A., S.S. and V.A.; Writing—Original Draft Preparation, N.A.A.; Writing—Review & Editing, N.A.A., S.S. and V.A.; Visualization: N.A.A., S.S. and V.A., Funding: N.A.A. All authors have read and agreed to the published version of the manuscript.

Funding: This research received no external funding.

Acknowledgments: We thank Susan Cuddy from CSIRO Land and Water branch for her revision of the paper, susan.cuddy@csiro.au.

Conflicts of Interest: The authors declare no conflict of interest.

References

1. Shroder, J.; Ahmadzai, S.J. *Transboundary Water Resources Climate Change and Land-Use Implications*; Elsevier: Amsterdam, The Netherlands, 2016.
2. Hagg, W.; Hoelzle, M.; Wagner, S.; Mayr, E.; Klose, Z. Glacier and Runoff Changes in the Rukhk Catchment, Upper Amu-Darya Basin until 2050. *Glob. Planet. Chang.* **2013**, *110*, 62–73. [CrossRef]
3. White, C.J.; Tanton, T.W.; Rycroft, D.W. The Impact of Climate Change on the Water Resources of the Amu Darya Basin in Central Asia. *Water Resour. Manag.* **2014**, *28*, 5267–5281. [CrossRef]
4. Glantz, M.H. Water, Climate, and Development Issues in the Amu Darya Basin. *Mitig. Adapt. Strateg. Glob. Chang.* **2005**, *10*, 23–50. [CrossRef]
5. International Federation of Red Cross and Red Crescent Societies. Emergency Plan of Action Operation Update Afghanistan: Drought and Flash Floods. 2020, Volume 2. Available online: https://www.ifrc.org (accessed on 21 September 2020).
6. McSweeney, C.; New, M.; Lizcano, G. UNDP Climate Change Country Profiles Documentation. 2007. Available online: http://country-profiles.geog.ox.ac.uk/ (accessed on 21 September 2020).

7. Savage, M.; Dougherty, B.; Hamza, M.; Butterfield, R.; Bharwani, S. *Socio-Economic Impacts of Climate Change in Afghanistan A Report to the Department for International Development*; Stockholm Environment Institute: Stockholm, Sweden, 2009; Available online: https://www.weadapt.org/sites/weadapt.org/files/legacy-new/placemarks/files/5345354491559sei-dfid-afghanistan-report-1-.pdf (accessed on 22 September 2020).
8. Ghulami, M. Assessment of Climate Change Impacts on Water Resources and Agriculture in Data-Scarce Kabul Basin. *Université Côte d'Azur* **2018**, *146*. Available online: https://tel.archives-ouvertes.fr/tel-01737052 (accessed on 22 September 2020).
9. Aich, V.; Akhundzadah, N.A.; Knuerr, A.; Khoshbeen, A.; Hattermann, F.; Paeth, H.; Scanlon, A.; Paton, E. Climate Change in Afghanistan Deduced from Reanalysis and Coordinated Regional Climate Downscaling Experiment (CORDEX)—South Asia Simulations. *Climate* **2017**, *5*, 38. [CrossRef]
10. NEPA. *Second National Communication Under the United Nations Framework Convention on Climate Change (UNFCCC)*; National Environmental Protection Agency (NEPA): Kabul, Afghanistan, 2017; Available online: https://postconflict.unep.ch/publications/Afghanistan/Second_National_Communication_Report2018.pdf (accessed on 22 September 2020).
11. UNEP. *Post-Conflict Environmental Assessment Afghanistan*; United Nation Environmental Program: Nairobi, Kenya, 2003; Available online: https://www.unenvironment.org/resources/assessment/afghanistan-post-conflict-environmental-assessment (accessed on 21 September 2020).
12. Ahmad, M.; Wasiq, M. *Water Resource Development in Northern Afghanistan and Its Implications for Amu Darya Basin*; World Bank: Washington, DC, USA, 2004. [CrossRef]
13. Parto, S.; Mihran, R. Climate Change and Food Security in Afghanistan: Evidence from Balkh, Herat, and Nangarhar. 2014. Available online: https://www.loc.gov/item/2014363278/ (accessed on 21 September 2020).
14. Akmuradov, M. Environment and Security. 2011. Available online: http://www.zaragoza.es/contenidos/medioambiente/onu/1171-eng.pdf (accessed on 21 September 2020).
15. Smith, T. Climate Vulnerability in Asia's High Mountains. *WWF USAID* **2014**, *116*. Available online: https://www.climatelinks.org/resources/climate-vulnerability-asias-high-mountains (accessed on 21 September 2020).
16. Wester, P.; Mishra, A.; Mukherji, A.; Shrestha, A.B. Summary of the Hindu Kush Himalaya Assessment Report. 2019. Available online: https://doi.org/doi:10.1007/978-3-319-92288-1 (accessed on 21 September 2020).
17. Chen, Y.; Li, W.; Deng, H.; Fang, G.; Li, Z. Changes in Central Asia's Water Tower: Past, Present and Future. *Sci. Rep.* **2016**, *6*. [CrossRef]
18. USGS. Conceptual Model of Water Resources in the Kabul Basin, Afghanistan. 2009. Available online: https://pubs.usgs.gov/sir/2009/5262/pdf/sir2009-5262_front_text_508_Pt1_i-32_rev111312.pdf (accessed on 21 September 2020).
19. Kamal, G.M. River Basins and Watersheds of Afghanistan. 2004, pp. 1–7. Available online: http://www.nzdl.org/gsdl/collect/areu/Upload/1710/Kamal_River%20basins%20and%20watersheds2004.pdf (accessed on 22 September 2020).
20. CSO. *Afghanistan Statistical Yearbook*; Afghanistan Islamic Republic Central Statistic Organization, 2018; Available online: https://doi.org/10.29171/azu_acku_musalsal_ha4570_6_alif2_seen22_v1391s (accessed on 21 September 2020).
21. Doebrich, J.L.; Wahl, R.R.; Ludington, S.D.; Chirico, P.G.; Wandrey, C.J.; Bohannon, R.G.; Orris, G.J.; Bliss, J.D.; Wasy, A.; Younusi, M.O. Geologic and Mineral Resource Map of Afghanistan. USGS, 2006. Available online: https://mrdata.usgs.gov/catalog/cite-view.php?cite=54 (accessed on 21 September 2020).
22. DABS. AFG: Multi Tranche Financing Facility for Energy Sector Development Investment Program (ESDIP)—Tranche 1. ADB, 2016. Available online: https://www.adb.org/projects/documents/multitranche-financing-facility-energy-sector-development-program-proposed-projec (accessed on 21 September 2020).
23. FAO. Land Cover Atlas of the Islamic Republic of Afghanistan. 2016. Available online: http://www.fao.org/3/a-i5457e.pdf. (accessed on 21 September 2020).
24. MEW. Afghanistan Water Resources Development (AWARD) Project—Technical and Implementation Support Consultancy (TISC) AWARD-TISC INVESTMENT PLAN FOR THE PANJ-AMU. 2013. Available online: https://afghanwaters.net/wp-content/uploads/2017/10/2013-KRB-Investment-Plan.pdf (accessed on 21 September 2020).

25. Ministry of Agriculture, Irrigation and Livestock. The Islamic Republic of Afghanistan Ministry of Agriculture, Irrigation and Livestock Climate Change Scenarios for Agriculture of Afghanistan. 2017. Available online: http://www.fao.org/documents/card/en/c/08636747-610a-4690-b588-97171b0cb4e5/ (accessed on 21 September 2020).
26. Afghanistan Meteorological Department. Available online: http://www.amd.gov.af/ (accessed on 13 February 2020).
27. Afghanistan, W.V. World Vision Afghanistan Annual Report 2019. Available online: https://reliefweb.int/report/afghanistan/world-vision-afghanistan-annual-report-2019 (accessed on 21 September 2020).
28. Favre, R.; Kamal, G.M. Watershed Atlas of Afghanistan, Working Document For Planners. 2004. Available online: http://www.cawater-info.net/afghanistan/pdf/afg_wat_atlas_part_1_2.pdf (accessed on 21 September 2020).
29. Cressey, G.B.; Michel, A.A. The Kabul, Kunduz, and Helmand Valleys and the National Economy of Afghanistan: A Study of Regional Resources and the Comparative Advantages of Development. *Geogr. Rev.* **1960**, *50*, 609. [CrossRef]
30. Ministry of Energy and Water. *Hydrological Year Book*; Islamic Republic Ministry of Energy and Water archive: Kabul, Afghanistan, 2018. Available online: https://www.loc.gov/item/lcwaN0015932/ (accessed on 21 September 2020).
31. U.S. Geological Survey. Landsat—Earth Observation Satellites. 2016. Available online: https://doi.org/10.3133/fs20153081 (accessed on 22 September 2020).
32. McLeod, A.I. Kendall Rank Correlation and Mann-Kendall Trend Test. 2011, p. 12. Available online: https://cran.r-project.org/web/packages/Kendall/Kendall.pdf (accessed on 22 September 2020).
33. Yue, S.; Pilon, P. A Comparison of the Power of the t Test, Mann-Kendall and Bootstrap Tests for Trend Detection. *Hydrol. Sci. J.* **2004**, *49*, 21–37. [CrossRef]
34. Mazza, A.; Punzo, A. R Package: DBKGrad. 2015, p. 12. Available online: https://rdrr.io/cran/DBKGrad/ (accessed on 22 September 2020).
35. Theil, H. A Rank-Invariant Method of Linear and Polynomial Regression Analysis. *Ned. Acad. Wetensch. Proc.* **1950**, *53*, 386–392.
36. Sen, P. Estimates of the Regression Coefficient Based on Kendall's Tau. *J. Am. Stat. Assoc.* **1968**, *63*, 1379–1389. [CrossRef]
37. Yue, S.; Pilon, P.; Phinney, B.; Cavadias, G. The Influence of Autocorrelation on the Ability to Detect Trend in Hydrological Series. *Hydrol. Process.* **2002**, *16*, 1807–1829. [CrossRef]
38. Bronaugh, D.; Werner, A. Package "Zyp". 2015, p. 9. Available online: http://cran.nexr.com/web/packages/zyp/zyp.pdf (accessed on 22 September 2020).
39. Boehmke, B.; Greenwell, B. Random Forests. 2001, 5–32. In *Data Science—Was Ist das Eigentlich?!* Springer: Berlin/Heidelberg, Germany, 2001; pp. 5–32. [CrossRef]
40. Gorelick, N.; Hancher, M.; Dixon, M.; Ilyushchenko, S.; Thau, D.; Moore, R. Google Earth Engine: Planetary-Scale Geospatial Analysis for Everyone. *Remote Sens. Environ.* **2017**. [CrossRef]
41. Jin, Y.; Liu, X.; Chen, Y.; Liang, X. Land-Cover Mapping Using Random Forest Classification and Incorporating NDVI Time-Series and Texture: A Case Study of Central Shandong. *Int. J. Remote Sens.* **2018**. [CrossRef]
42. Hall-Beyer, M. Practical Guidelines for Choosing GLCM Textures to Use in Landscape Classification Tasks over a Range of Moderate Spatial Scales. *Int. J. Remote Sens.* **2017**. [CrossRef]
43. Haralick, R.M.; Dinstein, I.; Shanmugam, K. Textural Features for Image Classification. *IEEE Trans. Syst. Man Cybern.* **1973**. [CrossRef]
44. Xu, H. Modification of Normalised Difference Water Index (NDWI) to Enhance Open Water Features in Remotely Sensed Imagery. *Int. J. Remote Sens.* **2006**. [CrossRef]
45. Liu, H.Q.; Huete, A. Feedback Based Modification of the NDVI to Minimize Canopy Background and Atmospheric Noise. *IEEE Trans. Geosci. Remote Sens.* **1995**. [CrossRef]
46. Goodwin, N.R.; Coops, N.C.; Wulder, M.A.; Gillanders, S.; Schroeder, T.A.; Nelson, T. Estimation of Insect Infestation Dynamics Using a Temporal Sequence of Landsat Data. *Remote Sens. Environ.* **2008**. [CrossRef]
47. IDB. Index: Green Optimized Soil Adjusted Vegetation Index. 2020, p. 2020. Available online: https://www.indexdatabase.de/db/i-single.php?id=29 (accessed on 21 September 2020).
48. Bouzekri, S.; Lasbet, A.A.; Lachehab, A. A New Spectral Index for Extraction of Built-Up Area Using Landsat-8 Data. *J. Indian Soc. Remote Sens.* **2015**. [CrossRef]

49. Zhao, H.; Chen, X. Use of Normalized Difference Bareness Index in Quickly Mapping Bare Areas from TM/ETM+. *Int. Geosci. Remote Sens. Symp.* **2005**, 3, 1666–1668. [CrossRef]
50. ESA. *Land Cover CCI Product User Guide 2.5*; UCL-Geomatics: Belgium, 2017; Available online: http://maps.elie.ucl.ac.be/CCI/viewer/download/ESACCI-LC-PUG-v2.5.pdf (accessed on 22 September 2020).
51. *Long-Term Climate Change: Projections, Commitments and Irreversibility*; Cambridge University Press: Cambridge, UK, 2013. [CrossRef]
52. WMO. *Guide to Climatological Practices*; WMO: Geneva, Switzerland, 2011. [CrossRef]
53. Unger-Shayesteh, K.; Vorogushyn, S.; Farinotti, D.; Gafurov, A.; Duethmann, D.; Mandychev, A.; Merz, B. What Do We Know about Past Changes in the Water Cycle of Central Asian Headwaters? A Review. *Glob. Planet. Chang.* **2013**, *110*, 4–25. [CrossRef]
54. Huss, M.; Hock, R. Global-Scale Hydrological Response to Future Glacier Mass Loss. *Nat. Clim. Chang.* **2018**, *8*, 135–140. [CrossRef]
55. *Afghanistan Development Update*; World Bank: Washington, DC, USA, 2018; Available online: http://documents1.worldbank.org/curated/en/985851533222840038/pdf/Afghanistan-development-update.pdf (accessed on 22 September 2020).
56. UNEP. *Building Adaptive Capacity and Resilience to Climate Change in Afghanistan (LDCF), Baseline Assessment Report*; UNEP: Kabul, Afghanistan, 2014; Available online: http://wedocs.unep.org/bitstream/handle/20.500.11822/22232/Adaptive_capacity_Afghanistan.pdf?sequence=1&isAllowed=y (accessed on 21 September 2020).
57. Jawid, A.; Khadjavi, M. Adaptation to Climate Change in Afghanistan: Evidence on the Impact of External Interventions. *Econ. Anal. Policy* **2019**, *64*, 64–82. [CrossRef]
58. Najmuddin, O.; Deng, X.; Siqi, J. Scenario Analysis of Land Use Change in Kabul River Basin—A River Basin with Rapid Socio-Economic Changes in Afghanistan. *Phys. Chem. Earth* **2017**, *101*, 121–136. [CrossRef]
59. Najmuddin, O.; Deng, X.; Bhattacharya, R. The Dynamics of Land Use/Cover and the Statistical Assessment of Cropland Change Drivers in the Kabul River Basin, Afghanistan. *Sustainability* **2018**, *10*, 423. [CrossRef]
60. Přívara, A.; Přívarová, M. Nexus between Climate Change, Displacement and Conflict: Afghanistan Case. *Sustainability* **2019**, *11*, 5586. [CrossRef]
61. Khairandish, E.; Mishra, S.K.; Lohani, A.K. Modeling on Long-Term Land Use Change Detection Analysis of Kabul River Basin, Afghanistan by Using Geospatial Techniques. *Model. Earth Syst. Environ.* **2020**. [CrossRef]
62. Global Water Partnership; International Network of Basin Organization. *A Handbook for Integrated Water Resources Management in Basins*; Global Water Partnership and International Network of Basin Organization, 2009; Volume 74. Available online: https://www.inbo-news.org/IMG/pdf/GWP-INBOHandbookForIWRMinBasins.pdf (accessed on 22 September 2020).

climate

MDPI

Case Report

Non-Conventional Agricultural Spaces and Climate Change: The Cases of Le Grenier boréal and Lufa Farms in Quebec, Canada

Mélanie Doyon * and Juan-Luis Klein

Department of Geography, Université du Québec à Montréal, Montréal, QC H2X 3R9, Canada; klein.juan-luis@uqam.ca
* Correspondence: doyon.melanie@uqam.ca

Abstract: The objective of this text is to present a reflection on the link between local initiatives to combat food insecurity and actions adapting to climate change. To this end, two case studies of ongoing experiments in the Canadian province of Quebec will be presented and compared. While these two cases are very different in terms of location, production and people involved, they share the objective of bringing fresh and healthy food, produced locally, to the population of their territory and of rethinking the relationship of the community to nature through food production. Despite their significant differences, each of these two cases features actions for responding to problems that have a common cause: an agro-industrial food system that, by decoupling the locations of production and consumption, in order to maximize the economic profitability of the capital invested, has compromised both the health of citizens and the ecological balance.

Keywords: climate change; food insecurity; local initiatives; food miles; ecological transition

Citation: Doyon, M.; Klein, J.-L. Non-Conventional Agricultural Spaces and Climate Change: The Cases of Le Grenier boréal and Lufa Farms in Quebec, Canada. *Climate* **2021**, *9*, 148. https://doi.org/10.3390/cli9100148

Academic Editors: Azzeddine Madani, Christopher Robin Bryant, Andrea Vitali and Nektarios Kourgialas

Received: 23 July 2021
Accepted: 25 September 2021
Published: 2 October 2021

1. Introduction

Climate change resulting from the production of greenhouse gases (GHGs) is certainly the most evident aspect of the environmental crisis facing the planet, even if it is not the only factor. As Swyngedouw [1] pointed out, this crisis is global and concerns a model of society that sees nature only as an unlimited pool of resources to be exploited for economic growth and financial profitability. The options for dealing with this crisis must target the various facets of human life, such as health, transportation, agriculture, finance and water, and how these interlock with nature [2]. Thus, social innovations aimed at transforming society's relationship with the environment must be deployed across several dimensions, with food production targeted as a main, if not a flagship, priority [3] (p. 8).

Climate change and other aspects of the crisis such as the nutritional crisis are intimately linked. We have seen that events caused by climate change, which are becoming more frequent, destabilize the world food system, which consequently diminishes food security. An example of this is the case of Russia, which in 2010, fearing that it would be unable to meet its domestic demand following a major heat wave, decided to stop exporting wheat. This caused an increase of more than 40% in the price of wheat, making it more difficult for citizens in several regions of the world to obtain this grain [4].

This growing insecurity signals the need to rethink the food system with a view to experiments that promote an alternative to the globalized option of the food industry, or even of the economy as a whole. The territory can thus become a framework for restructuring based on the local, supported by visions that mobilize the interaction of the various dimensions of human life while also taking non-human life into account [1]. This calls for a return to a territorialized vision of development that brings consumption and production closer together and that draws on a new post-capitalist model of development built on the basis of local experiments [5]. In such a model, territories are seen as life environments, which means they are used and valued primarily from a perspective of improving the quality of life of citizens [6].

For several years now, the global food system has been under pressure due to rising temperatures, changes in precipitation patterns and more frequent extreme events (heat waves, droughts, hail) [7]. These climate challenges are compounded by global population growth, non-food uses of food crops and a shift to an increasingly animal-based diet [8]. These major trends threaten the food security of populations. According to the FAO [9], considering the current food system, 50% more food will have to be produced to meet the growing needs of the world's population.

While tropical and subtropical regions are already feeling the negative impacts of climate change on their agricultural yields, more northern regions are benefiting from these changes and are experiencing increased productivity for certain crops such as corn, soybeans and wheat. However, the IPCC [7] warns that these positive impacts will be short-lived. Declining yields of the major cereals will increase their cost, and this increase will affect the price of food in general. Fruit and vegetable production is not left out and also remains vulnerable to climate change. Heat stress and extreme events affect plant growth and even destroy crops. In addition to these direct impacts on production, there are impacts on productive resources. Decreased water quantity and quality, soil degradation and the presence and proliferation of pests and diseases are also to be expected. Finally, it should be noted that extreme temperatures likewise have impacts on agricultural workers [7]. One third of the world's food production would no longer benefit from a "safe climate space" and would be threatened in the medium term [10].

The relationships between climate change and food systems are complex and have consequences on the four dimensions of food security identified by the FAO [11], namely, food availability, food access, utilization and stability of these three dimensions over time. Availability refers to food supply and is derived from production, productivity, provisioning and trade. Food access is both economic and physical. The economic dimension refers to income in relation to the price of food, while physical access refers to infrastructure and the organization of supply and distribution systems, as well as to non-market practices (home production, social and solidarity economy organizations) [12]. Food utilization concerns the attainment of nutritional well-being that satisfies all physiological needs. Finally, the fourth pillar relates to the stability of the three previous pillars and concerns the different temporalities (cyclical, seasonal, annual).

Several studies have focused on measures to adapt to and mitigate climate change in food systems. Some of these measures concern demand-side changes (e.g., changes in diets) [13]. Others address the supply side such as adapting food production systems [14] and maintaining traditional productive systems [15]. In this article, we would like to discuss the contribution of non-traditional agricultural production sites to food security in a context of climate change and the territorialization of production practices. Given the pressures of climate change on production areas and human and natural resources, we believe it is necessary to reflect on the alternative dimension of these experiments in terms of their potential effects on natural and social balances.

The objective of this text is to present a reflection on the link between actions to adapt to food insecurity. To this end, two case studies of ongoing experiments in the Canadian province of Quebec will be presented and compared. While these two cases are very different in terms of location, production and people involved, they share the objective of bringing fresh and healthy food, produced locally, to the population of their territory and of rethinking the relationship of the community to nature through food production.

2. Materials and Methods

This article is based on a meta-analysis of two case studies that were conducted separately: Le Grenier boréal initiative in the Côte-Nord region (the study of Le Grenier boréal initiative was carried out by Jessica Élie-Leonard as part of her master's degree under the direction of Mélanie Doyon [16]), and Lufa Farms in Montreal (the study of

Lufa Farms was carried out by Roufaï Ouro-Koura as part of a master's thesis under the supervision of Juan-Luis Klein [17]). Both applied the case study method that is appropriate for comprehensive inductive studies of complex initiatives that need to be seized in their territorial context (for the case study method, see Yin [18] and Crowe et al. [19]). Our meta-analysis is based on the pattern matching of both case studies (Figure 1). It is inspired by grounded theory [20]. Therefore, the resulting theoretical and strategic considerations and proposals about the potential effect of food security-oriented initiatives on the ecologic transition and, consequently, on climate dynamics are presented in the last section of the article.

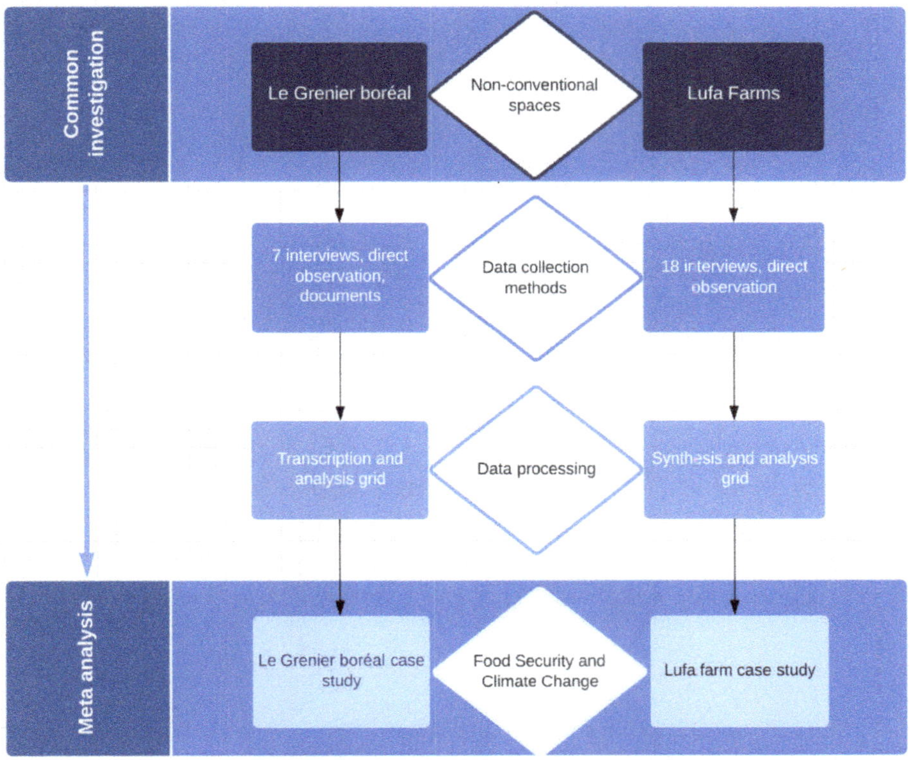

Figure 1. Meta-Analysis Chart.

The case study of Le Grenier boréal drew on a field survey which included direct observation as well as seven (7) semi-structured in-depth interviews with key project stakeholders (professionals, elected officials, project leaders, volunteers). These interviews, which lasted 90 min on average, were carried out in 2018. Information collected was recorded and transcribed and then analyzed using a reading grid. We also drew on government documents such as socio-territorial portraits of Minganie and Côte-Nord, a portrait and an action plan concerning poverty and social exclusion and the Minganie development plan, as well as maps, statistical information (Statistics Canada and Institut de la statistique du Québec) and graphs (Institut national de santé publique du Québec). Local newspapers and larger, more high-circulation newspapers, as well as radio and TV reports, were also mobilized and allowed us to determine the highlights of the project since its creation.

The data about Lufa Farms were drawn from eighteen (18) in-depth semi-structured interviews, conducted in 2018, with a variety of respondents, including two (2) main managers of the company, five (5) employees and nine (9) partners (agricultural producers, food artisans and distributors). These respondents were interviewed about the company's organizational practices and production methods. In addition, two (2) representatives of neighborhood organizations involved in food security and the social economy were interviewed about the company's community.

3. The Territorial Context of the Case Studies

The first case is located in an area on the outskirts of major urban centers, in the far east of the province. It is the agricultural initiative Le Grenier boréal (https://grenierboreal.coop/ accessed on 28 June 2021), an agroforestry solidarity cooperative located in the municipality of Longue-Pointe-de-Mingan in Quebec's Côte-Nord region. Founded as a result of citizen mobilization, this social economy enterprise, which has transformed an abandoned military site (tests have been conducted in order to ensure that there is no contamination) into agricultural land, relies on traditional resources and know-how as well as innovative methods in order to produce fruits, vegetables and herbs. The second study presents the case of a social enterprise (albeit private) of urban agriculture located in the metropolitan area of Montreal, thus in a central region and in a global city. This is the enterprise Les Fermes Lufa (https://montreal.lufa.com/fr/ accessed on 28 June 2021), or Lufa Farms, in English. In this project, actors mobilize innovative technologies to produce fresh food on a large scale in greenhouses located, in some cases, on the roofs of repurposed abandoned buildings.

Despite their significant differences, each of these two cases features actions showing autonomy-oriented responses to problems that, being different, have a common cause: an extractivist food system that, by decoupling the locations of production and consumption, in order to maximize the economic profitability of the capital invested, has compromised both the health of citizens and the ecological balance.

3.1. Case 1: Le Grenier boréal in Quebec's Côte-Nord Region—A Challenge with Regard to Remoteness and Climatic Conditions

Le Grenier boréal cooperative was established in 2013 by citizens of Longue-Pointe-de-Mingan, following a study on non-timber forest products (NTFPs) in Minganie and the obtaining of funding [21]. Two observations gave rise to the creation of this cooperative: the need to renew the economic base of the region, and the lack of fresh fruits and vegetables in the territory [22]. As a solidarity cooperative (A solidarity cooperative "is characterized by the diversification of its membership and its openness to partnership. It brings together, within the same enterprise, individuals or companies that have a common cause or interest, but varied needs" [23]. Entreprise Québec. La coopérative: un modèle d'affaires à découvrir. Available online: https://www2.gouv.qc.ca/portail/quebec/infosite?lang=fr&x=2371115552 accessed on 15 July 2021). Solidarity cooperatives are very present in home and health care services, professional services, business services and local services (grocery stores, gas stations, restaurants, etc.) as well as in arts and culture.), the mission of Le Grenier boréal is, on the one hand, to provide work for its members through the production, harvesting, processing and marketing of food and, on the other hand, to offer some agricultural and agroforestry products and services to the cooperative's members.

The enterprise has four components: (1) NTFP harvesting, which valorizes forest resources; (2) agrotourism and educational animation; (3) consulting services in agroforestry and the environment; and (4) market garden production [24]. The market garden component occupies approximately two hectares of leased public land. The site, which was a military wasteland, was cleaned up and reclaimed. Some development was required to address the challenges of the environment. Hedges were planted to protect the site from the strong winds that blow in Côte-Nord. Thanks in large part to local resources, the cooperative was also able to meet the challenge of poor soil. From the beginning, the cooperative relied on the residues from the local crab processing plant to produce compost. Over time, it has diversified its sources of supply, and fertilization is now mainly conducted with algae from the river, harvested by hand along the banks. The cooperative wishes to enhance the valorization of scallop residues of local companies and has set up a project for this purpose. Finally, shredded branch residues from the territory also contribute to fertilizing the gardens [25].

Although not "certified organic", the agriculture practiced by Le Grenier boréal respects organic farming principles [21]. The cooperative hopes that the agricultural and agroforestry expertise it develops will contribute to the social, economic and environmental progress of the communities located in the Minganie regional county municipality (RCM) [24].

3.1.1. The Territorial Framework: Minganie in Quebec's Côte-Nord Region

The municipality of Longue-Pointe-de-Mingan is located in the RCM of Minganie in the administrative region of Côte-Nord. This region is characterized by its vastness as well as its nature marked by the boreal forest and the St. Lawrence River. Located between Tadoussac and Blanc-Sablon, Côte-Nord is the second largest administrative region in Quebec, with a land area spanning 236,664 km^2 [26]. While the western part of the region is served by Route 138, the eastern part is not connected to the provincial road network and can only be reached by plane or boat. In 2020, the region had a population of 90,529 [27], a decrease of 4.5% from 2011 [28]. The region is subdivided into six RCMs, including Minganie.

Minganie is composed of eight municipalities (ten villages) and two Aboriginal (Innu) communities. It covers 55,355 km^2 of land, including the island of Anticosti [29]. In 2020, its population was 6437, with a density of 0.1 inhabitants/km^2 [30]. The municipality of Havre-Saint-Pierre, which provides the majority of services, had 3371 inhabitants [26]. Longue-Pointe-de-Mingan, where Le Grenier boréal is located, had a population of 423 in 2021 [29], a decrease of about 10% from 2011 [28]. The municipality is located nearly two hours by car east of Sept-Îles, the main city in the region (population 25,400 in 2016 [28]). The total median household income in 2015 was CAD 78,080, well above the Quebec average of CAD 59,822 [31]. However, in 2017, the percentage of low-income families was 12.4%, which is higher than the Quebec rate of 9.5% [32].

The majority of jobs in the RCM are in the sales and services and transportation and machinery sectors [33]. Since the early 2000s, the region has been developing ecotourism activities, including marine mammal watching tours and excursions to the islands of Anticosti and the Mingan Archipelago [34]. Nevertheless, the local economy depends on primary sector activities (fishing, mining, forestry), which are threatened by the depletion of resources and the lack of new workers. In addition, regional economic activity is driven by major extractive projects (e.g., the Romaine hydroelectric power complex, the Rio Tinto mine) [35]. However, only part of the Mingan population benefits from these projects, since many jobs are taken up by workers from outside the region, whose arrival puts pressure on local resources, particularly housing [35].

Agricultural activities, for their part, are rare in Minganie. In fact, in 2020, the RCM had no agricultural zone (In Quebec, provincial agricultural zoning was established in 1976 through the *Act respecting the preservation of agricultural land and agricultural activities*. The purpose of this act is to exclude non-agricultural uses in the agricultural zone. Essentially located along the St. Lawrence River and its main tributaries, the agricultural zone spans a total area of 63,000 km^2. It is present in 954 municipalities (out of 1132) and in all administrative regions of the province.). This can be explained by climatic conditions that are not very conducive to agriculture, notably average temperatures in July of 14.6 °C and an annual snowfall of 252 cm (data for Rivière-au-Tonnerre, between 1981 and 2010), compared to 21.2 °C and 158 cm on the island of Montreal (data for Rivière-des-Prairies, between 1981 and 2010) [36]. This results in an average growing season that is two months shorter than that of the Montreal region [37]. In 2011, the MAPAQ [38] counted only three agricultural enterprises in Minganie.

3.1.2. Food Insecurity in Minganie

Food insecurity in Minganie has various causes. First, the RCM, as with the entire region, relies on food deliveries from the large distribution centers located in the south of the province to meet most of its needs. Between two deliveries, fresh food can be scarce, indicating an instability in the food supply. In addition, transportation difficulties (e.g., accidents, weather) sometimes delay deliveries and disrupt supply, creating a sense of insecurity among the population [16].

Physical and economic access conditions are difficult for some segments of society. As it occurs elsewhere, poorer populations have less access to food [39], a problem that is exacerbated by high food prices. Indeed, in 2009, the cost per person per day of a nutritious food basket was CAD 7.84 in Minganie, while it was CAD 6.75 in Montreal [40].

The physical conditions of access are also an issue in Minganie. First, the food supply is limited by the scarcity of food stores. The only supermarket in the RCM is located in Havre-Saint-Pierre (30 min from Longue-Pointe-de-Mingan), and many villages have only small grocery stores, gas stations or convenience stores to obtain supplies. Furthermore, a car is almost always necessary for travel, due to the long distances involved; however, some people do not have one, and the region offers few alternatives to car transportation [35].

Finally, the quality of fruits and vegetables is regularly compromised in Minganie because of the thousands of food miles they have to travel. This is especially true in winter due to delivery difficulties, sometimes resulting in losses.

As with many rural communities in Quebec, Longue-Pointe-de-Mingan, and more generally the Minganie region, is characterized by a low demographic weight. Businesses are few and far between, and the cost of food is high, mainly due to transportation. While some foods are less affected by this circumstance, such as milk, for which a ceiling price was adopted (Although this price ceiling is still a little higher here than in the rest of Quebec, the price of milk is nonetheless much lower now than prior to the adoption of the price ceiling. Since 1 July 2016, the Régie des marchés agricoles et alimentaires has been regulating the price of milk across all of Quebec. In 2009, in Côte-Nord, the cost of four liters of 3.25% milk even reached CAD 15.96, while the maximum price in 2021 in Minganie was CAD 8.86 [40,41]. The price of other foods, however, is not regulated.), the price of other foods raises certain questions of equity.

3.1.3. The Marketing of Products

Since its inception, an increasing variety of fruits, vegetables and herbs have been produced by Le Grenier boréal, although the challenges remain significant. The cooperative has also had to adapt its production techniques. For example, it has installed plastic tunnels and greenhouses to extend the season, introduced winter covers to protect crops from the cold, chosen varieties adapted to the climate, made increasing use of local resources to fertilize the soil and planted willow hedges to protect the gardens from the wind [25]. Initially, its production was exclusively marketed through weekly baskets, the number of which grew from 10 baskets to 35 baskets between 2016 and 2017. In 2017, the cooperative successfully inaugurated a U-pick strawberry farm [42]. Since 2018, Le Grenier boréal has provided food to be sold at the Havre-Saint-Pierre grocery store [43] and, as of 2019, has supplied ingredients to the Puyjalon distillery in Havre-Saint-Pierre, to produce gin [44]. Although the 2020 season was complicated by pandemic-related delays in the construction of a greenhouse and the arrival of student interns, as well as redevelopment that took some acreage out of production, the harvest was very good. Le Grenier boréal also managed to set up a food stand in the villages of Minganie. Finally, market garden production is expected to increase over the next few years due to the cultivation of areas that had not been used prior to 2020 and the completion of the greenhouse [45].

3.2. Case 2: Lufa Farms in Montreal: Urban Agriculture and Socio-Technical Innovation

Lufa Farms is located in a territorial setting that is the polar opposite of the one that supports Le Grenier boréal. However, in relation to climate change and food autonomy, the two projects share certain objectives and techniques. Montreal is the largest city in the province of Quebec and the second largest in Canada in terms of population and economic importance. It is at the center of a metropolitan community of 4374 km^2 that includes more than four million inhabitants according to the 2016 census. As such, the agglomeration of Montreal had a population of 2,050,053, according to City of Montréal estimates in 2019 (The urban agglomeration of Montreal includes the City of Montréal and 15 autonomous cities located on the island of Montreal. The Communauté métropolitaine de Montréal includes the City of Montréal and 81 autonomous cities. For information on the City of Montréal, http://ville.montreal.qc.ca/portal/page?_pageid=6897,67633583&_dad=portal&_schema=PORTAL 30 September 2021. For information on the Communauté métropolitaine de Montréal, see https://cmm.qc.ca/ accessed on 25 September 2021). Montreal is one of North America's major metropolises serving as an industrial and service hub and offering high-level services. However, as with most large centers in our hyper-globalized and hyper-industrialized world [46,47], this city is fraught with social divides that separate the wealthy from the poor. Neighborhoods that are more affected by poverty are challenged with various forms of precariousness and socioeconomic vulnerability [48], including food insecurity [49,50].

As a result, several types of initiatives to fight for food security are emerging. These initiatives are divided into four sectors: production, processing, distribution and consumption. In terms of production, the most important activity is urban agriculture. It goes without saying that urban agriculture alone will never be able to feed the entire population of a city such as Montreal. Nevertheless, it remains a field of experimentation for complementary solutions to food problems in an environmental and social perspective [51]. Urban agriculture is spread over several sectors and is practiced in various types of areas at multiple scales.

Urban agriculture is on the rise all over the world. It has been practiced for a long time in the so-called countries of the South, where urban populations are more likely to have some of their needs met through food self-provisioning, often driven by poverty. On the other hand, in the cities of so-called developed countries, agricultural production in urban areas for the purpose of mitigating food insecurity is more recent. Historically, it was practiced for recreational, social or therapeutic purposes or was reserved for specific uses (e.g., vines, for homemade wine, in Montreal). Thus, in general, the food system, including food distribution, was expected to satisfy the needs of urban residents. Yet, the food system can no longer meet these demands. This is due to several factors including the deterioration of food quality [52] and the new aspirations of residents regarding the link between ecology, social justice and food production. In that context, for a low-income population with difficulties in obtaining adequate food supplies, urban agriculture can facilitate access to certain foods [53,54]. In addition, urban agriculture can contribute to residents' overall level of health [55] and education [54]. Finally, urban agriculture contributes to strengthening social ties [55,56]. Today, as environmental problems intensify, urban agriculture is increasingly called into play as a means to promote sustainable development [53,54,57]. It is part of the repertoire of collective actions oriented toward the co-construction of a "sustainable city" [58]. It is this aspect that we will address with the help of the Lufa Farms case.

3.2.1. Lufa Farms in Montreal

Lufa is one of Canada's leading urban agriculture production experiments. It is innovative at several levels. On the one hand, it was a pioneer in the implementation of commercial greenhouses on building roofs [59]. On the other hand, given that it involves the construction of rooftop greenhouses, it mobilizes high-level technologies for both production and management. In 2010, the company built its first greenhouse, with a surface area of 2973 m^2, on a disused building located in the borough of Ahuntsic-Cartierville.

Harvesting and delivery of products began in April 2011. In 2013, a second greenhouse of 3995 m^2 was built on a roof in Laval. In 2017, a third larger greenhouse (5853 m^2) was put into operation in the borough of Anjou. In August 2020, finally, Lufa inaugurated a fourth greenhouse, spanning 15,217 m^2, on the roof of a building located in the borough of Saint-Laurent, doubling the productive capacity that the company had reached by then. Altogether, Lufa Farms provides fresh food year-round to nearly 30,000 people. More than fifty varieties of vegetables are produced annually in its four greenhouses (many varieties of tomatoes, cucumbers, peppers, lettuce, eggplants, microgreens, basil and Swiss chard, among others). The production is conducted in accordance with the requirements of organic agriculture, that is, without pesticides, using biocontrol (the use of beneficial insects to combat pests) alongside rational use of water and electricity.

Lufa relies on a very complex computerized system run by the Argus Titan software, designed for facilities management in protected environments such as agricultural greenhouses (The company that markets this system is located in British Columbia. See http://arguscontrols.com/about-argus/system-applications/ accessed on 23 July 2021). This application manages parameters such as temperature, humidity, light, CO_2 level, air exchange and circulation, snow loads, rain and wind protection. In addition, the system takes into account the protection of buildings and crops.

3.2.2. The Creation and Putting into Operation of the Company

Lufa was founded in 2009 ("The name of the project is inspired by luffa, a climbing plant that grows in Lebanon, among other places, where it thrives in urban environments. It decorates the walls and fences it covers, provides shade under pergolas, and supplies squash that can be cooked or dried to make sponges" [60].). The project was the brainchild of Mohamed Hage, who was joined by Lauren Rathmell, a biochemist by training and researcher at McGill University's Macdonald campus, who is now the director of the greenhouses; Yahya Badran, director of engineering and a graduate of the Technical University of Construction in Bucharest; and Kurt Lynn, a Toronto-based contractor who acts as an advisor to the company.

A number of resources were needed to launch the business. Among these, the main ones were human. While experts, architects and engineers were hired, the initiators also learned a lot as they went along. Financial investments were likewise necessary. The construction of the pilot greenhouse, for example, was realized with private funding coming from only one partner, the senior manager. As this was a high-risk investment, it had been difficult to attract external private investors. By contrast, the company was able to attract several investors for the construction of the second greenhouse, in Laval. Cycle Capital Management (Cycle Capital invests in several areas including responsible agriculture, renewable energy and clean technology. See http://www.cyclecapital.com/ accessed on 23 July 2021 and http://www.cyclecapital.com/lufa-farms-inc-a-new-company-in-the-cycle-c3e-portfolio/ accessed on 23 July 2021), a venture capital fund that promotes sustainable technologies, was a main participant in this project. The construction of the third greenhouse, in Anjou, completed in 2016, was funded primarily by Solidarity Fund QFL, a fund created by the Fédération des travailleurs du Québec (FTQ, Montréal, QC, Canada) in 1983 to support job retention and creation in Quebec. The construction of the fourth greenhouse was supported by Sollio Groupe Coopératif (formerly Co-op fédérée), a large agricultural cooperative network in Quebec that saw Lufa Farms as a major ally. This partnership has strengthened the ties between Lufa Farms and agricultural producers located near Montreal.

The main goals of Lufa's creators are to increase the food autonomy of the city and to contribute to the improvement of the food distribution chain by bringing food production closer to the consumer. These goals are intended to address the dramatic growth (demographic and spatial) of cities and the attendant ever-increasing need for food products. The creators of the company also aim to offer an option in the face of the disappearance of farmland due to urbanization. In addition, by bringing food production closer to consumers, the company reduces the number of food miles traveled, thereby reducing the amount of energy required to distribute food products. In the beginning, the company only supplied consumers with products from its agricultural greenhouses. Today, Lufa's list of partners includes approximately 200 agricultural and food processing businesses, the majority of which are located on the island of Montreal or within a 25 km radius.

At Lufa, marketing is conducted exclusively through baskets. The distribution of products is divided into six steps, from the customer's registration to the reception of their basket. Subscription, orders and payment are all carried out online. The baskets are delivered to their respective pick-up points as soon as the order preparation is completed. The company's customers are referred to as "Lufavores," which promotes loyalty and a sense of belonging.

4. Cross Analysis: Le Grenier boréal and Lufa Farms at the Crossroads of Food Security and Ecological Transition

4.1. Contribution to Food Security

Le Grenier boréal contributes to the reduction in certain dimensions of food insecurity in Minganie, although it contributes, in particular, to the quality of the food, which is why the cooperative was created (Table 1). On the one hand, the cooperative's local production guarantees freshness, since it does not require traveling long distances for delivery or storage. On the other hand, the cooperative improves the diversity of the food offered. In addition to a better food quality, mobile food stands likewise contribute to food security, as they facilitate people's physical access to the fruits and vegetables produced. Indeed, in some places, the merchandise from the food stands was sold out in less than an hour [45].

Table 1. Contribution of Le Grenier boréal and Lufa Farms to food autonomy.

Themes	Le Grenier Boréal	Lufa Farms	Summary
Productive spaces	2 ha on abandoned military site in boreal forest space	2.8 ha on urban roofs	Agricultural use of non-conventional spaces for agriculture
Production chain	Collaborates with crab and scallop processing plants for compost production Algae harvested locally	Creation of a short food supply chain with a network of agricultural producers	Productive ecology perspective
Ecological production practices	Intensive organic farming practices, without certification Fertilization with local resources, greenhouses, tunnels, winter cover, windbreaks	Hydroponic greenhouse Use of an almost entirely organic management system (Lufa Farms is not certified organic, as the vegetables produced are not grown in the ground. However, the company, as with its partners, follows the same practices as organic farms for pest and disease control. See https://montreal.lufa.com/en/lufa-faq accessed on 23 August 2021) Heating of the greenhouses from solar energy and building losses[i]	Ecological methods
Distribution methods	Pre-determined base baskets Permanent and mobile food stands Local businesses Local distillery	Customized baskets Links with merchants for pick-up points	Promotes community-supported agriculture
Bringing production and consumption closer together	Food produced for the local population distributed in baskets and at food stands	Horticultural and other products delivered to the local population through pick-up points or home delivery	Promotes food autonomy
Challenges as a non-conventional location	Adapts to the climate with northern techniques and varieties	Ensures year-round production despite winter temperatures	Technological adaptation to climatic conditions and promotion of food autonomy

Table 1. *Cont.*

Themes	Le Grenier Boréal	Lufa Farms	Summary
Dimensions of food security to which the initiative contributes	Supply of fresh, locally grown fruits and vegetables Physical access through mobile food stands (launched in 2020)	Economic access for people in need through the direct donation program (launched in 2020) and 50% discount on fruits and vegetables from Lufa Physical access through pick-up points in areas underserved by fresh fruits and vegetables	Physical access to quality food products Contribution to food self-sufficiency
Challenges	Economic access Seasonality Expands production and production period	Access mainly for the middle class, to be expanded to vulnerable populations	Increases access and disseminates the model
Effect on environmental and climate protection	Climate-adapted crops Decreases food miles Promotes the productive use of waste Raises awareness of the virtues of organic production	Use of solar energy and decrease in food miles	Reduction in GHGs caused by distribution Decreases pollution caused by insecticides, herbicides and chemical fertilizers Promotes a cultural change in the population

Lufa Farms facilitates access to healthy food for a portion of the population, mainly young people or members of the middle classes who wish to have access to food produced according to ethical guidelines for social justice (fair trade, among others) and standards that guarantee respect for the environment. Thus, in terms of urban agriculture, Lufa is a pioneering, innovative and ecological company. Lufa's experience opens up various perspectives in the fight against food insecurity. Reclaiming space for year-round food production, bringing food products closer to consumers, establishing partnerships with local producers and saving energy used in production, processing, preservation and distribution are some of the avenues to pursue. In addition, Lufa and its partners are implementing short food supply chains, which promote local food systems and reduce the food miles required to make food accessible to consumers. Finally, Lufa's customer relationship model has a significant educational dimension.

Due to its prototypical nature, Lufa is a true laboratory for reflection, self-training, experimentation with new knowledge and collective learning. The vision of urban agriculture that Lufa cultivates has implications for the entire community. The production of food through a system of short food supply chains that allow for a circular economy integrating Lufa Farms and local producers builds bridges between the urban and the rural. The promotion of local products is recognized as a means of supporting local agriculture and a contribution to environmental protection due to the reduction in energy used for transportation. Its distribution system brings consumers closer to the producers of the food they consume, which promotes consumer awareness of food production. In this way, it responds to the concerns expressed by various specialists regarding the effects of the industrial and globalized food system regarding the origin of food and the conditions in which it is produced. As practiced by Lufa Farms, urban agriculture contributes to sustainable cities and shows a way to increase food self-sufficiency in cities.

4.2. Production Using Ecological Techniques

Lufa Farms uses a number of technologies for the construction of its greenhouses, the management of the company and the production of food. Firstly, it opts for biological management techniques for its production, recreating a balanced ecosystem of harmful and beneficial insects. Secondly, some greenhouses recover rainwater, which is then used in a closed circuit. Further, it applies various energy optimization techniques, including the recovery of heat loss from buildings located under the greenhouses. Finally, Lufa makes sure to reduce its waste by limiting its losses and by composting its green waste. While

Le Grenier boréal, for its part, was pest-free for the first few years of operation, this is no longer the case. Additionally, the enterprise uses natural products, in keeping with the principles of organic agriculture.

4.3. Local Marketing

Both Le Grenier boréal and Lufa Farms demonstrate the social and ecological value of community-supported agriculture. Le Grenier boréal distributes a significant share of its production through baskets, the contents of which are determined based on the products in season. For Lufa Farms, the entire production is marketed through baskets that are distributed mainly through pick-up points located in local businesses throughout the city and delivered by electric vehicles. Thanks to the quantity and variety of food produced in its own greenhouses, or in other farms and processing companies in its network, Lufa Farms is able to offer "Lufavores" the option of customizing their baskets. Le Grenier boréal, for its part, does not limit its marketing to baskets, offering a self-service counter on the cooperative's site and selling in selected local businesses. It also supplies a local company to produce a regional flavored gin. Finally, in 2020, it also began distributing its food via a mobile food stand [45].

4.4. Limitations of the Studied Cases

At present, however, both initiatives present some limitations. For instance, the production of Le Grenier boréal is not sufficient to meet demand, is seasonal and does not improve economic access to food by offering fruits and vegetables at a lower cost, although the cooperative strives to compete with supermarket prices. Still, the cases of La Clé des champs and Cultur'Innov, in the municipality of Saint-Camille, in the Eastern Townships (Quebec) [61], show that establishing gardens on land that is less suitable for growing vegetables is not an insurmountable obstacle. On the one hand, the improvement of a site (soil, redevelopment), the adoption of certain production techniques (greenhouses, varieties) and continuous learning on the part of the staff all increase productivity, at least in the first few years, as it can be observed at Le Grenier boréal. On the other hand, a community's attachment and commitment to the enterprise can also help to compensate for certain weaknesses, for example, by facilitating the establishment of markets or food stands, as is the case for La Clé des champs in Saint-Camille [61,62], or by volunteering, in the case of Le Grenier boréal. Thus, if Le Grenier boréal maintains its present core strengths, it can be expected to overcome some of the challenges as it matures, at which point it may have more impact on other dimensions of food insecurity in Minganie.

Of course, because of the costs of production, but also because of the differences in food emergencies felt by different social strata, Lufa Farms does not contribute directly to the fight against food insecurity for the poorest. Indeed, its products are inaccessible to the poorest segments of society. That said, while food emergencies have not yet led the poorest to focus on a quest for organic food but rather to improving access to food at affordable costs, the fact remains that the needs of this part of the population in terms of health are not different from those of the better-off classes. A partnership with the public sector should be designed to make these types of products accessible to people with low incomes who are likewise seeking access to quality products.

5. Discussion

Even though Le Grenier boréal and Lufa Farms are very different experiences, particularly in terms of their scope, the quantity of food produced and, therefore, the number of consumers served, the two cases have certain points in common and show the importance of promoting agricultural production models that contribute to bringing the sites of production and consumption closer together. Proximity agriculture promotes an awareness of the issues involved in food production, issues that are invisible to the dominant model. These issues concern, in particular, the link between food production and the geographical characteristics of places. It implements a change of scale insofar as the large-scale industry that dominates the global food system, driven by big capital, large companies and major distributors, is replaced by models implemented by small collective or socially oriented enterprises that make it possible for consumers to participate in production choices. This also serves to strengthen links with local communities. Our study of the two cases reveals the main analysis criteria, allowing for a better understanding of this model.

The two experiments operate on the periphery of the conventional agricultural system. On the one hand, the environments in which they are located (urban and northern environments) are not considered a priori as agricultural environments. While cities do have gardens and small developments producing some fruits and vegetables, Lufa Farms has launched a model that has no equal in Canada. This model could help reduce the dependence of large cities on the currently dominant extractivist food system. Le Grenier boréal, for its part, is located in a region that has had very little market garden production to date due to climatic and soil limitations. Its objective is also to reduce the double dependence of a remote region on extractivist production and on distribution centers located in large cities.

On the other hand, the areas they occupy—roofs in the case of Lufa, and, in the case of Le Grenier boréal, a boreal forest wasteland with soil that is very sandy and acidic and contains very little organic matter—are not known for their agronomic quality. However, the implementation of adapted and even innovative means of production has made it possible to overcome, at least partially, these limitations. Lufa Farms has opted for hydroponic greenhouse farming, which means that the company does not need land, and for a partnership approach with local agricultural producers while adopting an environmentally friendly distribution system (electric vehicles, for example).

Le Grenier boréal is also taking a partnership approach to productive ecology insofar as it fertilizes its gardens with residues from the fishing processing plant located in the municipality, thereby improving soil fertility. However, poor soil has remained a limiting factor for the cooperative, and in 2020, a portion of the fields was amended with green manure. While agricultural enterprises in the southern part of the province generally opt for manure spreading, this is not really an option for Le Grenier boréal, as there is no livestock in Côte-Nord. The manure would have to be trucked in over several hundred kilometers, which would be very expensive and would contribute to GHG emissions. To avoid this, other local resources such as branch residues and algae are also used [25]. In order to extend the season, Le Grenier Boréal has greenhouses and uses plastic tunnels. In addition, the cooperative makes a point of choosing its plant varieties so that they are well adapted to the northern climate [16]. The cooperative has also expressed an interest in collaborating with other crab and shellfish producers in the region and reusing their waste products, which would allow reducing the dependence on fertilizers from outside the region from an industrial ecology perspective.

In a context where the global food system is increasingly under pressure due to sudden crises and gradual but major transformations, such as those induced by climate change, there is a need to build more resilient food systems to ensure food security [63]. To do this, it is important to recognize and promote the diversity of production models [64]. Diversification improves the robustness of the system, in that it increases the chances that some links in the chain can take over if another link is weakened [65]. This diversification concerns production practices, the varieties used or the actors involved; yet, it also concerns, in our opinion, the sites and environments of production. Le Grenier boréal and Lufa are investing in sites and environments where agricultural production is little, if not very little, present, thus providing new possibilities for agricultural production.

Efforts must also be made to improve the food autonomy of territories [65]. This must, on the one hand, enable consumers to obtain supplies locally, as is the case for a part of the population of both the Les Grenier boréal and Lufa communities. On the other hand, the development of greater territorial autonomy must allow businesses to obtain the inputs they need locally (which is what Lufa is able to do to a large extent, and what Le Grenier boréal seeks to achieve by valorizing local resources).

Thus, in recent years, the development of agricultural initiatives of various sizes, both individual and collective, has led to the removal of local regulatory barriers that prohibited food production in some areas (in some Quebec municipalities, front yard gardens have recently been permitted; in others, greenhouses are now permitted in industrial zones). In addition, in 2021, the Ministère de l'Agriculture, des Pêcheries et de l'Alimentation du Québec (MAPAQ, Quebec, QC, Canada) launched a program to support the implementation of local food community development plans. These plans seek to promote "greater food autonomy and better resilience in the face of numerous challenges, including climate change" [66]. All this shows a progressive change in the understanding of the relationship between citizens and food production and testifies to the transformation of certain public policies.

Thus, while the cases studied have interesting impacts, although limited for the moment, in terms of both the food diversification and food autonomy of territories, they contribute to improving the resilience of the system, and, above all, they show possibilities. In this sense, they are beacons in their respective contexts and inspire other initiatives and projects implemented by local communities in their wake.

6. Conclusions

Climate change will gradually reduce the production capacity of many spaces. In order to maintain or even improve production levels, diversification of production environments is necessary, including the cultivation of traditionally non-productive areas. It is therefore important to diversify production systems and, above all, to integrate them in local processes where producers and consumers interact—which is essentially what Le Grenier boréal and Lufa Farms have been doing.

Climate change is accelerating. Let us recall that at the end of spring 2021, the media reported that early heat followed by late frosts had heavily impacted French viticulture and arboriculture [67]; that Australia was experiencing an invasion of mice that devoured crops, due to the years of drought that allowed them to proliferate [68]; that North Korea was in a "tense food situation" following typhoons and floods, occurring in 2020, that reduced domestic grain production [69]; and that the Western United States, including California, North America's vegetable garden, had entered a vicious cycle of drought, aridity and record high temperatures [70]. In Canada, in late June and early July 2021, the western provinces experienced the highest temperatures ever recorded in the country's history, breaking several records, especially in the Lytton area, where the temperature exceeded 49.6 degrees Celsius [71]. While some people such as the president of Brazil consider these extreme events to be deviations from the norm, which brings them to endorse the extractivist model, other people affirm that these events are part of a long-term trend and that the worst is yet to come, as stated in the IPCC report of 2021 [72]. These events are part of a context of globalized capitalism in which the food industry seeks to increase its profitability, which causes deterioration in product quality, leads to environmental degradation, especially due to transportation and the use of heavy machinery, and contributes to food insecurity.

The analysis of local initiatives aimed at food security points to ways of rethinking the relationship between food production, food consumption and a societal and ecological transition. Alternative models of action oriented toward social innovation are being implemented [3] and contribute to adaptation to climate change. We would do well to learn more about these models, especially in the context of a post-pandemic economic recovery, which many actors believe must be greener. For many, food is an area that calls on us to innovate in order to build a post-pandemic world that is more just and equitable and more respectful of nature [73,74]. In this perspective, the transformations to be made to food systems must have a strong territorial basis in relation to living environments [75] and must be conceived within a broad framework that implies a paradigm shift [76] in order to reduce dependence on inter-territorial imports. The cases of Le Grenier boréal and Lufa Farms point in this direction.

Funding: This research was funded by Fonds de recherche du Québec-Société et culture, grant number 253795.

Institutional Review Board Statement: The study was conducted according to the guidelines of the Declaration of Helsinki, and approved by the Ethics Committee of Université du Québec à Montréal (protocol code 1949, 14 June 2017 and 2229, 7 February 2018).

Informed Consent Statement: Informed consent was obtained from all subjects involved in the study.

Conflicts of Interest: The authors declare no conflict of interest.

References

1. Swyngedouw, E. The non-political politics of climate change. *ACME Int. J. Crit. Geogr.* **2013**, *12*, 1–8.
2. Fischer-Kowalski, M.; Haas, W.; Wiedenhofer, D.; Weisz, U.; Pallua, I.; Possanner, N.; Behrens, A.; Serio, G.; Alessi, M.; Weis, E. *Socio-Ecological Transitions: Definition, Dynamics and Related Global Scenarios*; Institute for Social Ecology: Brussels, Belgium; Centre for European Policy: Vienna, Austria, 2012.
3. Callorda, E.; Degrave, F.; Lévesque, B. Innovations sociales transformatrices et perspective multi-niveaux sur les transitions soutenables. Dialogue entre deux approches complémentaires du changement social. In Proceedings of the 7th EMES International Research Conference on Social Enterprise, Sustainable Development through Social Enterprise, Cooperative and Voluntary Action, Sheffield, UK, 24–27 June 2019.
4. Agence France Presse (AFP). *Suspension des Exportations de Céréales en Russie—Le FAO Fait une Mise en Garde Contre la Hausse du Prix du Blé;* Le Devoir 7 August 2010; Agence France Presse (AFP): Paris, France, 2010.
5. Gibson-Graham, J.K.; Cameron, J.; Healy, S.; McNeill, J. Roepke lecture in economic geography—Economic geography, manufacturing, and ethical action in the anthropocene. *Econ. Geogr.* **2019**, *95*, 1–21. [CrossRef]
6. Klein, J.-L.; Pecqueur, B. *Les Living Labs. Une Perspective Territoriale*; L'Harmattan: Paris, France, 2020.
7. Intergovernmental Panel on Climate Change (IPCC). *Climate Change and Land: An IPCC Special Report on Climate Change, Desertification, Land Degradation, Sustainable Land Management, Food Security, and Greenhouse Gas Fluxes in Terrestrial Ecosystems*; IPCC: Geneva, Switzerland, 2019.
8. Thornton, P.K. Livestock production: Recent trends, future prospects. *Philos. Trans. R. Soc. B Biol. Sci.* **2010**, *365*, 2853–2867. [CrossRef]
9. Food and Agricultural Organisation (FAO). *The Future of Food and Agriculture—Alternative Pathways to 2050;* FAO: Rome, Italy, 2018.
10. Kummu, M.; Heino, M.; Taka, M.; Varis, O.; Viviroli, D. Climate change risks pushing one-third of global food production outside the safe climatic space. *One Earth* **2021**, *4*, 720–729. [CrossRef] [PubMed]
11. Food and Agricultural Organisation (FAO). *Introduction Aux Concepts de la Sécurité Alimentaire*; FAO: Rome, Italy, 2008.
12. Touzard, J.-M.; Temple, L. Sécurisation alimentaire et innovations dans l'agriculture et l'agroalimentaire: Vers un nouvel agenda de recherche ? Une revue de la littérature. *Cah. Agric.* **2012**, *21*, 293–301. [CrossRef]
13. Herrero, M.; Henderson, B.; Havlík, P.; Thornton, P.K.; Conant, R.T.; Smith, P.; Wirsenius, S.; Hristov, A.N.; Gerber, P.; Gill, M.; et al. Greenhouse gas mitigation potentials in the livestock sector. *Nat. Clim. Chang.* **2016**, *6*, 452–461. [CrossRef]
14. Lipper, L.; Thornton, P.; Campbell, B.M.; Baedeker, T.; Braimoh, A.; Bwalya, M.; Caron, P.; Cattaneo, A.; Garrity, D.; Henry, K.; et al. Climate-smart agriculture for food security. *Nat. Clim. Chang.* **2014**, *4*, 1068–1072. [CrossRef]
15. Altieri, M.A.; Nicholls, C.I. The adaptation and mitigation potential of traditional agriculture in a changing climate. *Clim. Chang.* **2017**, *140*, 33–45. [CrossRef]
16. Élie-Léonard, J. La Sécurisation Alimentaire Dans Deux Milieux Ruraux Périphériques Québécois. Master's Thesis, Université du Québec à Montréal (UQAM), Montreal, QC, Canada, 2021.
17. Ouro-Koura, R. L'agriculture Urbaine Face à L'insécurité Alimentaire à Montréal: L'expérience des Fermes Lufa. Master's Thesis, Université du Québec à Montréal (UQAM), Montreal, QC, Canada, 2019.
18. Yin, R. *Case Study Research: Design and Methods;* Sage Publications: Los Angeles, CA, USA, 2014.
19. Crowe, S.; Cresswell, K.; Robertson, A.; Huby, G.; Avery, A.; Sheikh, A. The case study approach. *BMC Med. Res. Methodol.* **2011**, *11*, 100. [CrossRef] [PubMed]
20. Glaser, B. Conceptualization: On theory and theorizing using grounded theory. *Int. J. Qual. Methods* **2002**, *1*, 23–38. [CrossRef]
21. Bureau-Blouin, L. La Coopérative Le Grenier Boréal (Québec). Available online: https://www.resolis.org/initiatives/la-cooperative-le-grenier-boreal-quebec/438aa26d-6187-4fcb-bfc4-005add2e50de (accessed on 2 July 2021).
22. Lussier, C. Coop de solidarité agroforestière de Minganie: Le Grenier boréal. In Proceedings of the World Social Forum 2016. L'innovation Sociale: Le Défi de la Transformation Sociale. Atelier 3C: Sécurité et Souveraineté Alimentaires: Entre Besoins et Valeurs, Montreal, QC, Canada, 9–14 August 2016.
23. Entreprise Québec. La Coopérative: Un Modèle D'affaires à Découvrir. Available online: https://www2.gouv.qc.ca/portail/quebec/infosite?lang=fr&x=2371115552 (accessed on 1 April 2021).
24. Grenier Boréal. Produits et Services. Available online: https://grenierboreal.coop/produits-et-services/ (accessed on 22 June 2021).
25. Carmel, M. Sur la Côte-Nord, où le Sol Est Sableux et les Vents Sans Pitié, la Coop le Grenier Boréal Cultive DES Légumes Frais et Renforce la Sécurité Alimentaire de la Région, GES en Moins! Available online: https://unpointcinq.ca/ (accessed on 15 May 2020).

26. Ministère de l'Économie et de L'Innovation (MEI). Côte-Nord—Portrait Régional. Available online: https://www.economie.gouv.qc.ca/pages-regionales/cote-nord/portrait-regional/ (accessed on 30 March 2021).
27. Institut de la Statistique du Québec (ISQ). Principaux Indicateurs Sur le Québec et Ses Régions, Côte-Nord Available online: https://statistique.quebec.ca/fr/vitrine/region/09 (accessed on 1 March 2021).
28. Statistics Canada. Côte-Nord [Economic Region], Quebec and Quebec [Province] (Table). *Census Profile, 2016 Census, Statistics Canada Catalogue No. 98-316-X2016001.* Available online: https://www12.statcan.gc.ca/census-recensement/2016/dp-pd/prof/details/page.cfm?Lang=E&Geo1=ER&Code1=2480&Geo2=PR&Code2=24&SearchText=cote-nord&SearchType=Begins&SearchPR=01&B1=All&TABID=1&type=0 (accessed on 1 July 2021).
29. Ministère des Affaires Municipales et de L'habitation (MAMH). Répertoire des Municipalités. *Minganie.* Available online: https://www.mamh.gouv.qc.ca/fileadmin/cartes/mrc/981.pdf (accessed on 2 July 2021).
30. Institut de la Statistique du Québec (ISQ). Principaux Indicateurs Sur le Québec et Ses Régions, Minganie. Available online: https://statistique.quebec.ca/fr/vitrine/region/09/mrc/981 (accessed on 30 March 2021).
31. Statistics Canada. Longue-Pointe-de-Mingan, MÉ [Census Subdivision], Quebec and Quebec [Province] (Table). *Census Profile, 2016 Census, Statistics Canada Catalogue No. 98-316-X2016001.* Available online: https://www12.statcan.gc.ca/census-recensement/2016/dp-pd/prof/details/page.cfm?B1=All&Code1=2498045&Code2=24&Data=Count&Geo1=CSD&Geo2=PR&Lang=E&SearchPR=01&SearchText=Longue-Pointe-de-Mingan&SearchType=Begins&TABID=1 (accessed on 8 July 2021).
32. Institut de la Statistique du Québec (ISQ). Taux de Faible Revenu, Selon le Type de Famille, MRC et Ensemble du Québec. Available online: https://statistique.quebec.ca/fr/produit/tableau/taux-de-faible-revenu-selon-le-type-de-famille-mrc-et-ensemble-du-quebec#tri_coln1=5&tri_coln2=5 (accessed on 1 April 2021).
33. Statistics Canada. Minganie–Le Golfe-du-Saint-Laurent, CDR [Census Division], Quebec and Quebec [Province] (Table). *Census Profile, 2016 Census, Statistics Canada Catalogue Product No. 98-316-X2016001.* Available online: https://www12.statcan.gc.ca/census-recensement/2016/dp-pd/prof/details/page.cfm?B1=All&Code1=2498&Code2=24&Data=Count&Geo1=CD&Geo2=PR&Lang=E&SearchPR=01&SearchText=Minganie--Le+Golfe-du-Saint-Laurent&SearchType=Begins&TABID=1 (accessed on 8 July 2021).
34. Ministère des Affaires Municipales et de L'Habitation (MAMH). Organisation Territoriale. Côte-Nord (Région 09). Available online: https://www.mamh.gouv.qc.ca/organisation-municipale/organisation-territoriale/regions-administratives/cote-nord/ (accessed on 8 July 2021).
35. Municipalité Régionale de Comté (MRC) de Minganie. La Pauvreté et L'exclusion Sociale en Minganie—Portrait et Plan D'action, Havre-Saint-Pierre, MRC de Minganie. Available online: https://docplayer.fr/55809490-Mrc-deminganie-la-pauvrete-et-l-exclusion-sociale-en-minganie-portrait-et-plan-daction.html (accessed on 8 July 2021).
36. Ministère de L'environnement et de la Lutte Contre Les Changements Climatiques (MELCC). Normales Climatiques du Québec 1981–2010. Available online: https://www.environnement.gouv.qc.ca/climat/normales/index.asp (accessed on 20 March 2021).
37. Centre de Référence en Agriculture et en Agroalimentaire du Québec (CRAAQ). Saison de Croissance (Base 5.5 °C). Available online: https://www.agrometeo.org/atlas/category/saiscrois/therm (accessed on 1 April 2021).
38. Ministère de L'Agriculture Pêcherie et Alimentation du Québec (MAPAQ). Région 09 Côte-Nord: Secteur Bioalimentaire: Une Démarche vers une Vision Innovante. Available online: http://www.tablebioalimentairecotenord.ca/wp-content/uploads/2014/11/16-MAPAQ_Cote-Nord_Amenagement_des_secteurs_de_peche_et_agricole.pdf (accessed on 30 March 2021).
39. Doyon, M.; Klein, J.-L. *Pour la Sécurisation Alimentaire au Québec: Perspective Territoriale*; Presses de l'Université du Québec: Quebec, QC, Canada, 2019.
40. Duquette, M.-P.; Demmers, T.; Lacroix, A.; Scatliff, C.; Desrosiers-Choquette, J. *Étude sur la Détermination du Coût du Panier à Provisions Nutritif Dans Trois Régions du Québec*; Rapport du Projet—Région de la Côte-Nord; Dispensaire diététique de Montréal: Montreal, QC, Canada, 2011.
41. Québec. Règlement Sur les Prix du Lait de Consommation, 1 February 2017. Chapitre M-35.1, R. 206; Québec, QC, Canada, 1 February 2018; Available online: https://www.canlii.org/fr/qc/legis/regl/rlrq-c-m-35.1-r-206/derniere/rlrq-c-m-35.1-r-206.html (accessed on 2 July 2021).
42. Élie-Léonard, J.; Doyon, M. L'autonomisation alimentaire en milieu géographique adverse: Le cas du Grenier boréal en Minganie. In *Pour la Sécurisation Alimentaire au Québec; Perspective Territoriale*; Doyon, M., Klein, J.-L., Eds.; Presses de l'Université du Québec: Québec, QC, Canada, 2019; pp. 117–131.
43. Bélanger, J. Saison Estivale «Essouflante» Pour le GRENIER Boréal. Available online: https://ici.radio-canada.ca/ohdio/premiere/emissions/bonjour-la-cote/segments/entrevue/92655/bilan-saison-cooperative-grenier-boreal-longue-pointe-de-mingan-cote-nord (accessed on 2 July 2021).
44. Radio-Canada. *Le Grenier Boréal Cultive Son Printemps*; Radio-Canada: Montréal, QC, Canada, 2019.
45. Rioux-Berrouard, V. *De Bonnes Récoltes Pour le Grenier Boréal; Le Nord-Côtier*; Sept-Îles, QC, Canada, 9 September 2020. Available online: https://lenord-cotier.com/2020/09/09/de-bonnes-recoltes-pour-le-grenier-boreal/ (accessed on 2 July 2021).
46. Veltz, P. *La Société Hyper-Industrielle: Le Nouveau Capitalisme Productif*; Seuil: Paris, France, 2017.
47. Cary, P.; Fol, S. Les métropoles soumises à des transformations socio-spatiales. In *Ségrégation et Fragmentation dans les Métropoles: Perspectives Internationales*; Carrel, M., Cary, P., Washsberger, J.-M., Eds.; Presses universitaires du Septentrion: Villeneuve d'Ascq, France, 2013; pp. 11–22.
48. Klein, J.-L.; Shearmur, R. *Montréal: La Cité des Cités*; Presses de l'Université du Québec: Quebec, QC, Canada, 2017.

49. Enriquez, D. Les Initiatives Communautaires et de L'économie Sociale de Lutte Contre L'insécurité Alimentaire à Montréal: Le Cas Des Expérimentations Favorisées Par Parole D'exclues Dans Deux Quartiers de Montréal-Nord. Ph.D. Thesis, Université du Québec à Montréal, Montreal, QC, Canada, 2018.
50. Enriquez, D.; Klein, J.-L. La désertification alimentaire à Montréal: Logique économique et réponse sociale. In *Pour la Sécurisation Alimentaire au Québec: Perspective Territoriale*; Doyon, M., Klein, J.-L., Eds.; Presses de L'Université du Québec: Quebec, QC, Canada, 2019; pp. 21–41.
51. Audet, R.; Lefèvre, S.; El-Jed, M. *La Démarche D'innovation Des Marchés de Quartier de Montréal: Vers une Transition Socio-Écologique du Système Agroalimentaire*; Chaire de Responsabilité sociale et de Développement Durable (CRSDD): Montreal, QC, Canada, 2014; p. 44.
52. Moubarac, J.-C.; Batal, M. *La Consommation D'aliments Transformés et la Qualité de L'alimentation Chez les Québécois*; Transnut, Université de Montréal: Montreal, QC, Canada, 2016.
53. Duchemin, E.; Wegmuller, F.; Legault, A.M. Urban agriculture: Multi-dimensional tools for social development in poor neighbourhoods. *Field Actions Sci. Rep.* **2008**, *1*, 43–52. [CrossRef]
54. Wees, D. *L'agriculture Urbaine à Montréal: Un Projet Éducatif Multidisciplinaire*; Faculté des Sciences de L'agriculture et de L'environnement, McGill University: Montreal, QC, Canada, 2012.
55. Wegmuller, F.; Duchemin, É. Multifonctionnalité de l'agriculture urbaine à Montréal: Étude des discours au sein du programme des jardins communautaires. *Vertigo* **2010**, *10*. [CrossRef]
56. Renaud, G.; Royer, D. *Recension des Initiatives en Autonomie Alimentaire*; Cahier du Crises: Montréal, QC, Canada, 2012; Volume ES1204.
57. Granchamp-Florentino, L. L'agriculture urbaine; Un enjeu de la ville durable. *Rev. Des Sci. Soc.* **2012**, *47*, 142–152.
58. Forest, J.; Hamdouch, A. *Quand L'innovation Fait la Ville Durable*; Presses Polytechniques et Universitaires romandes: Lausanne, France, 2015.
59. Cockrall-King, J.; Eisenmann, M.; Galarneau, V. *La Révolution de L'agriculture Urbaine*; Écosociété: Montreal, QC, Canada, 2016.
60. Meyer, A.; Legaré, G. *Les Fermes Lufa: Une Nouvelle Façon D'utiliser L'espace Urbain et de Consommer Local*; Bulletin Oeconomia Humana: Montréal, QC, Canda, 2011; Volume 9.
61. Doyon, M.; Klein, J.-L.; Tremblay, P.-A. Community action against marginalization: The case of a rural social enterprise in the village of Saint-Camille, Quebec. In *Responses to Geographical Marginality and Marginalization*; Nel, E., Pelc, S., Eds.; Springer: New York, NY, USA, 2020.
62. Tremblay, P.-A.; Doyon, M.; Klein, J.-L.; Arsenault-Hétu, C. Le volet agricole d'une expérience d'innovation socioterritoriale: La Clé des champs de Saint-Camille. In *Pour la Sécurisation Alimentaire au Québec: Perspective territoriale*; Doyon, M., Klein, J.-L., Eds.; Presses de l'Université du Québec: Quebec, QC, Canada, 2019; pp. 151–168.
63. Tendall, D.M.; Joerin, J.; Kopanisky, B.; Edwards, P.; Shreck, A.; Le, Q.B.; Druetli, P.; Grant, M.; Six, J. Food system resilience; defining the concept. *Glob. Food Secur.* **2015**, *6*, 17–23. [CrossRef]
64. Touzard, J.-M.; Fournier, S. La complexité des systèmes alimentaires: Un atout pour la sécurité alimentaire? *VertigO* **2014**, *14*, 14840. [CrossRef]
65. Les Greniers de l'Abondance. *Vers la Résilience Alimentaire: Reair Face Aux Mensaces Globales à L'échelle Des Territoires*; Éditions Yves Michel: Gap, France, 2020.
66. MAPAQ. Plan de Développement D'une Communauté Nourricière. Available online: https://www.mapaq.gouv.qc.ca/SiteCollectionDocuments/Formulaires/Appui_elaboration_plan_developpement_communaute_nourriciere.pdf (accessed on 2 July 2021).
67. Orjollet, S. Le Réchauffement Climatique, Suspect Numéro 1. *La Presse*, 14 June 2021.
68. Leeson, A. Après les Catastrophes Climatiques, les Souris Dévastatrices. In *La Presse*; 14 June 2021.
69. Agence France Presse (AFP). *Kim Jong-Un Reconnaît Une «Situation Alimentaire Tendue»*; La Presse 16 June 2021; Agence France Presse (AFP): Paris, France, 2021.
70. Agence France Presse (AFP). *Une Vague de Chaleur Précoce Pourrait Faire Tomber des Records*; La Presse 15 June 2021; Agence France Presse (AFP): Paris, France, 2021.
71. La Presse Canadienne, C.-B. *Lytton bat un nouveau record de chaleur pour le Canada*; Radio-Canada: 28 June 2021; La Presse Canadienne: Toronto, ON, Canada, 2021.
72. Agence France Presse (AFP). *L'humanité à L'aube de Retombées Climatiques Cataclysmiques*; Radio-Canada 23 June 2021; Agence France Presse (AFP): Paris, France, 2021.
73. Benach, J. We must take advantage of this pandemic to make a radical social change: The Coronavirus as a global health, inequality, and eco-social problem. *Int. J. Health Serv.* **2021**, *51*, 50–54. [CrossRef] [PubMed]
74. Desa, G.; Jia, X. Sustainability transitions in the context of pandemic: An introduction to the focused issue on social innovation and systemic impact. *Agric. Hum. Values* **2020**, *37*, 1207–1215. [CrossRef] [PubMed]
75. Nurhasanah, I.S.; Medina Garcia, C.; Nyagwalla Otieno, J.; Balcha, W.G.; Paidakaki, A.; Van den Broeck, P. Social innovation in the face of the COVID-19 pandemic. *INSIST Cahier 4*, Leuven, Belgium, 2020.
76. Svampa, M. Reflexiones para un mundo post-coronavirus. *Nueva Sociedad*. Available online: https://nuso.org/articulo/reflexiones-para-un-mundo-post-coronavirus/ (accessed on 19 July 2021).

climate

Article

Determinants of Smallholder Livestock Farmers' Household Resilience to Food Insecurity in South Africa

Vuyiseka A. Myeki and Yonas T. Bahta *

Department of Agricultural Economics, University of the Free State, P.O. Box 339, Bloemfontein 9300, South Africa; 2014110122@ufs4life.ac.za
* Correspondence: Bahtay@ufs.ac.za; Tel.: +27-514-019-050

Abstract: This study identified factors affecting livestock farmers' agricultural drought resilience to food insecurity in Northern Cape Province, South Africa. Data of 217 smallholder livestock farmers were used in a principal component analysis to estimate the agricultural drought resilience index. The structural equation approach was then applied to assess smallholder livestock farmers' resilience to food insecurity. The study found that most smallholder livestock farmers (81%) were not resilient to agricultural drought. Assets ($\beta = 0.150$), social safety nets ($\beta = 0.001$), and adaptive capacity ($\beta = 0.171$) indicators positively impacted households' resilience to food insecurity with 5% significance. Climate change indicators negatively impacted households' resilience to food insecurity. Two variables were included under climate change, focusing on drought, namely drought occurrence ($\beta = -0.118$) and drought intensity ($\beta = -0.021$), which had a negative impact on household resilience to food insecurity with 10% significance. The study suggests that smallholder livestock farmers need assistance from the government and various stakeholders to minimize vulnerability and boost their resilience to food insecurity.

Keywords: agricultural drought; resilience; food insecurity; assets; adaptive capacity; social safety net

Citation: Myeki, V.A.; Bahta, Y.T. Determinants of Smallholder Livestock Farmers' Household Resilience to Food Insecurity in South Africa. *Climate* **2021**, *9*, 117. https://doi.org/10.3390/cli9070117

Academic Editors: Christopher Robin Bryant, Andrea Vitali and Azzeddine Madani

Received: 12 April 2021
Accepted: 2 July 2021
Published: 13 July 2021

1. Introduction

Agriculture, including the livestock sector, is one of the industries on which drought has a major influence, causing loss of agricultural production. The impact of agricultural drought on livestock production is becoming a significant physical stressor in temperate and humid regions, including South Africa [1]. Agricultural drought impacts livestock production and quality, which is dependent on several factors, such as intensity, recurrent agricultural drought, vulnerability, water stress, and socio-economic factors [2]. Globally, agricultural drought is the costliest natural disaster compared to other natural disasters such as floods, hurricanes, tornadoes, and earthquakes. Loss caused by agricultural drought is estimated to cost from USD 6 to 8 billion annually [3]. An estimated 40 million people have been affected by drought in southern Africa, with a cereal deficit of 9.3 million tonnes recorded at the end of the 2015/16 cropping season [4,5]. The high regional deficit raised staple food prices and constrained the already limited purchasing power of vulnerable families. More than 643,000 livestock deaths were reported in five countries alone due to lack of feed and water and disease outbreaks in southern Africa. In addition, the income sources of many households were diminished due to the loss of crops, livestock, labor, trading, and self-employment activities [4,5]. During 2015, agricultural production in South Africa declined by 8.4% due to drought. The livestock industry, for example, had a 15% reduction in the national herd stock due to the drought [6].

Smallholders are characterized by labor-intensive farming, adoption of traditional production techniques, and inadequate institutional capacity and support [7]. Smallholder agriculture in general and the smallholder livestock sector in particular remain at the center of rural development policy discussions in Africa [8]. Smallholder agriculture plays a significant role and will contribute to feeding approximately 9 billion people worldwide by 2050, although there are still debates about the role of smallholder agriculture. The contribution of agriculture to poverty reduction depends on its own growth performance, its indirect impact on growth in other sectors, the extent to which poor people participate in the sector, and the sector's size in the overall economy. Agriculture is significantly more effective than non-agriculture in reducing poverty among the poorest of the poor (as reflected by the USD 1/day squared poverty gap). It is also up to 3.2 times better at reducing USD 1/day headcount poverty in low-income and resource-rich countries (including those in Sub-Saharan Africa), at least when societies are fundamentally equal. However, when it comes to the better-off poor (reflected in the USD 2/day measure), non-agriculture has the edge [9].

Smallholder livestock farming contributes to improving the livelihoods of the rural poor in South Africa and plays a vital role by providing food and has the potential to strengthen households' economy. Livestock production plays multiple roles in the lives of the poor and meets the various objectives desired by resource-poor farmers [10]. Smallholder agriculture, including the livestock sector in South Africa, has been identified as a notable vehicle to foster poverty reduction, solve household food insecurity, and enhance resilient livelihoods.

Even though smallholder agriculture has the potential to enhance resilience, the decline in average rainfall and rapid population growth have resulted in food insecurity [11]. In Sub-Saharan Africa, smallholder livestock farmers do not produce output beyond household consumption. Their output does not generate enough income nor do they engage in off-farm or non-farm income-generating activities, even in export. The insufficient production is further undermined by factors such as a lack of assets (resources), a lack of adaptive capacity, climate change (agricultural drought), a lack of social safety nets, increasing farm input prices, a lack of information, and inadequate institutional infrastructure [12,13].

International and national studies, such as those of Boukary et al. [11], Melketo et al. [14], Ogunniyi et al. [15], Chamdimba et al. [16], and Galarza [17], focus on the impact of Jatropha cultivation for resilience in food insecurity, pastoral households' resilience, rural households' resilience to food insecurity, drought impact, coping and adaptation, and socio-economic drivers of food security. However, none of the studies empirically assess smallholder livestock farmers' resilience to food insecurity in the livestock sector.

To our knowledge, no studies have specifically focused on smallholder livestock farming households' resilience to food insecurity. Therefore, this study identified factors affecting livestock farmers' agricultural drought resilience to food insecurity in Northern Cape Province, South Africa, using a survey, principal component analysis, and structural equation approach. The findings of this study could help government and policymakers to develop suitable policies and mitigation strategies to build and boost smallholder livestock farmers' resilience to agricultural drought with the alignment of the National Development Plan (NDP) of South Africa and the Sustainable Development Goal of ending hunger and poverty. The NDP considers small livestock producers as a strategy given the role of the livestock sector in food security. This work is original academic research carried out by the authors and part of an MSc dissertation by Vuyiseka A. Myeki [18] entitled "Factors affecting smallholder livestock farmers' agricultural drought resilience to food insecurity in the Northern Cape, South Africa". The University of the Free State, Bloemfontein, South Africa.

2. Literature Review and Conceptual Framework

The definitions and conceptual framework used to identify factors affecting livestock farmers' agricultural drought resilience to food insecurity in Northern Cape Province, South Africa, were adopted from international and national studies/literature.

There are different definitions for resilience with shared characteristics [19–22]. However, nearly all definitions stress the common elements of resilience: ability, mitigation, adaptation, coping, recovery, withstanding shocks, resistance, and bouncing back against shocks. Resilience in this study is considered to be the ability of a household to "bounce back" after exposure to livelihood threats, shocks, or stressors (such as agricultural drought and vulnerability to food insecurity).

Household resilience to food insecurity is defined as the ability of a household to maintain a certain level of well-being (food security) when faced with agricultural drought, and depends on the options available to make a living, and on the ability to handle agricultural drought. Therefore, it refers to ex ante actions aimed at reducing or mitigating agricultural drought and ex post actions to cope with agricultural drought. Thus, the options available for a household to make a living and cope with agricultural drought will determine the resilience of the household [23]. In scenarios where the ecosystems that communities depend on during shocks are vulnerable and exhibit eroding resilience, it is evident that the coping and adaptive strategies tend to overlap. Therefore, the concept of resilience stresses the dynamic nature of agricultural drought and usefully categorizes resilience into absorptive, adaptive, and transformative capacities. Absorptive capacity highlights the ability to show an initial "persisting" response to cope with agricultural drought. Adaptive capacity reflects the ability to function consistently as before in the face of incremental changes in climate change shocks, including agricultural drought. Transformative capacity reflects the ability to show a substantial changing response to agricultural drought or prolonged disturbance, including value systems, regimes, financial systems, technological systems, and biological systems [24,25].

Further, it might involve improving infrastructure, supporting social protection mechanisms, providing basic social services, or developing institutional capacity. These changes might be voluntarily chosen or forced (such as conflict forcing people to flee their country). To be successful, these transformational changes typically require shifts in economic and social policies, land use legislation, and resource management practices, as well as inclusion of various institutions and social practices [24,25].

Food insecurity is defined as a household's inability to meet target consumption levels in the face of shocks, such as agricultural drought [14]. In this paper, the concept of resilience to food insecurity refers to the adaptive capacity of smallholder livestock farmers in Northern Cape Province of South Africa.

Rockstrom [26] highlighted that social, economic, situational, and institutional preparedness to cope with stresses and shocks as well as their effects are the core mechanisms of household resilience to food insecurity. In addition, numerous studies have documented several factors determining the means and processes of achieving household resilience [27–31].

Various resilience analysis frameworks have been suggested [32]. However, Hoddinott [33] argues that the plethora of frameworks for resilience analysis have similar components. These include highlighting the broader environment in which a household (or individual or some other unit of observation) resides; the resources available to that household; how that household uses those resources; how the economic returns on those uses are affected by shocks that the household experiences; and how the outcomes of those uses lead to consumption of food and other goods and services, savings, health, nutritional status, and other such outcomes.

Therefore, resilience frameworks commonly guide studies on household resilience to food insecurity [14,34,35]. This study adopted an updated framework developed by Alinovi et al. [22,36]. Figure 1 presents the conceptual framework applied in this study. The selection of the framework is justified, because it is mainly proposed for analysis (Equations (5)–(7)) of households' resilience to food security shocks such as agricultural drought. This framework elicits the extent of resilience-building variation from one household to another and that the variation is determined by diverse factors. Factors include assets (herd/flock size (HFS); agricultural assets (AA), non-agricultural assets (NAA)), adaptive capacity (perception; source of income (Incsource); migration; credit), social safety nets (cash; training; food support; water rights; garden equipment; sanitary latrines, farm input), climate change (occurrence and intensity of drought). The factors are associated with the outcome variable of the agricultural drought resilience index (ADRI) as illustrated in Figure 1 and Equations (5)–(7).

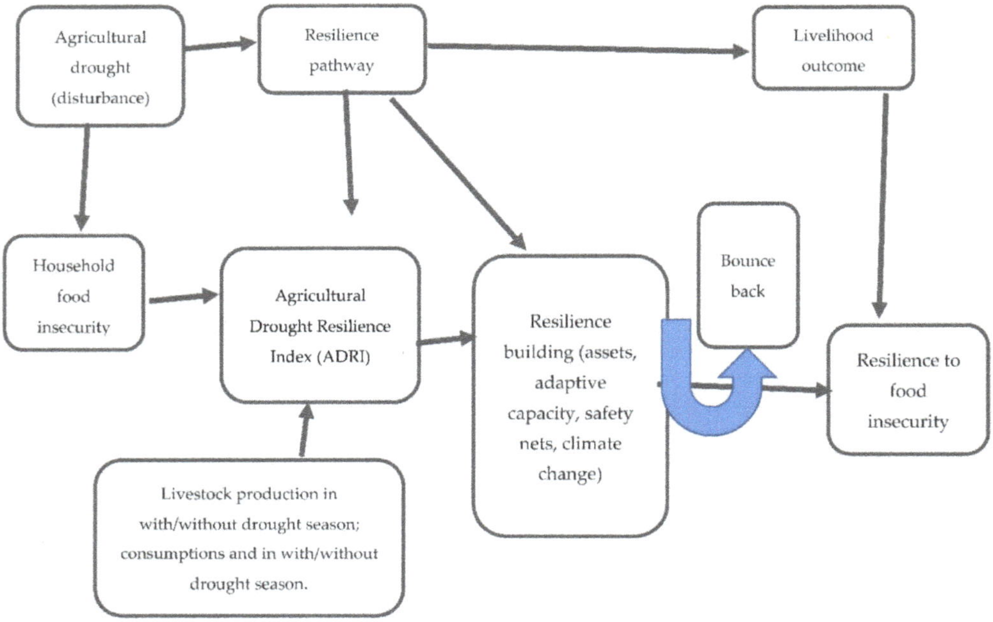

Figure 1. Conceptual framework of the study. Source: Authors' work from observations of various studies.

As shown in Figure 1 and Equation (5), the ADRI is calculated using principal component analysis (PCA) and variables related to livestock production and consumption with or without a drought season. Furthermore, as demonstrated in Figure 1 and Equations (6) and (7), the ADRI is determined using a structural equation model against independent variables as aggregate and disaggregate variables of assets, adaptive capacity, social safety nets, and climate change.

3. Materials and Methods

3.1. Description of the Study Area

This study was conducted in Northern Cape Province of South Africa, in the Frances Baard District Municipality. The municipality's total geographical area is 12,384 km^2 and accounts for 3.4% of the total area of Northern Cape Province [37]. The study was conducted in the following four local municipalities: Sol Plaatje, Dikgatlong, Magareng, and Phokwane (Figure 2).

3.2. Sampling Procedure and Sample Size Determination

A multiple-stage sampling procedure was employed. Firstly, Northern Cape Province was chosen from the nine provinces of South Africa, because most households were involved in livestock production, and the province was declared a disaster zone by the South African Government due to agricultural drought. Secondly, four district municipalities in the province were randomly selected using balloting and included Dikgatlong, Magareng, Sol Plaatje, and Phokwane.

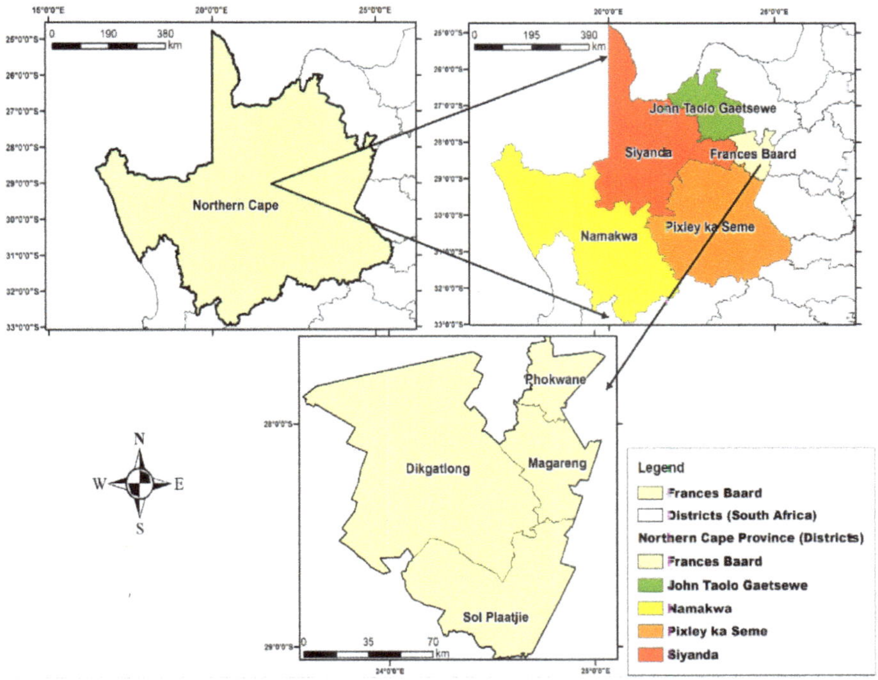

Figure 2. Maps of South Africa highlighting Northern Cape Province, district municipalities of the Northern Cape, and the four local municipalities of Frances Baard District Municipality. Source: FBDM [38].

Appropriate sample sizes were calculated using the simple random sampling formulae of Cochran [39] and Bartlett et al. [40]. Subsequently, 217 smallholder livestock farmers were selected from 878 farmers registered for government and local government assistance during the 2015/2016 production season (Table 1); this season was the worst drought season in South African history [41]. The assistance from the government included feed and medication for livestock, strengthening access to agricultural credit and farm input, and enhancing smallholder farmers' involvement in agricultural drought resilience activities by giving training and disseminating information.

Table 1. Number of farmers who received assistance from the government and sampling procedure.

Local Municipality	Number of Smallholder Farmers	Share of Farmers (Number of Farmers/Total) %	Number of Samples (Percentage × Total Sample Size [217])
Dikgatlong	351	40	87
Magareng	120	14	30
Sol Plaatje	141	16	35
Phokwane	266	30	65
Total	876	100	217

Note: The "×" represents multiplication. Sources: Northern Cape Department of Agriculture, Forestry and Fisheries (NDAFF) [41], beneficiaries of drought relief program.

A sample of 217 smallholder livestock farmers were interviewed. Cochran's [40] sample size formula was applied to determine the correct sample size (Equation (1)):

$$\text{Sample size} = \frac{(y)^2 * (f)(g)}{(z)^2} \tag{1}$$

where y is the level of confidence/alpha level, f and g are the estimates of the variance of the population, and z is the margin of error (5% (0.05)). Therefore (Equation (2)):

$$\text{Sample size} = \frac{(1.65)^2 * (0.515)(0.515)}{(0.05)^2} \text{Sample size} = 288.3 \quad (2)$$

Resulting in a sample size of 288.3. Note that, if the sample size exceeds 5% of the population, the Cochran's correctional formula should be applied (Equations (3) and (4)):

$$\text{N1} = \frac{\text{Sample size}}{1 + (\text{N0}/\text{population})} \quad (3)$$

$$\text{N1} = \frac{288.83}{1 + (288.83/878)} \text{N1} = 217 \quad (4)$$

3.3. Data Collection

This research was qualitative and quantitative. Data were collected using a structured questionnaire and face-to-face interviews. The questionnaire included continuous and categorical data, which comprised socio-economic characteristics, livestock production, assets, adaptive capacity, climate change, social safety net, and other variables. Face-to-face interviews were conducted from October to December 2020 using a structured questionnaire (part of the questionnaire is available in Appendix A). Ethical clearance was obtained from the University of the Free State.

3.4. Analytical Procedures

3.4.1. Agricultural Drought Resilience Index (ADRI)

Principal component analysis (PCA) was used to construct the agricultural drought resilience index (ADRI). Production of livestock in a normal year (W_nP_n), production of livestock with agricultural drought (W_dP_d), the number of months a household consumes food produced by the household in a normal year ($W_{cn}M_n$), and the number of months a household consumes food produced by the household in agricultural drought ($W_{cd}M_d$) were aggregated in PCA to develop the ADRI. The ADRI formula is expressed as (Equation (5)):

$$\text{ADRI} = W_nP_n + W_dP_d + W_{cn}M_n + W_{cd}M_d \quad (5)$$

where W represents each component as a weighted linear combination of the variables and is determined from the component loadings from principal components with a zero mean and unit variance.

The four variables, production of livestock produced in a normal year (Pn), livestock produced in a year with agricultural drought (Pd), the number of months a household consumes food produced by the household in a normal year (Mn), and the number of months a household consumes food produced by the household in drought (Md), were analyzed using Bartlett's test of sphericity and Kaiser–Meyer–Olkin (KMO) using SPSS software.

3.4.2. Structural Equation Modeling

A structural equation model was applied to the determinants of smallholder livestock farming households' resilience to food insecurity (Table 2). The model applies a factor analysis-type model to measure the latent variables via observed variables, simultaneously using a regression-type model for the relationship among the latent variables [36,42]. The structural equation model for a household *i* is illustrated as (Equation (6)):

$$\text{ADRI}_i = f(\text{ASS}_i, \text{ADC}_i, \text{SSF}_i, \text{CH}) \quad (6)$$

Table 2. Description of variables used in structural equation modeling.

Variables	**Descriptions**
Dependent variable	
Agricultural drought resilience index (ADRI)	
Explanatory variables	Sub-variables and description
Assets (ASS)	Herd/flock size (HFS) (cattle, sheep, and goats), agricultural assets (AAs) (tractors, feeding equipment, livestock trailers, water tanks, and corral systems), non-agricultural assets (NAAs) (house, television, chairs, radio, and bed)
Adaptive capacity (ADC)	Perception, source of income (Incsource), migration, credit
Social safety nets (SSF)	Cash, training, food support, water rights, garden equipment, sanitary latrines, farm input
Climate change (CH)	Drought occurrence and intensity

Source: Authors' observation (2020).

Equation (6) is disaggregated in detail as (Equation (7)):

$$\begin{aligned} ADRI_i = f(&ASS_i(\text{herd/flock size (HFS); agricultural assets (AA); non-agricultural assets (NAA)}),\\ &ADC_i\ (\text{perception; source of income (Incsource); migration; credit}),\\ &SSF_i\ (\text{cash; training; food support; water rights; garden equipment; sanitary latrines, farm input), herd/flock size (HFS)};\\ &CH\ (\text{drought occurrence and intensity})) \end{aligned} \tag{7}$$

4. Results

4.1. Socio-Economic Characteristics of the Respondents

Table 3 depicts the socio-economic characteristics of the respondents. When it comes to farming, age is a debatable topic; the average age of the farmers was 52 years. The average formal education of smallholder livestock farmers was eight years (Table 3). The results show that 54% of the respondents had primary education, 42% secondary education, and only 4% had tertiary education.

An average of 11 years of farming experience was observed. As indicated in Table 3, the minimum length of farming experience was half a year, and the maximum was 60 years. The average number of household members was five, with a minimum of one and a maximum of 25 members. From the study's findings, 61 (28%) of the respondents were women, while 156 (72%) were men.

The majority of the respondents were married (57%), 27% single, 9% widowed, 2% divorced, 1% separated, and the remaining respondents (4%) noted other (Table 3). The findings indicated that 51% of the respondents used their family savings to support their farming business, while 8% borrowed money and 41% used other ways of supporting their farming business. Farming is considered a business entity, and thus the majority of the smallholder livestock farmers (86%) depended solely on farming, and 14% owned additional businesses. In addition, only 5% of the respondents owned additional property as a source of income besides livestock farming.

4.2. Respondents' Agricultural Drought Resilience Profile

As indicated, a PCA was applied to construct the outcome variable of the ADRI. Table 4 shows the communalities, component factors, and correlations of variables utilized when constructing the ADRI. All the initial commonalities were above 0.30, which was good. The component variance explained 94% of the total variance. The variables used in PCA were not inter-correlated, and Bartlett's test of sphericity and Kaiser–Meyer–Olkin (KMO) were applied. Bartlett's test of sphericity was significant (p-value = 0.000 with chi-square = 2224.837). As a result, the variables were suitability correlated, warranting the application of PCA, because the inter-correlation matrix did not derive from a population. The KMO was 0.64, which was above 0.5, showing that KMO was suitable for PCA. Therefore, the data set met both KMO and Bartlett's sphericity test requirements and was considered suitable for dimension reduction using PCA.

Table 3. Socio-economic characteristics of the respondents (n = 217).

Variables		Frequency	Percentage	Average	Min	Max	St.dev
Age	21–50	102	47	51.66	21.00	85.00	14.16
	51–85	115	53				
Education	Primary	118	54.38	8.01	0.00	16.00	4.31
	Secondary	91	41.94				
	Tertiary	8	3.68				
Farming experience	0.5–20	196	90.32	10.96	0.50	60.00	8.85
	21–60	21	9.68				
Household members	1–10	204	94	5.19	1.00	25.00	2.88
	11–25	13	6				
Gender	Female	61	28.1	0.72	0.00	1.00	0.45
	Male	156	71.9				
Marital status	Single	59	27.2	2.05	1.00	6.00	1.09
	Married	123	56.7				
	Widowed	19	8.8				
	Divorced	4	1.8				
	Separated	2	0.9				
	Other	9	4.1				
Source of farm funding	Family Savings	111	51.2	1.92	1.00	3.00	0.96
	Borrowings	18	8.3				
	Other Sources	88	40.6				
Other businesses	No	187	86.2	0.14	0.00	1.00	0.35
	Yes	30	13.8				
Property owned	No	207	95.4	0.05	0.00	1.00	0.21
	Yes	10	4.6				

Source: Authors' compilation based on survey (2020).

Table 4. Correlation matrix used for agricultural drought resilience index (ADRI).

Variables	Communalities		Component Factors	Corr. ADRI
	Initial	Extraction		
PN	1	0.935	0.967	0.894
PD	1	0.958	0.979	0.995
Mn	1	0.280	0.963	0.890
MD	1	0.955	0.977	0.984
Eigenvalue variances (%) = 94.402				
Cumulative (%) = 94.402				
KMO test of sampling adequacy = 0.636				
Bartlett's test of sphericity is significant at p = 0.0000; chi-square = 2224.837				

Source: Authors' compilation based on survey (2020).

As a result, Equation (5) is rewritten to estimate the ADRI (Equation (8)):

$$\text{ADRI} = \text{PN} * 0.967 + \text{PD} * 0.979 + \text{Mn} * 0.963 + \text{Md} * 0.977 \quad (8)$$

Based on the findings using Equations (5) and (8), Table 5 presents the ADRI of the study area. An ADRI greater than zero represents households that were resilient to drought, while ADRI less than zero represents households that were not resilient. An estimated 81% (176) of the farming households were not resilient to agricultural drought.

Table 5. Agricultural drought resilience index (ADRI) of Northern Cape Province of South Africa.

	Number	Percentage
ADRI > 0	41	19
ADRI < 0	176	81
Total	217	100

Source: Authors' estimation (2020).

4.3. Econometric Results (Structural Equation Modeling)

The ADRI as an outcome variable was regressed using Equation (6) at the aggregate level (general) and Equation (7) at disaggregate level against the explanatory variables to the determinants of smallholder livestock farmers' household resilience to food insecurity. A structural equation modeling approach was applied to empirically assess smallholder livestock farmers' resilience to food insecurity in Northern Cape Province of South Africa. The results in Table 6 (aggregated) and Table 7 (disaggregated) show that assets, adaptive capacity, safety nets, and climate change indicators significantly impacted households' resilience to food insecurity. ADC (β = 0.171), ASS (β = 0.150), CH (β = 0.053), and SSF (β = 0.001) contributed to the regression model. Asset, SSF, and adaptive capacity indicators positively impacted households' resilience to food insecurity and were significant at 5%. The variance inflation factor (VIF) statistics indicated that there was no multicollinearity problem in the analysis.

Table 6. Structural equation modeling results (aggregated).

Variables	Unstandardized Coefficients		Standardized Coefficients	Sig.	VIF
	B	Std. Error	B		
Constant	11.366	2.086			
Assets (ASS)	0.007	0.003	0.150	0.036 **	1.86
Social safety nets (SSF)	−0.005	0.319	0.001	0.987	1.46
Adaptive capacity (ADC)	0.910	0.360	0.171	0.012 **	1.72
Climate change (CH)	0.095	0.127	−0.053	0.454	1.65

** Significant at 5%. Source: Authors' estimation based on survey (2020).

Households' resilience to food insecurity in the Northern Cape was empirically assessed in detail (Table 7). The results indicated that HFS (β = 0.333), AA (β = 0.089), and NAA (β = −0.019) influenced households' resilience to food insecurity. Herd/flock size (HFS) and AA indicators positively impacted households' resilience to food insecurity. The HFS was the most crucial dimension compared to the other components of assets. Smallholder farmers used livestock as a coping and adaptation mechanism, because they sold livestock during agricultural drought to enhance their resilience.

Four dummy variables were used to estimate the resilience impact of adaptive capacity on food insecurity. The results in Table 7 showed that migration indicators positively impacted households' resilience to food insecurity. Migration (β = 0.037), credit (β = −0.250), perception (β = −0.181), and income source (β = −0.122) contributed to the regression model.

The results in Table 7 showed that all the social safety net indicators had a positive and significant impact on households' resilience to food insecurity. Cash (β = 0.044), training (β = 0.124), food support (β = 0.075), water rights (β = 0.111), garden equipment (β = 0.195), sanitary latrines (β = 0.037), and farm input (β = 0.145) contributed to the regression model.

The two variables that were included under climate change, focusing on drought, namely, drought occurrence and drought intensity, had a negative and significant impact at 10% on household resilience to food insecurity (Table 7). Drought occurrence (β = −0.118) and drought intensity (β = −0.021) contributed to the regression model.

Table 7. Structural equation modeling results (disaggregated).

Variables	Unstandardized Coefficients	Standardized Coefficients	Sig.	Variables
	B	Std. Error	B	
Constant	11.366	2.086		
Assets (ASS)				
Herd/flock size (HFS)	3.435	0.676	0.333	0.000 ***
Agricultural assets (AA)	37.494	27.567	0.089	0.175
Non-agricultural assets (NAA)	−2.795	9.997	−0.019	0.780
Social safety nets (SSF)				
Cash	0.038	0.059	0.044	0.524
Training	0.096	0.057	0.124	0.092 *
Food support	0.060	0.057	0.075	0.297
Water rights	0.114	0.079	0.111	0.147
Garden equipment	0.271	0.106	0.195	0.012 **
Sanitary latrines	0.040	0.077	0.037	0.607
Farm input	0.118	0.055	0.145	0.032 **
Adaptive capacity (ADC)				
Perception	−0.154	0.057	−0.181	0.007 ***
Source of income (Incsource)	−0.235	0.132	−0.122	0.077 *
Credit	−0.541	0.155	−0.250	0.001 ***
Migration	0.059	0.113	0.037	0.603
Climate change (CH)				
Drought occurrence	−0.052	0.030	−0.118	0.090 *
Intensity	−0.007	0.032	−0.021	0.825

*** Significant at 1%; ** significant at 5%; * significant at 10%. Source: Authors' estimation (2020).

5. Discussion

The socio-economic variables, such as age, gender, sex, marital status, access to credit, and assets, were the main factors determining the enhancement of resilience to agricultural drought. It is concerning that the average age of farmers was relatively high. It meant that fewer young people were farming and mostly joined other industries. This could be due to a lack of funding for start-up farmers and the negative stigmas surrounding agriculture as a career choice. This finding is supported by Meterlerkamp et al. [43], who found that one-third of young people show a positive attitude towards farming and choose agriculture as a career.

The male household heads spent more years in school than their female counterparts. This implied that the more educated and higher-skilled individuals were likely to be the least vulnerable to climate shocks such as agricultural drought. This is consistent with the finding of Brenda [44], who highlighted that, commonly, the more educated and higher-skilled individuals of a household are likely to be the least vulnerable to climate shocks such as agricultural drought and have more adaptive capacity than less-educated farmers, because they could obtain information about climate change to assess their situation.

Gender and its impact on social and economic aspects are essential for decision making. It is clear that there is a gender imbalance in farming, agreeing with the study of Matlou and Bahta [45]. Marital status is critical in the determination of the level of involvement in farming. Married household heads can make better decisions during agricultural drought with the assistance of their partners. This finding is in line with a study by Ngeywo et al. [46], who found that the youth who dedicate their energy to farming as a business are denied a chance to do so, because they believe they are not responsible enough if they are not married.

Access to credit or funding is the main challenge for smallholder farmers in Africa, including South Africa. The findings indicated that only a few respondents had access to credit. This finding is in line with the study of Bahta et al. [47]. They highlight that access to credit enhances the working capital of households and resilience to agricultural drought. The majority of the respondents depended on farming. Diversification of income helps to enhance the resilience of smallholder farmers when shocks (such as agricultural drought) occur. However, a minority of farmers owned additional property; this indicated that most smallholder farmers were not resilient to shocks such as agricultural drought. These findings concurred with the findings of Maltou and Bahta [45].

Results from the ADRI indicated that the majority of the respondents were not resilient to agricultural drought. This suggests that smallholder livestock farmers need assistance from the government and different stakeholders in industry to enhance their resilience. The assistance could be feed for livestock (fodder), medication for livestock, strengthening access to agricultural credit and farm input, as well as enhancing smallholder farmers' involvement in agricultural drought resilience activities by giving training and disseminating information. This finding is in line with the study of Matlou and Bahta [45].

The structural equation modeling result indicated that assets, adaptive capacity, safety nets, and climate change indicators significantly impacted households' resilience to food insecurity. This implied that the more assets a farming household owned, the higher the resilience to agricultural drought. These findings are consistent with literature stating that having more assets may increase a household's resilience to food insecurity [11,13–16,48]. Further, the literature also indicates that resilience is the key to enhancing adaptive capacity [49].

The social safety net refers to benefits and protects vulnerable households from the risk of food insecurity. All the social safety net indicators (cash, training, food support, water rights, garden equipment, sanitary latrines, and farm input) had a positive and significant impact on households' resilience to food insecurity. The finding indicates that benefiting from the social safety net provides support for individual households. Our findings concurred with Mane et al. [50], Boukary et al. [11], Szabo et al. [43], and Shah and Dulal [51].

Climate change (drought occurrence and intensity) had a negative and significant impact on household resilience to food insecurity. Indeed, the Northern Cape climate is characterized by hot summers (between 34 °C and 40 °C) and cold winters (below zero nightfall temperatures and frost). Coupled with low rainfall (mean annual precipitation of 200 mm), the climate is consistently dry, which leads to the reduction of livestock production. The findings concur with Shah and Dulal [51], who indicated that a climate shock such as agricultural drought affects food production.

6. Conclusions

This study identified factors affecting livestock farmers' agricultural drought resilience to food insecurity in Northern Cape Province, South Africa. A principal component analysis was applied to estimate the agricultural drought resilience index. A structural equation model was then applied using a survey of 217 smallholder livestock farmers.

The study found that most (81%) smallholder livestock farmers were not resilient to agricultural drought. The study also showed that asset, social safety net, and adaptive capacity indicators positively and significantly impacted households' resilience to food insecurity. However, climate change indicators had a negative and significant impact on households' resilience to food insecurity. This implied that the more assets a farming household owned, the higher the resilience to agricultural drought. The findings further indicated that benefiting from the social safety net provided support for individual households. Indeed, the Northern Cape climate is characterized by hot summers (between 34 °C and 40 °C) coupled with low rainfall (mean annual precipitation of 200 mm). The climate is consistently dry, which leads the reduction of livestock production. As a result, the government needs to strengthen the drought relief program for affected smallholder farmers by supplying fodder, medication, and farming inputs, and strengthening access to agricultural credit.

The study suggests that smallholder livestock farmers need assistance from the government and various stakeholders to minimize vulnerability and boost their resilience to food insecurity. They should target disadvantaged smallholder farmers to build their resilience by enhancing their persistence and adaptability. The government may help smallholder livestock farmers to gather resources to acquire more assets and reduce vulnerability to food insecurity via strengthening access to agricultural credit and farm input. Additionally, the government should address viable off-farm employment as a source of income, and strengthen social safety nets, which include smallholder farmers' involvement in agricultural drought resilience activities by giving training and disseminating information.

Furthermore, the government could improve water rights and access to boost the resilience of smallholder farmers to agricultural drought. This could be achieved through collaboration and coordination among all stakeholders. This includes coordination between monitoring agencies in terms of reliable early warning data, communicated in a comprehensive way to decision makers, farmers' organizations such as the African Farmers' Association of South Africa (AFASA; AFASA is very active in Northern Cape Province of South Africa), and the private sector, such as banks, to strengthen the resilience of farmers against shocks.

Collaboration with national and provincial governmental departments should also be strengthened. This includes collaboration with the Department of Agriculture, Forestry and Fisheries (DAFF), provincial Departments of Agriculture, National and Provincial Disaster Management Centres (NDMC and PDMC), the Department of Water Affairs (DWA), and the South African Weather Service (SAWS).

Author Contributions: All authors made a significant contribution to the manuscript preparation. V.A.M. was involved in data collection, analysis, and writing the first draft. Y.T.B. was a supervisor of the first author, and aided in the study design and conceptualization, review, and writing the final draft. All authors have read and agreed to the published version of the manuscript.

Funding: National Research Foundation (NRF) of South Africa funded this research, grant number TTK170510230380.

Institutional Review Board Statement: The study obtained an ethical clearance certificate from the University of the Free State Research Ethics Committee, and the reference number is UFS-HSD2020/0359/2704.

Informed Consent Statement: Informed consent was obtained from all subjects involved in the study.

Data Availability Statement: Data will be available on request from the corresponding author (Y.T.B.).

Acknowledgments: We acknowledge and thank the anonymous reviewers for their valuable comments and suggestions. We acknowledge Liesl van der Westhuizen (Science Writer) for language editing in this manuscript.

Conflicts of Interest: The authors declare no conflict of interest.

Appendix A

1. Socio-economic characteristics of the respondents

1.1 How old is the household head (age):	
1.2 Gender:	Female = 0 and Male = 1
1.3 Marital Status:	Single = 1, Married = 2, Widowed = 3, Divorced = 4, Separated = 5 and Other = 6
1.4 Educational level (years spent at school)	
1.5 How long have you been farming/farm experience?	
1.6 Where do you get funding for your farm business?	Family savings = 1, Borrowings = 2, and Other = 3
1.7 How many household members are staying in the household?	

2. Assets at household home and farm.

Do you own any of the following assets? How many of the following assets do you own (specify the number)?

Asset		Number of Assets
2.1 Herd/Flock Size (HFS)		
2.1.1 Cattle	No = 0 and Yes = 1	
2.1.2 Sheep	No = 0 and Yes = 1	
2.1.3 Goat	No = 0 and Yes = 1	
2.1.4 Chicken	No = 0 and Yes = 1	
2.1.5 Pig	No = 0 and Yes = 1	
2.1.6 Others		
2.2 Agricultural Assets (AA)		
2.2.1 Tractor	No = 0 and Yes = 1	
2.2.2 Feeding equipment (feed mixer)	No = 0 and Yes = 1	
2.2.3 Livestock trailer	No = 0 and Yes = 1	
2.2.4 Water tank	No = 0 and Yes = 1	
2.2.5 Corral system	No = 0 and Yes = 1	
2.2.6 Others	No = 0 and Yes = 1	
2.3 Non-Agricultural Assets (NAA)		
2.3.1 House	No = 0 and Yes = 1	
2.3.2 Television	No = 0 and Yes = 1	
2.3.3 Chairs	No = 0 and Yes = 1	
2.3.4 Radio	No = 0 and Yes = 1	
2.3.4 Bed	No = 0 and Yes = 1	
2.3.5 Others	No = 0 and Yes = 1	

3. Social Safety Net

Do you or did you receive any of the following benefits?

Support Type	Response		Support from where?
3.1 Cash	No = 0 and Yes = 1		
3.2 Training	No = 0 and Yes = 1		
3.3 Support for food	No = 0 and Yes = 1		
3.4 Vegetable gardening equipment	No = 0 and Yes = 1		
3.5 Sanitary latrine (toilet)	No = 0 and Yes = 1		
3.6 Farm inputs (feed, medication, etc.)		No = 0 and Yes = 1	
3.7 Water rights		No = 0 and Yes = 1	
3.8 Others			

4. Adaptive Capacity

Adaptive capacity 4.1 Credit	Questions	Response
4.1.1 Institution (financial institution)	4.1.1.1 Do you have access to credit? If yes, how effective is the support from the institutions?	No = 0 and Yes = 1
4.2 Perception		
4.2.1 Perception of risk	4.2.1.1 Do you believe that the climate is changing to the extent that it will affect your livestock production?	No = 0 and Yes = 1
4.3 Income source		
4.3.1 Employment	4.3.1.1 How many members of your household are employed? 4.3.1.2 How do they contribute during the drought?	
4.3.2 Business	4.3.2.1 Is there any other business the household is doing besides farming? If yes, please specify 4.3.2.2 How does the business contribute to your farm during drought year?	No = 0 and Yes = 1
4.4 Migration		

Adaptive capacity 4.1 Credit	Questions	Response
4.4.1 Migration 4.2.2 Other options	4.1.1.1 Is migration is an adaptive option during the drought? 4.4.1.2 If no, do you have any other options available? What are they?	N0 = 0 and Yes 1

5. Climate change

Do you usually experience agricultural drought in your community? (Yes/No), if yes.

Climate change	Questions	Response
5.1 Drought occurrence	When was the last time drought occurred? (less than 12 months = 1, less than 5 years = 2, and more than 5 years = 3)	
5.2 Drought intensity	Do you think the intensity of this drought is: (worse than the previous droughts = 1; similar to the previous droughts = 2; better than previous droughts = 3)	

References

1. Wilhite, D.A. Preparedness and coping strategies for agricultural drought risk management: Recent progress and trends. In *Managing Weather and Climate Risk in Agriculture*; Sivakumar, M.V.K., Motha, R., Eds.; Springer: Berlin/Heidelberg, Germany, 2007; pp. 21–38.
2. Benton, T.; Gallani, B.; Jones, C.; Lewis, K.; Tiffin, R.; Donohoe, T. *Severe Weather and UK Food Chain Resilience*; Food Research Partnership: Resilience of the UK Food System Subgroup: Swindon, UK, 2012.
3. Food and Agricultural Organization of the United Nations (FAO). UN Lays Foundations for More Drought Resilient Societies. Meeting Urges Disaster Risk Reduction Instead of Crisis Management. 2013. Available online: https://www.fao.org/news/story/en/item/172030/icode/ (accessed on 26 May 2021).
4. Food and Agricultural Organization of the United Nations (FAO). *Southern Africa—Situation Report October 2016*; FAO: Roma, Italy, 2016.
5. Sithole, V.B. *El Nino Impacts in Southern Africa: Highlights from the 2015/16 Season*; Paper Presented at the Southern Africa Regional Climate Services Workshop (29 November to 2 December 2016); Global Framework for Climate Services Office: Harare, Zimbabwe, 2016.
6. Agricultural Organizations in South Africa (Agri SA). *A Raindrop in the Drought; Report to the Multi-Stakeholder Task on the Drought*; Agri SA's Status Report on the Current Drought Crises; AGRI: Pretoria, South Africa, 2016.
7. Louw, A. Sustainable policy support for smallholder agriculture in South Africa: Key issues and options for consideration. In *Smallholder and Agro-Food Value Chains in South Africa*; Greenburg, S., Ed.; PLAAS: Cape Town, South Africa, 2013; pp. 23–28.
8. Larson, D.F.; Muraoka, R.; Otsuka, K. Why African rural development strategies must depend on small farms. *Global Food Secur.* **2016**, *10*, 39–51. [CrossRef]
9. Christiaensen, L.; Demery, L.; Kuhl, J. The (evolving) role of agriculture in poverty reduction: An empirical perspective. *J. Dev. Econ.* **2010**, *96*, 239–254. [CrossRef]
10. Cheteni, P.; Mokhele, X. Small-scale livestock farmers' participation in markets: Evidence from the land reform beneficiaries in the central Karoo, western Cape, South Africa. *S. Afr. J. Agric. Exten.* **2019**, *47*, 118–136. [CrossRef]
11. Boukary, A.G.; Diaw, A.; Wünscher, T. Factors affecting rural households' resilience to food insecurity in Niger. *Sustainability* **2016**, *8*, 181. [CrossRef]
12. Maponya, S.; Mpandelo, P. Constraints and challenges facing the small scale farmers in Limpopo province, South Africa. *J. Agric. Sci.* **2014**, *6*, 135–143.
13. Matlou, R.; Bahta, Y.T.; Owusu-Sekyere, E.; Jordaan, H. Impact of Agricultural Drought Resilience on the Welfare of Smallholder Livestock Farming Households in the Northern Cape Province of South Africa. *Land* **2021**, *10*, 562. [CrossRef]
14. Melketo, T.; Schmidt, M.; Bonatti, M.; Sieber, S.; Müller, K.; Lana, M. Determinants of pastoral household resilience to food insecurity in Afar region, northeast Ethiopia. *J. Arid Environ.* **2021**, *188*, 104454. [CrossRef]
15. Ogunniyi, A.I.; Omotoso, S.O.; Salman, K.K.; Omotayo, A.O.; Olagunju, K.O.; Aremu, A.O. Socio-economic drivers of food security among rural households in Nigeria: Evidence from smallholder maize farmers. *Soc. Indic. Res.* **2021**, *155*, 583–599. [CrossRef]
16. Chamdimba, O.; Ortmann, G.; Wale, E. Resilience to shocks and food insecurity: Determinants, and the impact of smallholder jatropha curcas cultivation in Southern Malawi. *Afr. J. Sci. Technol. Innov. Dev.* **2020**, 1–15. [CrossRef]
17. Galarza, M.X.L. Resilience to food insecurity: Theory and empirical evidence from international food assistance in Malawi. *J. Agric. Econ.* **2020**, *71*, 936–961. [CrossRef]
18. Myeki, V.A. Factors Affecting Smallholder Livestock Farmers' Agricultural Drought Resilience to Food Insecurity in the Northern Cape, South Africa. Master's Thesis, The University of the Free State, Bloemfontein, South Africa, 2021.
19. Luthar, S.S.; Cicchetti, D. The construct of resilience: Implications for interventions and social policies. *Dev. Psychopathol.* **2000**, *12*, 857–885. [CrossRef]

20. Keil, A.; Zeller, M.; Wida, A.; Sanim, B.; Birner, R. What determines farmers' resilience towards ENSO-related drought? An empirical assessment in central Sulawesi, Indonesia. *Clim. Chang.* **2007**, *86*, 291–307. [CrossRef]
21. Gwaka, L.; Dubihlela, J. The resilience of smallholder livestock farmers in sub-saharan Africa and the risks imbedded in rural livestock systems. *Agriculture* **2020**, *10*, 270. [CrossRef]
22. Alinovi, L.; Mane, E.; Romano, D. Towards the measurement of household resilience to food insecurity: Applying a model to Palestinian household data. In *Deriving Food Security Information from National Household Budget Surveys*; Sibrian, R., Ed.; Experiences, Achievement, Challenges; FAO: Rome, Italy, 2008; pp. 137–152.
23. Grist, N.; Mosello, B.; Roberts, R.; Hilker, L. Resilience. In The Sahel: Building Better Lives With Children. ODI. Available online: https://cdn.odi.org/media/documents/9369.pdf (accessed on 10 March 2021).
24. Bickel, G.; Mark, N.; Cristofer, P.; William, H.; John, C. Guide to Measuring Household Food Security, Revised; Measuring Food Security in the United States; United States Department of Agriculture (USDA). Available online: https://www.ers.usda.gov/briefing/foodsecurity (accessed on 25 March 2021).
25. Kegan, R.; Lahey, L.L. *Immunity to Change: How to Overcome It and Unlock the Potential in Yourself and Your Organization*; Harvard Business Press: Boston, MA, USA, 2009; p. 341.
26. Rockstrom, J. Resilience building and water demand management for drought mitigation. *Phys. Chem. Earth* **2003**, *28*, 869–877. [CrossRef]
27. Abate, T. Indigenous ecological knowledge and pastoralist perception on rangeland management and degradation in Guji zone of South Ethiopia. *J. Sustain. Dev.* **2016**, *15*, 192–218.
28. Birhanu, Z.; Ambelu, A.; Tesfaye, A.; Woldemichael, K. Understanding resilience dimensions and adaptive strategies to the impact of recurrent droughts in Borana zone, Oromia region, Ethiopia: A grounded theory approach. *Int. J. Environ. Res. Public Health* **2017**, *14*, 118. [CrossRef]
29. Kwadwo, O.; Ayalew, D.; Zerfu, E. Determinants of food security in selected agro-pastoral communities of Somali and Oromia regions, Ethiopia. *J. Food Sci. Eng.* **2013**, *3*, 453–471.
30. Magersa, B.; Markemann, A.; Angassa, A.; Anne, V.Z. The role of livestock diversification in ensuring household food security under a changing climate in Borana, Ethiopia. *Food Sec.* **2014**, *6*, 15–28. [CrossRef]
31. Food and Agricultural Organization of the United Nations (FAO). *El Nino Response Plan 2016*; Food and Agricultural Organization of the United Nations (FAO): Addis Ababa, Ethiopia, 2015.
32. Fan, S.; Pandya-Lorch, R.; Yosef, S. *Resilience for Food and Nutrition Security*; IFPRI: Washington, DC, USA, 2014.
33. Hoddinott, J. Looking at development through a resilience lens. In *Resilience for Food and Nutrition Security*; Part of an IFPRI 2020, Book; Fan, S., Pandya-Lorch, R., Yosef, S., Eds.; IFPRI: Washington, DC, USA, 2014; pp. 19–26.
34. Mulugeta, H. Green Famine in Ethiopia: Understanding the Cause of Increasing Vulnerability to Food Insecurity and Policy Response in the Southern Ethiopia Highlands. Ph.D. Thesis, University of Sussex, Brighton, UK, 2014.
35. Guyu, D.; Muluneh, W. Household resilience to seasonal food insecurity: Dimensions and magnitude in the "green famine" belt of Ethiopia. *Appl. Sci. Rep.* **2015**, *11*, 125–143.
36. Alinovi, L.; D'Errico, M.; Mane, E.; Romano, D. Livelihoods strategies and household resilience to food insecurity: An empirical analysis to Kenya. In *The European Report of Development in Dakar, Senegal, 28–30 June*; FAO: Rome, Italy, 2010; pp. 1–52.
37. Stats, S.A. Frances Baard District Municipality Annual Budget 2019/2020. Available online: http://francesbaard.gov.za/wp-content/uploads/2015/04/Annual-budget-2019.20.pdf (accessed on 1 April 2021).
38. Frances Baard District Municipality (FBDM). Map of the Four Local Municipalities of the Frances Baard Municipal District in the Northern Cape. 2019. Available online: https://municipalities.co.za/map/134/frances-baard-district-municipality (accessed on 21 February 2021).
39. Cochran, W.G. *Sampling Techniques*, 3rd ed.; John Wiley and Sons: New York, NY, USA, 1977.
40. Bartlett, J.E.; Kotrlik, J.W.; Higgins, C.C. Organizational research: Determining the appropriate sample size in survey research. *Inf. Techn. Learn. Perfor. J.* **2001**, *19*, 43–50.
41. Northern Cape Department of Agriculture, Forestry, and Fisheries (NDAFF). *Beneficiaries of an Agricultural Drought Relief Program*; NDAFF: Kimberly, South Africa, 2020.
42. Bollen, K.A. *Structural Equations with Latent Variables*; Wiley and Sons: New York, NY, USA, 1989.
43. Metelerkamp, L.; Drimie, S.; Biggs, R. We're ready, the system's not-youth perspectives on agricultural careers in South Africa. *Agrekon* **2019**, *58*, 154–179. [CrossRef]
44. Brenda, B.L. Resilience in agriculture through crop diversification: Adaptive management for environmental change. *Bioscience* **2011**, *61*, 183–193.
45. Matlou, R.; Bahta, Y.T. Factors influencing the resilience of smallholder livestock farmers to agricultural drought in South Africa: Implication for adaptive capabilities. *J. Disast. Risk Stud.* **2019**, *11*, a805.
46. Ngeywo, J.; Basweti, E.; Shitandi, A. Influence of gender, age, marital status and farm size on coffee production: A case of Kisii county, Kenya. *Asian J. Agric. Exten. Econ. Soc.* **2015**, *5*, 117–125. [CrossRef]
47. Bahta, Y.T.; Strydom, D.B.; Donkor, E. Microcredit and gender empowerment: Policy implications for sustainable agricultural development in Eritrea. *Dev. Pract.* **2017**, *27*, 90–102. [CrossRef]
48. Szabo, S.; Hossain, M.S.; Adger, W.N.; Matthews, Z.; Ahmed, S.; Lazar, A.N.; Ahmad, S. *Soil Salinity, Household Wealth and Food Insecurity in Tropical Deltas: Evidence from South-West Coast of Bangladesh*; Springer: Tokyo, Japan, 2015.

49. Folke, C.; Colding, J.; Berkes, F. Building resilience for adaptive capacity in social-ecological systems. In *Navigating Social-Ecological Systems: Building Resilience for Complexity and Change*; Berkes, F., Colding, J., Folke, C., Eds.; Cambridge University Press: Cambridge, UK, 2002.
50. Mane, E.; Rocca, M.; Conforti, P. *Social Protection and Food Security Indicators: An Inquiry through Data from 10 Household Budget Surveys*; FAO Statistics Division—Working Paper Series. 9/2015; FAO: Rome, Italy, 2015.
51. Shah, K.U.; Dulal, H.B. Household capacity to adapt to climate change and implications for food security in Trinidad and Tobago. *Reg. Environ. Chang.* **2015**, *9*, 83–100. [CrossRef]

climate MDPI

Article

Impact of Climate Change on Crop Production and Potential Adaptive Measures in the Olifants Catchment, South Africa

Mary Funke Olabanji *, Thando Ndarana and Nerhene Davis

Department of Geography, Geo-informatics and Meteorology, University of Pretoria, Private Bag x 20, Hatfield 0028, South Africa; thando.ndarana@up.ac.za (T.N.); nerhene.davis@up.ac.za (N.D.)
* Correspondence: u18098267@tuks.co.za; Tel.: +27-718779860

Abstract: Climate change is expected to substantially reduce future crop yields in South Africa, thus affecting food security and livelihood. Adaptation strategies need to be implemented to mitigate the effect of climate change-induced yield losses. In this paper, we used the WEAP-MABIA model, driven by six CORDEX climate change data for representative concentration pathways (RCPs) 4.5 and 8.5, to quantify the effect of climate change on several key crops, namely maize, soya beans, dry beans, and sunflower, in the Olifants catchment. The study further investigated climate change adaptation such as the effects of changing planting dates with the application of full irrigation, rainwater harvesting, deficit irrigation method, and the application of efficient irrigation devices on reducing the impact of climate change on crop production. The results show that average monthly temperature is expected to increase by 1 °C to 5 °C while a reduction in precipitation ranging between 2.5% to 58.7% is projected for both RCP 4.5 and RCP 8.5 relative to the baseline climate for 1976–2005, respectively. The results also reveal that increased temperature and decreased precipitation during planting seasons are expected to increase crop water requirements. A steady decline in crop yield ranging between 19–65%, 11–38%, 16–42%, and 5–30% for maize, soya beans, dry beans, and sunflower, respectively, is also projected under both RCPs climate change scenarios. The study concludes that adaptation measures such as the integration of changing planting dates with full irrigation application and the use of rainwater harvest will help improve current and future crop production under the impact of climate change.

Keywords: climate change; crop yield; adaptation strategies; water requirement; WEAP-MABIA model

Citation: Olabanji, M.F.; Ndarana, T.; Davis, N. Impact of Climate Change on Crop Production and Potential Adaptive Measures in the Olifants Catchment, South Africa. *Climate* **2021**, 9, 6. https://doi.org/10.3390/cli9010006

Received: 20 October 2020
Accepted: 28 December 2020
Published: 31 December 2020

1. Introduction

Increased greenhouse gas emission caused by human-induced activities such as the burning of fossil fuels and deforestation has accelerated the impacts of climate change in the 20th century. Recently, the intergovernmental panel on climate change report [1] indicated that natural and human systems have been significantly impacted by increased temperature, and increased frequencies of extreme weather events such as droughts and floods that are caused by changes in the climate system. The impacts of climate change on agriculture can no longer be ignored as agricultural production is largely dependent on the amount of water available and dry-land farmers who rely on rain-fed farming for their livelihood are particularly vulnerable [2]. Climate change impacts on agriculture resulting in the decline in crop yield may increase food insecurity globally [3,4]. This is largely because agricultural crops relevant to food security such as maize, wheat, and rice require significant amounts of water for production [5]. The amount of water required to produce 1 kg of these crops is estimated at 1.5 m^3, 1.0 m^3, and 2.5 m^3 respectively [6]. It is therefore anticipated that areas with limited water availability due to climate change impacts will experience significant crop yield losses which would compromise food security in the long term.

Climate change impacts on agricultural production are increasingly becoming a major area of scientific interest [7]. Such impacts are significant in arid and semi-arid areas which include a country like South Africa, a semi-arid country, with about two-thirds of its land area receiving a mean annual rainfall of less than 500 mm [8] during the summer months over the eastern parts of the country [8], where most of the agricultural activity takes place. The share of agriculture in the country's gross domestic product (GDP) is barely 4 percent [9]. However, despite this seemingly negligible contribution towards the country's economy, the agricultural sector accounts for almost 10 percent of the total employment in the country and about one-third of the country's total crop production is exported with considerable financial returns [10]. In addition, more than a million people in the country are indirectly dependent on the agricultural sector for their livelihood [11] and 94 percent of the agricultural products such as wheat, maize, and rice are consumed within the country [12]. The strategic importance of the agricultural sector in South Africa is therefore evident.

Increased rainfall variability and high temperatures are currently the key factors expected to have significant impacts on agricultural production in South Africa [8,13]. For instance, climate projection studies have illustrated that the frequency and intensity of droughts coupled with higher variability in rainfall will have negative implications on crop production [1]. A study undertaken by Erasmus et al. [14], projected a decline in precipitation in the Western Cape Province which they predicted would result in less water available for agriculture with related socio-economic impacts for farmers in the area. The anticipated increase in temperature of 1.2 °C in 2020, 2.4 °C in 2050, and 4.2 °C by the year 2080 and a projected rainfall decrease of about 5–10% in the next 50 years [8], thus presents a significant risk to South Africa's food security and socio-economic stability. Considering the socio-economic importance of agriculture and food security, it is therefore imperative to assess how future climate change will affect crop yield. Adaptation is an important factor that will minimize the severity of the impact of climate change on future crop production [15,16]. Potential adaptation strategies should thus be developed and consistently evaluated to effectively cope with climate risk.

Numerous studies have assessed the impact of climate change on crop production in South Africa [17–22], their findings indicate a decrease in crop yield as a result of a changing climate. Mayowa [23] examined the impact of climate variability on maize yield in South Africa using satellite-derived data and a neutral framework. The result of the study indicated that maize phenology could be impacted by climate variability, especially if the impacts are most severely experienced during the vegetative and reproductive period of plants. Studies by Gbetibouo and Hassan [24] and Deraasa et al. [25] used the Ricardian model to investigate the economic impact of climate change on major South African field crops and found that the production of field crops was more sensitive to changes in marginal temperature as compared to changes in precipitation. Results from their study implied that an increase in temperature somehow positively affected net revenue whereas the effect of precipitation decrease was negative. The study went further to highlight the importance of season and location in dealing with climate change, indicating that the spatial distribution of climate change impact and consequently the needed adaptation strategies vary across the different agro-ecological regions of South Africa.

Despite the fairly extensive research undertaken towards assessing the potential impacts of climate change on crop production in South Africa, to date, no study has evaluated the efficiency of adaptation strategies in order to provide farmers and decision-makers with clear guidance on the best practices to be implemented. Further, many of the existing studies were either conducted at a national, regional, or provincial level which fails to capture climate change-related dynamics and its implications at a catchment level. It is important to understand how climate change affects crop production at a catchment scale since vulnerability and the intensity of climate change are location-specific and the formulation of adaptation strategies depends on the level of impacts. Specifically, the assessment of climate change and adaptive measures in terms of crop production have not been investigated in the Olifants catchment where crop yield could be more sensitive to climate change due to the vulnerability of the catchment to global change.

This study, therefore, aims to assess the potential impact of climate change on crop yield in the Olifants catchment using a high-resolution climate change model and it evaluates the efficiency of the adaptive strategies deployed by farmers to improve crop yield under current and future climate change scenarios for the catchment. The specific objectives of this study are: (1) to evaluate the impacts of climate change on the catchment reference evapotranspiration and crop water requirement; (2) to assess the influence of soil texture on crop yield under current climate condition; (3) to evaluate the impact of current and projected climate change on crop yield, and lastly, (4) to assess crop yield response to different adaptation measures in the context of changing climate.

The findings from this study are intended to provide relevant information on the expected changes in climate and its impact on crop yield at a catchment-scale as well as to guide policy-makers on the most suitable adaptation options to be implemented in order to improve future crop production and ensure food security.

The rest of the paper consists of Section 2 which details the material and methods used for analysis in the study, followed by a presentation of the results in Section 3. The discussion and concluding remarks are provided in Sections 4 and 5 of the paper.

2. Materials and Methods

2.1. Study Area and Data

The Olifants River Basin is one of the nineteen water management areas in South Africa. It is a principal sub-catchment of the Limpopo River. It originates in the north of South Africa in the province of Mpumalanga and flows northeast through the northern province before joining in Mozambique and emptying into the Indian Ocean (Figure 1). An estimated 3.2 million people live within the catchment area with two-thirds of this population living in the rural community [26]. The Olifants River Basin is recognized as one of the most important basins in South Africa as it contributes largely to the country's economic hub, with an annual contribution of six percent to the Gross Domestic Product arising from agricultural, mining, and industrial activities [27]. The catchment consists of both large and medium-scale agricultural farms that consume a lot of water for irrigation (540 Mm^3 per year) with approximately 130,000 ha irrigated (i.e., 11% of the total cultivated area in the catchment), primarily in the commercial farming sector. The water used for irrigation is obtained from both dams and groundwater in the catchment [28]. Precipitation in the catchment occurs during the summer months from October to April, with average annual rainfall ranging between 500 mm to 800 mm in most parts of the catchment and surpasses 1000 mm along the escarpment which separates the Highveld from the Lowveld. Evaporation varies across the catchment with high levels occurring in the north and west, and lower levels of evaporation recorded in the southeast. Elevations range from 300 m to over 2300 m above sea level, which explains the relatively cool winter and annual wide-range of temperature variations of −4 to 35 °C [29]. Runoff from the catchment reflects the temporal and spatial distribution of the rainfall with the greatest volumes occurring in the south and along the escarpment. The average annual runoff from the catchment is 37.5 mm (i.e., 6% of the average annual rainfall), which equates to 2040 million cubic meters (Mm^3). However, there is considerable inter-annual variation and consecutive years where the flow is below the mean annual discharge [30].

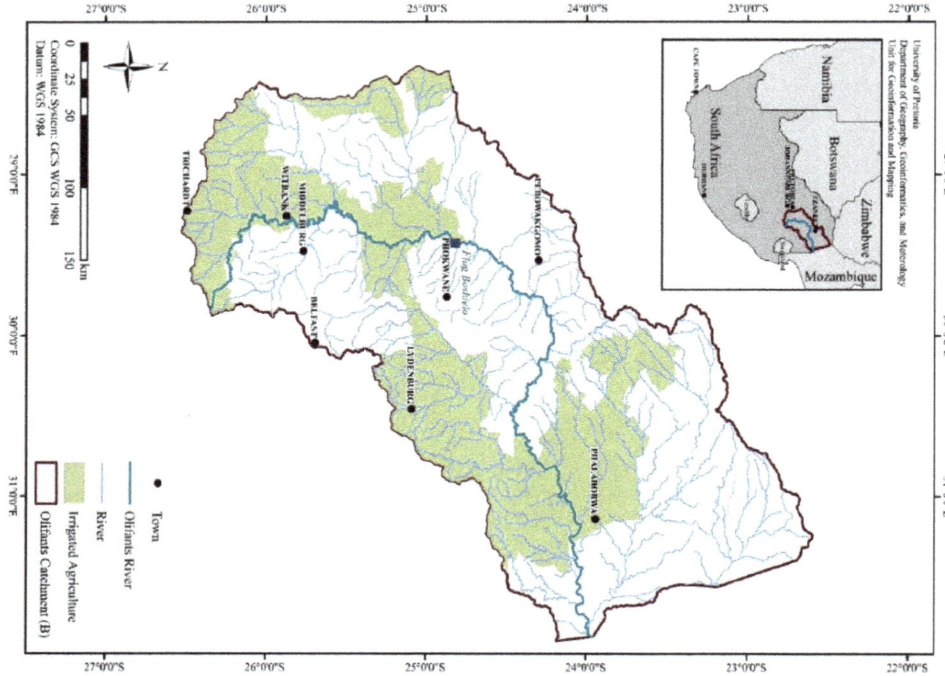

Figure 1. Map of Olifants River Basin showing towns, rivers, and irrigated areas.

This study uses climate simulated data (daily rainfall, minimum, and maximum temperature, average relative humidity, 2 m wind speed, and solar radiation) from the Coordinated Regional Climate Downscaling Experiment (CORDEX) database. The data were obtained from http://cordexesg.dmi.dk/esgf-web-fe/. The output from CORDEX RCMs are quality controlled and can be used according to the terms of use (http://wcrp-cordex.ipsl.jussieu.fr/). It should be noted that all Coordinated Regional Downscaling Experiment-Regional Climate Models (CORDEX RCMs) are set to 0.44° by 0.44°spatial resolutions, which corresponds to 50 km by 50 km. The area-weighted average method [31] was used to calculate the average climate variables from the CORDEX RCMs over the entire Olifants River catchment (latitude 24° and 26° and longitude 29° and 32°). Daily climate variables listed above were obtained from a single RCM driven by six Global Climate Models (GCMs) namely, Commonwealth Scientific and Industrial (CSIRO), National Centre for Meteorological Research (CNRM), Canadian Centre for Modelling and Analysis (CCMA), Institut Pierre Simon Laplace (IPSL), Model for Interdisciplinary Research on Climate (MIROC), and Max Planck institute for Meteorological Earth System Model (MPI-ESM) for a period of 30 years (1976–2005). Two projected climate change scenarios, representative concentration pathways (RCP 4.5 and 8.5) were used. The former represents an intermediate stabilization of radioactive forcing by 2100, without surpassing 4.5 W/m^2 (~650 ppm CO_2), which constitutes a high mitigation scenario [32]. Whilst the RCP 8.5 scenario assumes that the radioactive forcing pathway reaches above 8.5 W/m^2 (~1370 ppm CO_2) by 2100 [33].

Using CORDEX-RCM climate change data, three time periods 2010–2039, 2040–2069, 2070–2099 were considered in this study to project future climate for both greenhouse gas concentration scenarios. These time periods were then compared to the baseline period, 1976–2005. CORDEX climate variables were biased corrected using a linear scaling bias correction method. Observed climate variables obtained from the South African weather service were used to bias correct the current and projected CORDEX climate data [34]. It was necessary to bias correct the simulated climate data in order to compensate for any over or underestimation of the downscaled variables. The linear scaling bias correction is based on the average difference between daily observed time series data. These differences were then applied to the simulated climate data to obtain bias-corrected variables [34]. The biased-corrected climate variables were then integrated into a decision support system (Water Evaluation and Planning) model to evaluate current and future crop yield and adaptation scenarios using the WEAP-MABIA method.

The WEAP-MABIA model used in this study has a soil profile library functionality that provides typical values for water content at saturation, field capacity, wilting point, and the available water capacity for 12 textural classes. It uses a pedotransfer function to estimate the average soil water capacity. In this study, we assumed scenarios of three textural classes of soil to evaluate its impacts on crop yield under current climate conditions. The three textural classes of soil were (S1-sandy loam, S2-loamy sand, and S3-Sandy clay loam) presented in Table 1.

Table 1. Classification soil type used in this study (S1 = Sandy loam, S2 = Loamy sand, and S3 = Sandy clay loam).

Texture Classes Soil		Properties as A Percentage (%) of Volume		
Soil Type	**Total Available Water (TAW)**	**Saturation (SAT)**	**Field Capacity (FC)**	**Wilt Point (WP)**
S1	15.72	41.20	23.74	8.02
S2	10.36	40.10	14.90	4.54
S3	12.97	33.00	25.13	12.16

Crop parameters were also obtained from the crop library functionality within the WEAP-MABIA. The "Crop Library" provides the required parameters for over 100 crops, some with multiple entries for different climates or regions of the world. The end-user can add, edit, remove, copy, export, import, or search the "Crop Library" for a particular crop. This study selected four crops namely: maize, soya beans, dry beans, and sunflower from the crop library using the crop scheduling wizard. The crop parameter used in this study is presented in Table 2.

Table 2. Database of crop parameters used in this study.

Crop	Maize	Soya Beans	Dry Beans	Sunflower
Planting area (in thousand hectare)	180	72	41	20
Planting date	10/24	11/05	11/17	11/25
Stage length initial [days]	25	15	20	25
Stage length dev [days]	40	30	30	35
Stage length mid-season [days]	45	60	40	45
Stage length end-season [days]	30	25	30	25
Stage length total [days]	140	130	120	130
Kcb: initial	0.15	0.15	0.15	0.15
Kcb: mid-season	1.15	1.10	1.00	1.10
Kcb: end-season	0.30	0.30	0.80	0.25
Depletion factor initial	0.55	0.50	0.45	0.45
Depletion factor mid-season	0.55	0.35	0.45	0.45
Depletion factor: end-season	0.55	0.50	0.45	0.45
Yield response factor [ky] initial	0.40	0.40	0.20	0.40
Yield response factor dev	0.40	0.80	0.60	0.60
Yield response factor mid-season	1.30	1.00	1.00	0.80
Yield response factor end-season	0.50	0.40	0.40	0.80
Yield response factor total	1.25	0.85	1.15	0.95
Maximum Height [m]	2.00	0.75	0.40	2.00
Rooting Depth [m] Minimum	0.15	0.15	0.15	0.15
Rooting Depth [m] Maximum	1.35	0.95	0.60	1.15

The lengths of growth stages (Lini, Ldev, Lmid, Llate) were computed according to the FAO-56 method as a function of vegetation cover (fc). The initial stage (Lini) runs from the sowing date to when the fc reaches a value of 0.1, the development stage (Ldev) runs from a fc of 0.1 to full vegetation cover (fc of 0.9). The mid-season stage (Lmid) runs from the end of the development stage until canopy cover (fc) drops back to the same value it had at the end of the development stage and the beginning of the mid-season period (fc = 0.9). The late-season stage (Llate) runs from the end of the mid-season stage until the end of the growing season.

The basal crop coefficient (Kcb) is defined as the ratio of the crop evapotranspiration ETc over the reference evapotranspiration ETref when the soil surface is dry but transpiration is occurring at a potential rate. Therefore, Kcb represents primarily the transpiration component of ETc. The Kcb coefficient serves as a lumped parameter for the physical and physiological differences between crops. Variation in Kcb between the growth stages is mainly dependent on how the crop canopy develops. The values given in the "crop library" represent a standard climate having mean daily minimum relative humidity (RHmin) equal to 45% and mean daily wind speed measured at 2 m (u2) equal to 2 m s^{-1}.

The depletion factor (p) is the fraction of the total available water (TAW) that can be depleted from the root zone before moisture stress occurs. Different values can be defined to express the variation of the crop sensitivity to water shortage over the different crop stages.

The yield response factor (Ky) is a factor that describes the reduction in relative yield according to the reduction in the crop evapotranspiration (ETc) caused by soil water shortage. Ky values are crop-specific and may vary over the growing season. The values for Ky are given for the individual growth periods as well as for the complete growing season.

The rooting depth for annual crops has three growth stages. The rooting depth is held constant at the minimum depth (Zr min) throughout the initial crop growth stage. The root zone increases linearly to a maximum depth (Zr max) throughout the vegetative growth and development stages

The maximum root depth is attained at the beginning of the mid-season stage (peak growth) and is maintained throughout the mid and late season stage [35].

2.2. Description of WEAP Model

The water evaluation and planning model developed by the Stockholm Environment Institute (SEI) is a decision support system (DSS) used for the integration of water resources management and planning. It is easy to use for water planning and scenario assessment. WEAP simulates water balance for water demand, supply, and storage on a monthly basis and it allows the assessment of water resource management policies between different sectors (agriculture, industry, tourism). It can be applied at a catchment level as well as other more complex levels such as regional and country levels [36].

Within the WEAP model, different agricultural catchment calculation methods can be used. In this study, we used the WEAP-MABIA method version 1.0.1 [35] to simulate crop water requirement, crop yield as well as agricultural management plans under different climate conditions. The selection of this method was based on the fact that it has been applied by scientists, engineers, and resource managers to simulate runoffs, infiltration, and percolation processes resulting from natural rainfall, irrigation scheduling, and crop yield reduction [35,37–39]. The WEAP-MABIA method calculate evapotranspiration using the 'dual' crop coefficient kc method (Kc = Ke + Ks Kcb), as described in Allen et al. [40], whereby the Kc value is divided into a 'basal' crop coefficient, Kcb, and a separate component, with Ke, representing evaporation from a shallow soil surface layer. The basal crop coefficient represents actual ET conditions when the soil surface is dry but sufficient root zone moisture is present to support full transpiration.

The reference evapotranspiration (ET_0) for the Olifants catchment was calculated using the modified Penman-Monteith equation recommended by Allen et al. [40]. The equation utilizes some assumed constant parameters for a clipped grass reference crop. It was assumed that the definition for the reference crop was a hypothetical reference crop with a crop height of 0.12 m, a fixed surface resistance of 70 s m^{-1} and an albedo value (i.e., a portion of light reflected by the leaf surface) of 0.23 [41]. The equation used for calculating ET_0 is given below:

$$ET_0 = \frac{0.408\Delta(R_N - G) + \lambda\frac{900}{T+273}\,U_2\,(e_s - e_a)}{\Delta + \gamma(1 + 0.34\,U_2)}$$

where R_N is the net radiation at the crop surface, G is the soil heat flux density, T represents the mean daily air temperature at 2 m height, U_2 is the wind speed at 2 m height, e_s is the saturation vapor pressure, e_a is the actual vapor pressure, (e_s–e_a) represents the vapor pressure deficit of the air, Δ is the slope vapor pressure curve, and γ represents the psychometric constant.

The following climate parameters such as daily temperature (minimum and maximum), average relative humidity, 2-m wind speed, and solar radiation were used to estimate the current and projected reference crop evapotranspiration.

The performance of the WEAP-MABIA was verified by calibrating and validating observed crop yield data for Mpumalanga province where the Olifants catchment is situated, as there was no recorded crop data for the catchment. The data was obtained from the Department of Agriculture, Fisheries, and Forestry (DAFF). The efficiency of the model performance was determined by comparing the observed against the simulated crop yield using two verification statistics such as Coefficient of Determination (R^2) and Nash-Sutcliffe Efficiency (NSE). The values of R^2 ranges between 0–1, values higher than 0.5 are considered acceptable. While NSE ranges between $-\infty$ and 1.0, where NSE = 1 indicates a perfect match of simulated and observed yield. An efficiency of 0 shows that the model prediction is as accurate as the mean of the observed data, while an efficiency less than 0 shows that the observed mean is a better predictor than the model. For more detail on the procedure and statistical equation used for the calibration and validation of the model, readers should consult Olabanji et al. [29]. We calibrated and validated the WEAP-MABIA crop model using observed crop yield data for the period of 1995–2000 for calibration and 2001–2004 for validation. The results presented in Table 3 show that the simulated crop yield perfectly agrees with the observed yield with NSE ranging between 0.97 to 0.99 during calibration and 0.87 to 0.96 during validation. The coefficient of determination (R^2) ranged between 0.98 to 1.0 for calibration and 0.95 to 0.98 during the validation process. The agreement between the simulated and observed crop yield indicates the capability of the crop model to simulate future crop yields.

Table 3. Model calibration and validation result using yearly simulated and observed crop yield.

Crops	Crop Yield (t/ha) (Calibration)				Crop Yield (t/ha) (Validation)			
	Sim	Obs	R^2	NSE	Sim	Obs	R^2	NSE
Maize	4.30	4.62	0.98	0.97	4.32	5.17	0.95	0.92
Soya beans	0.40	0.45	0.99	0.98	0.65	0.72	0.96	0.94
Dry beans	0.35	0.37	1.0	0.99	0.36	0.45	0.92	0.87
Sunflower	0.20	0.23	1.0	0.98	0.21	0.25	0.98	0.96

2.3. Experimental Design

We conducted five experiments to assess crop yield response to climate change impacts and adaptation strategies in the Olifants catchment (Figure 2). The adaptation scenarios used for this analysis were derived from a comprehensive literature review [42–46] and augmented with the results from a field survey that was completed by seventy-three smallholder farmers who provided information about the adaptive strategies they are deploying in the area. During the course of the study, four adaptation measures were evaluated using the WEAP-MABIA model.

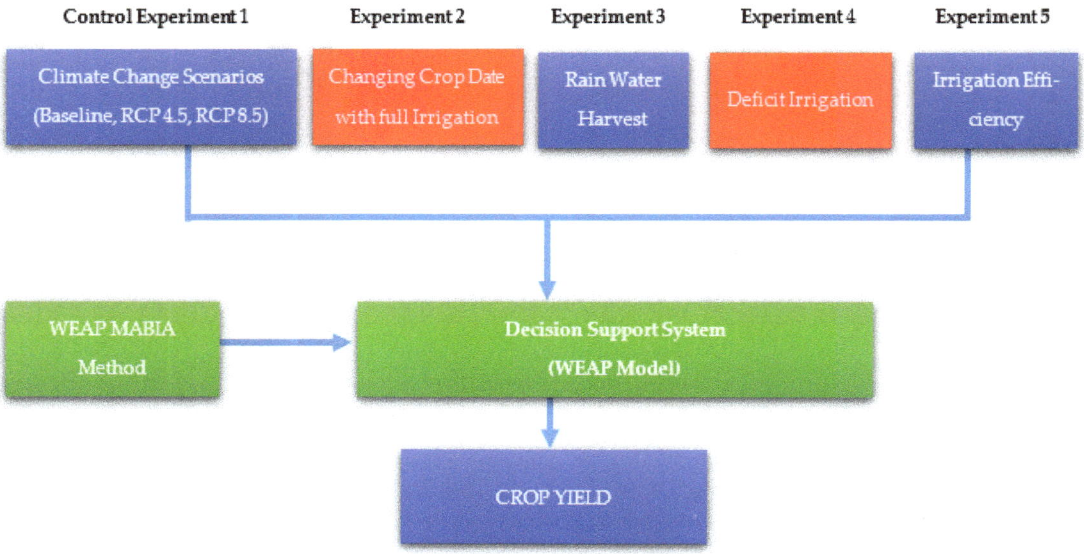

Figure 2. Diagrammatic representation of the conceptual framework of the study.

- Experiment 1, which is the control experiment, presents crop yield under baseline and projected climate scenarios (RCP 4.5 and RCP 8.5) without the implementation of adaptation strategies.
- Experiment 2, presents crop yield with the adoption of a shifting plantation date and the application of full irrigation water. This scenario assumes a delay in the planting of crops. It is expected that crops in the catchment, particularly those investigated in this study, are to be planted from early October to mid-December. However, due to the delay in rainfall, most farmers have adjusted their farming practice by shifting their planting dates to wait for rain. We, therefore, assumed a delay of 25 days from the initial date planted. The strategy also assumed the application of full irrigation to supplement rainfall should there be any shortfall during the cropping cycle. The scenario assumed that irrigation would be applied at 100% of the readily available water. This application implied that soil depletion would never drop below the readily available water level, and as such, water stress is not expected to occur in this scenario.
- Experiment 3, presents crop yield with the application of rainwater harvesting. This scenario assumes the harvesting of rainwater from runoffs during the period of heavy rainfall within the cropping cycle to irrigate crops. This strategy is expected to augment limited irrigation water from the system, it also serves as an agricultural water management measure as it helps to restrain the over-exploitation of freshwater during crop production.
- Experiment 4, presents crop yield with the adoption of the deficit irrigation method. This strategy involves the application of limited water during crop production. The formulation of this strategy is as a result of the increased demand for water resources by other water use sectors which may cause shortages during irrigation for farmers in the near future [29]. According to the findings of Geerts and Raes [47], the deficit irrigation scheme is a promising and tested irrigation technique, especially in periods

of low rainfall. The scenario assumes allowing shortage of 50% of the readily available water (RAW) before irrigation.

- Finally, experiment 5 presents crop yield with the use of efficient irrigation devices such as the drip irrigation technique, or the sprinkler and furrow. These devices are deemed more efficient compared to traditional irrigation techniques such as the center pivot system, or the buckets and pipes approach which consumes a lot more water during irrigation. Specifically, for this analysis, the scenario assumes the use of a drip irrigation technique with an application efficiency of 95% to improve crop yield.

The results obtained from experiments 2, 3, 4, and 5 were compared against the control experiment (crop yield without adaptation strategies adopted). The assessment of each experiment enabled us to determine which adaptation measure performed the best in terms of crop productivity.

3. Results

3.1. Climate Model Validation

This section validates the performance of CORDEX-RCM current climate data for the period 1976–2005. Results presented in Figure 3a,b shows that ensemble monthly historical RCM after bias correction mostly agreed with the observed precipitation and temperature in the catchment. The agreement between the simulated RCM climate and the observed indicated that CORDEX-RCM data after bias correction were capable of projecting future climate.

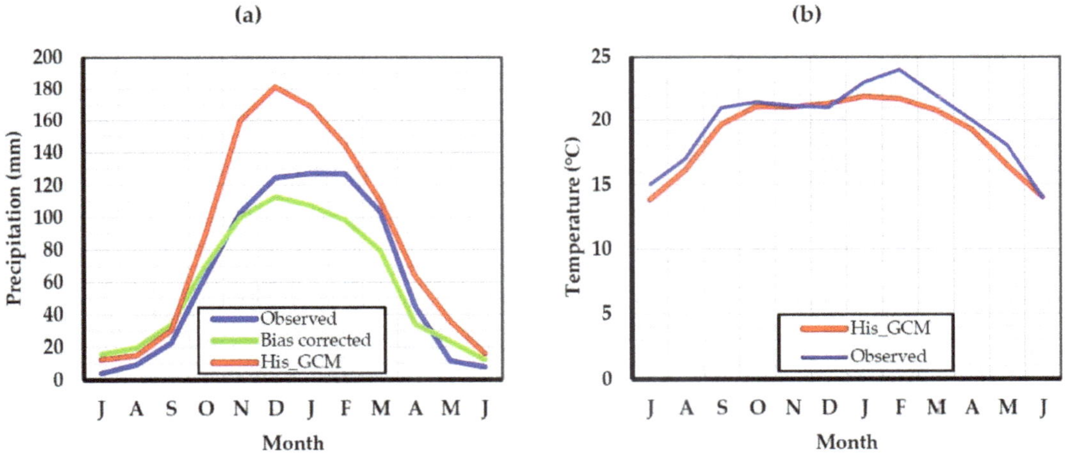

Figure 3. Olifants catchment current monthly climate (**a**) showing observed monthly South African Weather service (SAWS) precipitation plotted against the uncorrected historical ensemble precipitation from the Regional Climate Model (RCM) and the bias-corrected ensemble RCM for the period 1976–2005; (**b**) ensemble historical average monthly temperature against SAWS average monthly temperature.

3.2. Future Climate Projections

The intra-annual changes in temperature and precipitation for the study area under the two future projected climate change (RCP 4.5 and RCP 8.5) scenarios are presented in Figure 4a–d. From the results obtained, it was clear that elevated CO_2 concentration would significantly increase temperatures in the future. The monthly average temperature was projected to increase by 1.0, 1.6, and 2.9 °C for the 2010–2039, 2040–2069, and 2070–2099 periods respectively, for RCP 4.5 scenario relative to baseline. Under the RCP 8.5 scenario, the average monthly temperature increased by 2.3, 3.0, and 5.0 °C for the three future time periods as shown in Figure 4a,b. The highest temperature increase was expected for the RCP 8.5 scenario towards the end of the century. Results further revealed that summer months were likely to experience increased temperature particularly the month of October during the mid and end of the century. In addition, the results presented in Figure 4c,d show the variations in intra-annual predicted precipitation for the Olifants catchment for RCP 4.5 and RCP 8.5 respectively. As indicated in Figure 4c, a slight increase in precipitation for the RCP 4.5 scenario was expected for the early-term in the month of August and September and a decrease was expected for other months. For the mid-term and far-term, an average increase in precipitation was expected for the months of July and August, with the remaining part of the months expected to be dry. A precipitation decline between 3.2 to 51.4% was predicted for RCP4.5 with the mid-term expected to be the driest. The results presented in Figure 4d show a decreasing trend in most of the months in the mid and end-term for the RCP 8.5 climate change scenario. However, a slight increase was projected for the months of July, August, and September in the early term. The end-term was projected to be the driest period for the RCP 8.5 scenario, a decrease in precipitation ranging between 2.5 to 58.7% was anticipated. The summer months of October to February for both climate change scenarios were expected to experience a greater decline in precipitation during the early-term, mid-term, and end-term respectively. Based on the outcome of the climate analysis, the decline in the projected precipitation during the summer months and an increasing trend in temperature will have implications for future crop production, as most of the crops in the catchment are planted during this period.

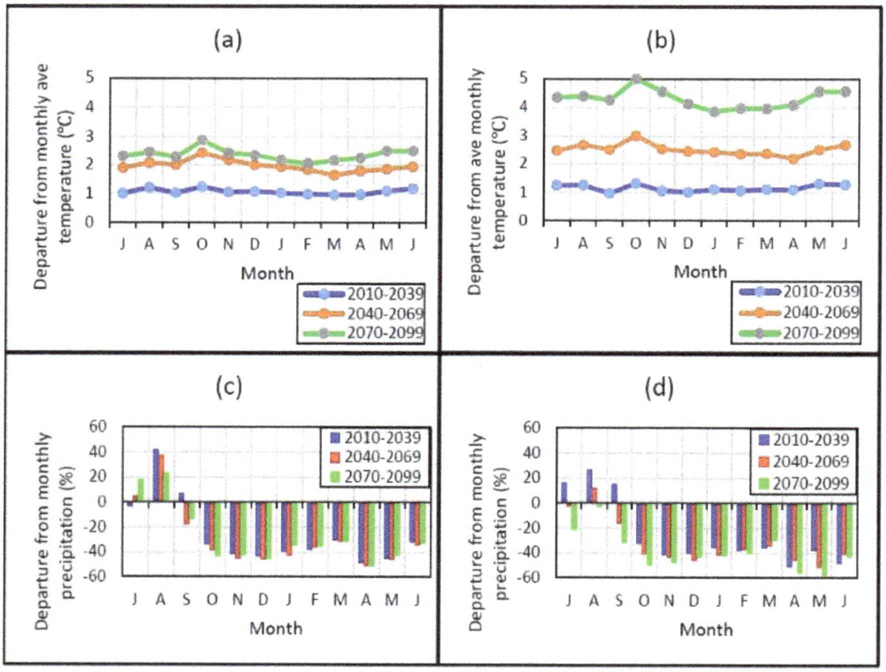

Figure 4. Olifants catchment future monthly climate departure from observed climate: (**a**) temperature anomaly for representative concentration pathways (RCP) 4.5 and (**b**) for RCP 8.5; (**c**) and (**d**) showing precipitation anomaly for RCP 4.5 and RCP 8.5 for the period 2010–2039, 2040–2069, and 2070–2099 respectively.

3.3. *Analysis of Reference Crop Evapotranspiration*

This study estimated reference crop evapotranspiration (ET_0) using the Penman-Montieth equation. Projected changes in ET_0 were calculated as the difference between the average monthly and annual ET_0 for RCP 4.5 and RCP 8.5 relative to baseline (1976–2005). Results from the calculation show an increasing trend in total annual ET_0 of 770.9 mm, 810.0 mm, and 817.8 mm for RCP 4.5 and 778.0 mm, 829.3 mm, and 904.5 mm for RCP 8.5 for the early-term, mid-term, and end-term relative to 729.1 mm of baseline climate (1976–2005). The intra-annual ET_0 variation indicated a higher increase in the months of September, August, and October while April had the smallest increase for both climate change scenarios as illustrated in Figure 5. The increase in ET_0 aligned with the increasing trend in temperature, implying that a change in temperature would have a significant impact on the amount of soil evaporation and crop transpiration.

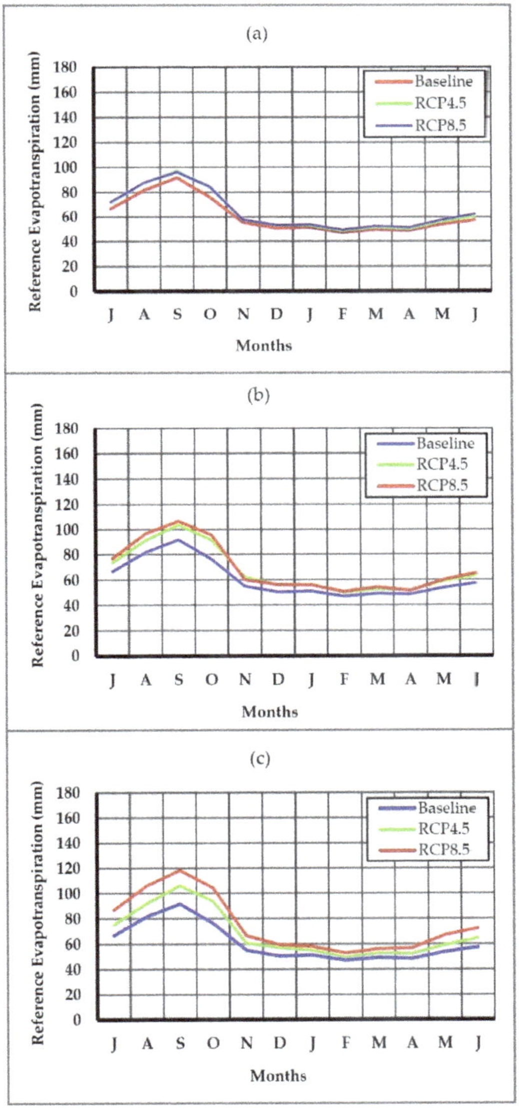

Figure 5. Predicted changes in the reference evapotranspiration for RCP 4.5 and RCP 8.5 relative to baseline climate (1976–2006) for the period (**a**) 2010–2039, (**b**) 2040–2069, and (**c**) 2070–2099.

3.4. Climate Change Impact on Crop Water Requirement

This study assessed the impacts of climate change on crop water requirements by integrating climate parameters into the WEAP-MABIA crop model. Table 4 shows the projected changes in crop water requirements for RCP 4.5 and RCP 8.5 climate change scenarios. Results revealed a steady increase in crop water requirements for both RCPs. Water requirements for maize crops increased from 7.6×10^8 m^3 for the baseline climate to 9.1×10^8 m^3 and 1.0×10^9 m^3, while soya beans increased from 2.2×10^8 m^3 to 3.1×10^8 m^3 and 3.6×10^8 m^3 for RCP 4.5 and RCP 8.5 climate change scenarios. An increase of 1.4×10^8 m^3 and 7.3×10^7 m^3, 1.6×10^8 m^3, and 9.1×10^7 m^3 was anticipated for dry beans and sunflower for both RCPs relative to 9.3×10^7 m^3 and 4.3×10^7 m^3 of baseline climate. Comparing both RCPs, a slight decrease in water requirements for maize and soya beans was expected under RCP 8.5 in the early term, with the highest water requirements for all crops expected towards the end of the century. The increased water requirements for all crops modeled in the WEAP-MABIA crop model is due to the high evapotranspiration rate resulting from an increased temperature and a decline in precipitation. Water stress can only occur when rainfall or irrigation do not meet crop water requirements. Therefore, the simulated results for crop water requirements show that crops in the Olifants catchment will likely face severe water stress in the future which may lead to a decline in crop yield.

Table 4. Future changes in crop water requirements (in m^3) for RCP 4.5 and RCP 8.5 climate change scenarios relative to baseline climate.

	Baseline		RCP 4.5			RCP 8.5	
	1976–2005	2010–2039	2040–2069	2070–2099	2010–2039	2040–2069	2070–2099
Crops	**CWR**	**CWR**	**CWR**	**CWR**	**CWR**	**CWR**	**CWR**
Maize	7.6×10^8	8.7×10^8	8.9×10^8	9.1×10^8	8.6×10^8	9.3×10^8	1.0×10^9
Soya beans	2.2×10^8	2.6×10^8	3.0×10^8	3.1×10^8	2.5×10^8	3.0×10^8	3.6×10^8
Dry beans	9.3×10^7	1.1×10^8	1.3×10^8	1.4×10^8	1.2×10^8	1.4×10^8	1.6×10^8
Sunflower	4.3×10^7	5.7×10^7	7.0×10^7	7.3×10^7	6.0×10^7	7.8×10^7	9.1×10^7

3.5. Influence of Soil Texture on Crop Yield

We analyzed the effect of soil texture on crop yield under the current climatic conditions (1976–2005). Simulated results of crop yield (maize soya beans, dry beans, and sunflower) were higher in sandy loam and sandy clay loam, compared to loam sand, which is ascribed to high water retention in sandy loam and sandy clay loam soil. The grain yield in sandy loam and sandy clay loam was 5.6 t/ha and 4.5 t/ha respectively, as presented in Figure 6. Similarly, a study by Jalota et al. [46] found an increased grain yield of maize and wheat in sandy loam soil which the study attributed to the high water holding capacity of the soil. In our current study, it is evident that soil with low water retention capacity will have larger percolation loss which would increase the amount of water required for crop production. Adapting to this issue will require farmers to irrigate more often, particularly during periods of limited rainfall. However, this situation can be rectified through the application of organic matter to increase the density of the soil as suggested by Jalota et al. [46].

3.6. Crop Yield Analysis

This study analyzed the impact of climate change on crop yield in the Olifants catchment. The WEAP-MABIA model was used to simulate crop yield under RCP 4.5 and RCP 8.5 emission scenarios. Results presented in Table 5 show the changes in crop yield for both RCPs climate change scenarios relative to the 30 years baseline average. Based on the analysis, an average annual decrease in crop yield was expected for both projected climate change scenarios (RCP 4.5 and RCP 8.5) for the early, mid, and end-term periods with RCP

8.5 showing the highest rate of decrease towards the end of the century. A decrease of 19 to 40%, 12 to 25%, 19 to 32%, and 5 to 20% was anticipated for maize, soya beans, dry beans, and sunflower under the RCP 4.5 climate change scenario. For the RCP 8.5 scenario, maize and soya beans were expected to decline by 20 to 65% and 11 to 38% respectively, while a decrease of 16 to 42% and 10 to 30% was expected for dry beans and sunflower. A slight increase in the yield of soya beans in the early term for RCP 8.5 was anticipated. The general decrease in crop yield in the Olifants catchment is attributed to the decreasing trend in precipitation coupled with an increase in temperature. The anticipated decline in crop yield in the catchment suggests the need for the development and implementation of plausible adaptation measures.

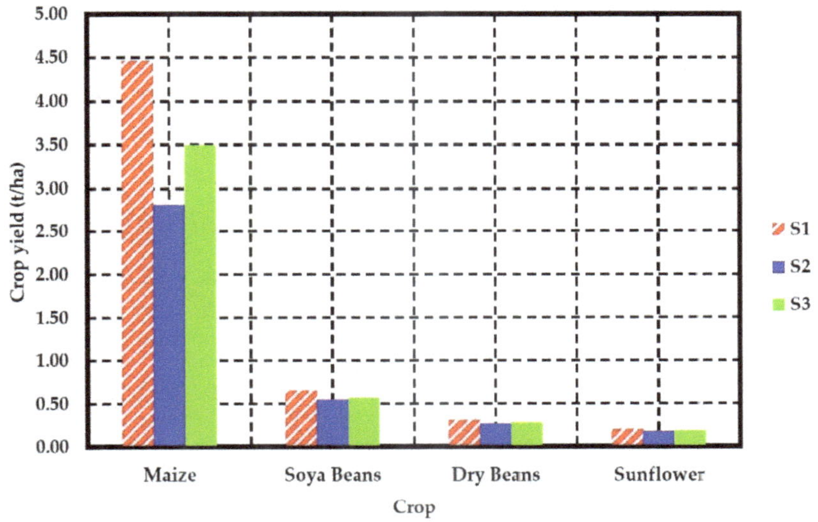

Figure 6. Influence of soil texture on crop yield, a plot showing crop yield response to different soil water holding capacity resulting from soil characteristics.

Table 5. Changes in crop yield (in t/ha) for RCP 4.5 and RCP 8.5 emission scenarios relative to the average 30 years historical baseline (1976–2005).

Crops Yield (t/ha)	1976–2005	2010–2039		2040–2069		2070–2099	
	Mean	Mean	% Change	Mean	% Change	Mean	% Change
Maize	4.46	3.57	−19	2.93	−34	2.69	−40
		3.55	−20	2.54	−43	1.65	−63
Soya beans	0.65	0.57	−12	0.51	−22	0.49	−25
		0.58	−11	0.47	−28	0.40	−38
Dry beans	0.31	0.25	−19	0.23	−26	0.21	−32
		0.26	−16	0.21	−32	0.18	−42
Sunflower	0.20	0.19	−5	0.17	−15	0.16	−20
		0.18	−10	0.16	−20	0.14	−30

3.7. Evaluation of Adaptation Strategies

Considering the negative effect of climate change on crop yield arising from increased temperature and decreased precipitation, this study evaluated the capabilities of adaptation strategies to improve current and future crop yield. Four agricultural management strategies were simulated in the WEAP-MABIA crop model. These strategies included a combination of changing planting date with the application of full irrigation, use of rainwater harvesting, application of the deficit irrigation method, and the use of efficient irrigation devices. Table 6 presents changes in crop yield with the adoption of changing planting date and full irrigation technique for both baseline and projected climate (RCP 4.5 and RCP 8.5) scenarios for the period 2010–2039, 2040–2069, and 2070–2099 respectively. Results revealed that maize and soya bean yields will increase by 39 to 270% and by 52 to 138% while dry bean and sunflower was expected to increase by 45 to 144% and 15 to 57% respectively.

Table 6. Changes in crop yield with the adoption of a change in planting date along with full irrigation technique for the baseline and projected climate scenarios for the periods 1976–2005, 2010–2039, 2040–2069, and 2070–2099.

			RCP 4.5						RCP 8.5					
	1976–2005		2010–2039		2040–2069		2070–2099		2010–2039		2040–2069		2070–2099	
Crops	**Y(t/h)**	**%**	**Y(t/h)**	**%**	**Y(t/h)**	**%**	**Y(t/h)**	**%**	**Y(t/h)**	**%**	**Y(t/h)**	**%**	**Y(t/h)**	**%**
Maize	6.18	39	6.13	72	6.12	108	6.13	128	6.13	72	6.13	141	6.10	270
Soya beans	1.00	52	0.99	74	0.98	92	0.98	100	0.99	71	0.98	104	0.95	138
Dry beans	0.45	45	0.44	69	0.44	91	0.44	100	0.45	73	0.44	110	0.44	144
Sunflower	0.23	15	0.23	21	0.22	29	0.22	38	0.23	28	0.22	38	0.22	57

For the application of rainwater harvest, results from the analysis revealed an increase in the mean potential yield ranging between 14 to 21% and 5 to 8% for maize and soya beans respectively. For dry beans, an increase of between 4 to 10% was expected while an increase of 5 to 13% was expected for sunflower yield, as presented in Table 7. The analysis of the results also revealed that the ability of RWH to improve crop yield in the Olifants catchment was expected to continue toward the mid and end-term for both RCPs.

Table 7. Changes in crop yield with the adoption of rainwater harvesting for the baseline and projected climate scenarios for the periods 1976–2005, 2010–2039, 2040–2069, and 2070–2099.

			RCP 4.5						RCP 8.5					
	1976–2005		2010–2039		2040–2069		2070–2099		2010–2039		2040–2069		2070–2099	
Crops	**Y(t/h)**	**%**	**Y(t/h)**	**%**	**Y(t/h)**	**%**	**Y(t/h)**	**%**	**Y(t/h)**	**%**	**Y(t/h)**	**%**	**Y(t/h)**	**%**
Maize	5.10	14	4.17	17	3.43	17	3.16	17	4.14	16	3.06	20	2.00	21
Soya beans	0.69	5	0.60	5	0.55	8	0.52	6	0.61	5	0.51	6	0.43	8
Dry beans	0.33	6	0.27	4	0.24	4	0.23	5	0.28	8	0.23	10	0.19	6
Sunflower	0.21	5	0.20	5	0.18	6	0.18	13	0.19	6	0.17	6	0.15	7

With regards to the application of the deficit irrigation method as an adaptive strategy. The results illustrated in Table 8 show that the yield of maize was expected to decrease by 1 to 3% for both the baseline and projected climate scenarios. While an increase of 2 to 4% was anticipated for the soya beans yield. For dry beans, a different situation was observed as there were no changes in the yield for both baseline and projected climate scenarios with the exception of RCP8.5 which showed an increase of 4 and 5% for the early term (2010–2039) and mid-term (2040–2069) respectively. Similar to dry beans, the yield of sunflower remains unchanged.

Table 8. Changes in crop yield with the adoption of the deficit irrigation method for the baseline and projected climate scenarios for the periods 1976–2005, 2010–2039, 2040–2069, and 2070–2099.

					RCP 4.5						RCP 8.5			
	1976–2005		2010–2039		2040–2069		2070–2099		2010–2039		2040–2069		2070–2099	
Crops	**Y(t/h)**	**%**	**Y(t/h)**	**%**	**Y(t/h)**	**%**	**Y(t/h)**	**%**	**Y(t/h)**	**%**	**Y(t/h)**	**%**	**Y(t/h)**	**%**
Maize	4.37	−2	3.48	−3	2.85	−3	2.63	−2	3.46	−3	2.48	−2	1.63	−1
Soya beans	0.67	2	0.58	2	0.53	4	0.50	2	0.60	3	0.49	2	0.41	3
Dry beans	0.31	0	0.26	0	0.23	0	0.22	0	0.27	4	0.22	5	0.18	0
Sunflower	0.20	0	0.19	0	0.17	0	0.16	0	0.18	0	0.16	0	0.14	0

Under the application of the efficient irrigation device, results demonstrated in Table 9 show an increase in maize yield for both baseline and projected climate, while there were no changes in the yield of soya beans, dry beans, and sunflower.

Table 9. Changes in crop yield with the adoption of an irrigation efficiency device for the baseline and projected climate scenarios for the periods 1976–2005, 2010–2039, 2040–2069, and 2070–2099.

					RCP 4.5						RCP 8.5			
	1976–2005		2010–2039		2040–2069		2070–2099		2010–2039		2040–2069		2070–2099	
Crops	**Y(t/h)**	**%**	**Y(t/h)**	**%**	**Y(t/h)**	**%**	**Y(t/h)**	**%**	**Y(t/h)**	**%**	**Y(t/h)**	**%**	**Y(t/h)**	**%**
Maize	4.47	0.2	3.59	0.5	2.94	0.3	2.70	0.4	3.57	0.3	2.55	0.4	1.66	0.6
Soya beans	0.66	0	0.57	0	0.51	0	0.49	0	0.58	0	0.48	0	0.40	0
Dry beans	0.31	0	0.26	0	0.23	0	0.22	0	0.26	0	0.21	0	0.18	0
Sunflower	0.20	0	0.19	0	0.17	0	0.16	0	0.18	0	0.16	0	0.14	0

The evaluation of adaptation measures in this study show that crops respond to adaptive measures differently. It is, however, important to note that not all adaptive strategies can improve crop yields. Among the different adaptation strategies evaluated, the combination of changing the planting date with full irrigation technique and the application of rainwater harvesting were found to be the most suitable measures for all crops studied.

4. Discussion

There is great uncertainty about the future effects of climate change on crop production. This analysis explores the potential implications of climate change on catchment crop yield based on CORDEX-RCM driven by six GCMs for two IPCC emission scenarios (RCP 4.5 and RCP 8.5). Furthermore, four possible adaptation strategies to cope with climate change impact in the Olifants catchment were considered: the combination of changing sowing date and application of full irrigation, application of rainwater harvesting, adoption of the deficit irrigation method, and the use of efficient irrigation devices.

Our analysis revealed an increased warming trend of between 1 °C to 5 °C for both climate change scenarios (RCP 4.5 and RCP 8.5) relative to the baseline climate of 1976–2005. This analysis aligns with the predictions of Durand [8]. On the other hand, average monthly precipitation is expected to decrease in the future for both climate change scenarios with the exception of the month of August and September for RCP 4.5 during the early term. This finding aligns with Kusangaya et al. [48] who also predicted decreased precipitation over Southern Africa. The changes in precipitation and temperature, particularly in the summer months when crops are being planted, have caused significant declines in crop yield. Such decreases are mostly attributed to increased temperature [17,20,21,49]. Contrary to this finding, a study by Cazadilla et al. [9] found an increase in crop production for South Africa under the MIROC AIB scenario. However, this contradiction may be due to the uncertainty associated with global climate models.

Considering the negative impacts of climate change on crop production in this study, we evaluated different adaptation strategies to improve crop production. Among the adaptation strategies assessed, the integration of changing sowing date with full irrigation application had the highest crop yield under both current and projected climate change scenarios as compared to other adaptation strategies evaluated. The adoption of changing sowing date alone might not be an effective measure to cope with climate change considering the significant decline in the catchment precipitation. However, few studies [50,51] have observed an increase in crop yield with the adoption of this strategy.

Combining changing sowing date with the application of full irrigation is seen to be an effective measure to cope with the long-term impact of climate change, but the application of full irrigation might prove to be challenging, as the Olifants catchment is already experiencing water stress arising from increased demands. The adoption of rainwater harvesting is therefore seen as an effective measure towards resolving this challenge as it involves harvesting rainwater from runoffs during periods of heavy rainfall. This approach would thus reduce the over-exploitation of fresh water from the system during irrigation. Our analysis has shown that the application of rainwater harvesting would improve the yield of crops under current and projected climate change which is consistent with the findings of Lebel et al. [52] and Rasuiba [53]. Contrary to the findings of the study conducted by Chimonyo et al. [54], where the application of the deficit irrigation method improved the yield of sorghum and cowpea plant, our experiments provided a different result. However, our findings aligned with the findings of a study by Shrestha L and Shrestha N [55] who also reported a decline in winter wheat yield with the adoption of the deficit irrigation approach. The application of an efficient irrigation device improved the yield of maize crops while the yield of other studied crops remained unchanged under current and projected climate change.

Findings from this study have shown that not all adaptive measures are capable of improving crop yield under the impact of climate change. It is therefore important to evaluate crop response to different adaptation measures before implementation in order to determine the most suitable and appropriate strategies to be adopted.

5. Conclusions

The analysis of climate change impacts has shown that crop yield may be declined by as much as 65% by the end of the century in the Olifants catchment of South Africa. Yet, studies have also suggested that much of the yield loss can be mitigated using adaptation measures. In this paper, we used an ensemble of six biased corrected GCMs downscaled with one regional climate model to assess crop yield response to projected climate change (RCP 4.5 and RCP 8.5) for the Olifants catchment. The WEAP-MABIA processed-based crop model was used for yield prediction and to investigate the effect of adaptation strategies. The findings from the study revealed that surface temperature in the Olifants catchment will increase in the future while precipitation, on the other hand, is expected to decrease, which will consequently decline crop production. The analysis also revealed that soil with high water holding capacity tends to retain more water for crop use and thus is able to improve crop yield under limited rainfall.

Based on the adaptation strategies evaluated to cope with the impacts of climate change, the combination of changing sowing date with full irrigation application as well as the adoption of rainwater harvesting resulted in a significant positive yield change.

Finally, the effect of climate change on crop yield is considerable and poses serious threats not just to farmers but also to regional food security, especially given the rapidly growing population of South Africa which necessitates the production of more food. Ultimately, the solution to climate change lies in the effective deployment of adaptive strategies that could mitigate the impacts of climate change. This study thus provides actionable knowledge and insights that could be used to avoid yield losses in the future. Adopting the measures proposed in this study are also well within the ability of policy-makers and the majority of the smallholder farmers. The implications of the analysis and findings of this study are to pave the way towards a more proactive agricultural management planning with regards to climate change and its impending impacts on food security in the region.

Author Contributions: Conceptualization, M.F.O. and T.N.: Methodology, M.F.C.; Software, M.F.O.; Validation, M.F.O., T.N. and N.D.; Formal Analysis, M.F.O.; Investigation, M.F.O., Resources, M.F.O.; Data Curation, T.N. and M.F.O.; Writing—Original Draft Preparation, M.F.O.; Writing—Review & Editing, T.N. and N.D.; Visualization, M.F.O.; Supervision, T.N. and N.D.; Funding Acquisition, M.F.O. All authors have read and agreed to the published version of the manuscript.

Funding: This research was supported by the Nation Research foundation-The World Academy of Science (NRF-TWAS) Grant No. 110823.

Institutional Review Board Statement: The study was conducted according to the guidelines of the Declaration of Mary Olabanji, and approved by the Institutional Review Ethics Committee of the University of Pretoria (protocol code NAS133/2019 and approval date 21-08-2019).

Informed Consent Statement: Informed consent was obtained from all subjects involved in the study.

Data Availability Statement: Data available in a publicly accessible repository that does not issue DOIs. Publicly available datasets were analyzed in this study. This data can be found here: http://cordexesg.dmi.dk/esgf-web-fe/.

Acknowledgments: The authors wish to express their gratitude to the South African weather service for the provision of observed climate data used in this study.

Conflicts of Interest: Authors declare no conflict of interest.

References

1. Intergovernmental Panel on Climate Change (IPCC). *Climate Change 2014: Synthesis Report. Contribution of Working Groups I, II and III to the Fifth Assessment Report of the Intergovernmental Panel on Climate Change*; IPCC: Geneva, Switzerland, 2014.
2. Shrestha, S.; Thin, N.M.M.; Deb, P. Assessment of Climate Change Impacts on Irrigation Water Requirement and Rice Yield for Ngamoeyeik Irrigation Project in Myanmar. *J. Water Clim. Chang.* **2014**, *5*, 427–442. [CrossRef]
3. Malla, G. Climate change and its impact on Nepalese agriculture. *J. Agric. Environ.* **2008**, *9*, 62–71. [CrossRef]
4. Bhatt, D.; Maskey, S.; Babel, M.S.; Uhlenbrook, S.; Prasad, K.C. Climate Trends and Impacts on Crop Production in the Koshi River Basin of Nepal. *Reg. Environ. Chang.* **2014**, *14*, 1291–1301. [CrossRef]
5. Nana, E.; Corbari, C.; Bocchiola, D. A Model for Crop Yield and Water Footprint Assessment: Study of Maize in the Po Valley. *Agric. Syst.* **2014**, *127*, 139–149. [CrossRef]
6. Bouman, B. How much water does rice use. *Management* **2009**, *69*, 115–133.
7. Kang, Y.; Khan, S.; Ma, X. Climate Change Impacts on Crop Yield, Crop Water Productivity and Food Security—A Review. *Prog. Nat. Sci.* **2009**, *19*, 1665–1674. [CrossRef]
8. Durand, W. *Assessing the Impact of Climate Change on Crop Water Use in South Africa*; CEEPA DP28; University of Pretoria: Pretoria, South Africa, 2006.
9. Calzadilla, A.; Zhu, T.; Rehdanz, K.; Tol, R.S.; Ringler, C. Climate change and agriculture: Impacts and adaptation options in South Africa. *Water Resour. Econ.* **2014**, *1*, 24–48. [CrossRef]
10. Source OECD. *OECD Review of Agricultural Policies—South Africa*; OECD: Paris, France, 2006. [CrossRef]
11. Adisa, O.M.; Botai, C.M.; Botai, J.O.; Hassen, A.; Darkey, D.; Tesfamariam, E.; Adisa, A.F.; Adeola, A.M.; Ncongwane, K.P. Analysis of Agro-Climatic Parameters and Their Influence on Maize Production in South Africa. *Theor. Appl. Climatol.* **2018**, *134*, 991–1004. [CrossRef]
12. Muroyiwa, B.; Mushunje, A. Price discovery in the South African white and yellow maize futures market. *Int. J. Sci. Res.* **2017**, *73*, 73. [CrossRef]
13. Botai, C.M.; Botai, J.O.; Dlamini, L.C.; Zwane, N.S.; Phaduli, E. Characteristics of Droughts in South Africa: A Case Study of Free State and North West Provinces. *Water* **2016**, *8*, 439. [CrossRef]
14. Erasmus, B.; Van Zyl, J. The Effects of Climate Change on the Farm Sector in the Western Cape. *Agrekon* **2000**, *39*, 559–573. [CrossRef]
15. Intergovernmental Panel on Climate Change (IPCC). *Climate Change 2007: The Physical Science Basis*; Cambridge University Press: Cambridge, UK, 2007.
16. Lobell, D.B.; Burke, M.B.; Tebaldi, C.; Mastrandrea, M.D.; Falcon, W.P.; Naylor, R.L. Prioritizing Climate Change Adaptation Needs for Food Security in 2030. *Science* **2008**, *319*, 607–610. [CrossRef] [PubMed]

17. Mangani, R.; Tesfamariam, E.H.; Engelbrecht, C.J.; Bellocchi, G.; Hassen, A.; Mangani, T. Potential Impacts of Extreme Weather Events in Main Maize (*Zea Mays* L.) Producing Areas of South Africa Under Rain-fed Conditions. *Reg. Environ. Chang.* **2019**, *19*, 1441–1452. [CrossRef]
18. Thornton, P.K.; Jones, P.G.; Ericksen, P.J.; Challinor, A.J. Agriculture and Food Systems in Sub-Saharan Africa in a 4 °C + World. Philosophical Transactions: Mathematical. *Phys. Eng. Sci.* **2011**, *369*, 117–136.
19. Du Toit, A.S.; Prinsloo, M.A.; Durand, W.; Kiker, G. *Report on Vulnerability of Maize Production to Climate Change and Adaptation Assessment in South Africa*; Agricultural Research Council, Grain Crops Institute: Potchefstroom, South Africa, 2001.
20. Schulze, R.E.; Kiker, G.A.; Kunz, R.P. Global Climate Change and Agricultural Productivity in Southern Africa. *Glob. Environ. Chang.* **1993**, *3*, 330–349. [CrossRef]
21. Akpalu, W.; Hassan, R.M.; Ringler, C. Climate variability and maize yield in South Africa. *Environ. Prod. Technol. Div. IFPRI Pap.* **2008**.
22. Matji, O. The Impact of Climate Change on Agricultural Crop Distribution in South Africa. Master's Thesis, University of Witwatersrand, Johannesburg, South Africa, 2015.
23. Mayowa, A.O. The Use of Satellite-Derived Data and Neural-Network Analysis to Examine Variation in Maize Yield Under Changing Climate. Ph.D. Thesis, University of Pretoria, Pretoria, South Africa, 2019.
24. Gbetibouo, G.A.; Hassan, R.M. Measuring the Economic Impact of Climate Change on Major South African Field Crops: A Ricardian Approach. *Glob. Planet. Chang.* **2005**, *47*, 143–152. [CrossRef]
25. Deressa, T.; Hassan, R.; Poonyth, D. Measuring the Impact of Climate Change on South African Agriculture: The Case of Sugarcane Growing Regions. *Agrekon* **2005**, *44*, 524–542. [CrossRef]
26. Magagula, T.F.; van Koppen, B.; Sally, H. Water access and poverty in the Olifants Basin: A spatial analysis of population distribution, poverty prevalence and trends. In Proceedings of the 7th WaterNet/WARFSA/GWP Annual Symposium, Lilongwe, Malawi, 1–3 November 2006.
27. Nkhonjera, G.K.; Dinka, M.O. Significance of Direct and Indirect Impacts of Climate Change on Groundwater Resources in the Olifants River Basin: A Review. *Glob. Planet. Chang.* **2017**, *158*, 72–82. [CrossRef]
28. Richard, T.; Michael, S. *Olifants River Irrigation Schemes: 1. Crop and Irrigation Data for Four Separate Irrigation Schemes: 2. Irrigation Management Structures for Four Separate Irrigation Schemes*; South Africa working paper no. 5; International Water Management Institute (IWMI): Colombo, Sri Lanka, 2000.
29. Olabanji, M.F.; Ndarana, T.; Davis, N.; Archer, E. Climate Change Impact on Water Availability in the Olifants Catchment (south Africa) with Potential Adaptation Strategies. *Phys. Chem. Earth* **2020**, *120*, 102939. [CrossRef]
30. McCartney, M.P.; Yawson, D.K.; Magagula, T.F.; Seshoka, J. *Hydrology and Water Resources Development in the Olifants River Catchment*; Working Paper 76; International Water Management Institute (IWMI): Colombo, Sri Lanka, 2004.
31. Ly, S.; Degre, A.; Charles, C. Geostatistical Interpolation of Daily Rainfall at Catchment Scale: The Use of Several Variogram Models in the Ourthe and Ambleve Catchments, Belgium. *Hydrol. Earth Syst. Sci.* **2011**, *15*, 2259–2274. [CrossRef]
32. Amin, A.; Iqbal, J.; Asghar, A.; Ribbe, L. Analysis of Current and Future Water Demands in the Upper Indus Basin Under Ipcc Climate and Socio-Economic Scenarios Using a Hydro-Economic WEAP Model. *Water* **2018**, *10*, 537. [CrossRef]
33. Riahi, K.; Grübler, A.; Nakicenovic, N. Scenarios of Long-Term Socio-Economic and Environmental Development under Climate Stabilization. *Technol. Forecast. Soc. Chang.* **2007**, *74*, 887–935. [CrossRef]
34. Shrestha, M. *Data Analysis Relied on Linear Scaling Bias Correction (V.1.0), Microsoft Excel File*; ResearchGate GmbH: Berlin, Germany, 2015. [CrossRef]
35. Jabloun, M.; Sahli, A. *WEAP-MABIA Tutorial: A Collection of Stand-Alone Chapters to Aid in Learning the WEAP-MABIA Module*; Federal Institute for Geosciences and Natural Resources: Hannover, Germany, 2012.
36. Sieber, J.; Purkey, D. *WEAP: Water Evaluation and Planning System. User Guide*; Stockholm Environment Institute, US Center: Somerville, MA, USA, 2011; Available online: http://weap21.org/downloads/WEAP_User_Guide.pdf (accessed on 22 March 2020).
37. Esteve, P.; Varela-Ortega, C.; Blanco-Gutiérrez, I.; Downing, T.E. A hydro-economic model for the assessment of climate change impacts and adaptation in irrigated agriculture. *Ecol. Econ.* **2015**, *1*, 49–58. [CrossRef]
38. Rajendran, M.; Gunawardena, E.R.; Dayawansa, N.D. Runoff Prediction in an Ungauged Catchment of Upper Deduru Oya Basin, Sri Lanka: A Comparison of HEC-HMS and WEAP Models. *Int. J. Progress. Sci. Technol.* **2020**, *18*, 121–129.
39. Allani, M.; Mezzi, R.; Abdallah, W.; Gharbi, A.; Zouabi, A.; Hedhli, K.; Beji, R.; Jemli, A.; Joumade-Mansouri, F.; Affli, E.; et al. A contribution to an advisory plan for integrated irrigation water management at Nebhana dam system: From research to operational support. *EPiC Ser. Eng.* **2018**, *3*, 26–35.
40. Allen, R.G.; Pereira, L.S.; Raes, D.; Smith, M. Crop evapotranspiration-Guidelines for computing crop water requirements-FAO Irrigation and drainage paper 56. *FAO Rome* **1998**, *300*, D05109.
41. Smith, M.; Allen, R.G.; Monteith, J.L.; Pereira, L.S.; Perrier, A.; Pruitt, W.O. *Report on the Expert Consultation on Procedures for Revision of FAO Guidelines for Prediction of Crop Water Requirements*; Land and Water Development Division, United Nations Food and Agriculture Service: Rome, Italy, 1992.
42. Maponya, P.; Mpandeli, S. Climate change adaptation strategies used by Limpopo Province farmers in South Africa. *J. Agric. Sci.* **2012**, *4*, 39. [CrossRef]

43. Parry, M.L.; Rosenzweig, C.; Iglesias, A.; Livermore, M.; Fischer, G. Effects of Climate Change on Global Food Production Under Sres Emissions and Socio-Economic Scenarios. *Glob. Environ. Chang.* **2004**, *14*, 53–67. [CrossRef]
44. Mpandeli, S.; Nesamvuni, E.; Maponya, P. Adapting to the impacts of drought by smallholder farmers in Sekhukhune District in Limpopo Province, South Africa. *J. Agric. Sci.* **2015**, *7*, 115. [CrossRef]
45. Sani, S.; Chalchisa, T. Farmers' perception, impact and adaptation strategies to climate change among smallholder farmers in sub-Saharan Africa: A systematic review. *J. Resour. Dev. Manag.* **2016**, *26*, 1–8.
46. Jalota, S.K.; Singh, S.; Chahal, G.B.; Ray, S.S.; Panigraghy, S.; Singh, K.B. Soil texture, climate and management effects on plant growth, grain yield and water use by rainfed maize–wheat cropping system: Field and simulation study. *Agric. Water Manag.* **2010**, *97*, 83–90. [CrossRef]
47. Geerts, S.; Raes, D. Deficit Irrigation as an On-Farm Strategy to Maximize Crop Water Productivity in Dry Areas. *Agric. Water Manag.* **2009**, *96*, 1275–1284. [CrossRef]
48. Kusangaya, S.; Warburton, M.L.; van Garderen, E.A.; Jewitt, G.P.W. Impacts of Climate Change on Water Resources in Southern Africa: A Review. *Phys. Chem. Earth* **2014**, *67–69*, 47–54. [CrossRef]
49. Özdoğan, M. Modelling the impacts of climate change on wheat yields in North-western Turkey. *Agric. Ecosyst. Environ.* **2011**, *1*, 1–12. [CrossRef]
50. Liangzhi, J.Z.D.G.; Chunlin, S. Food Crop Production Strategies for Adaptation to Global Climate Change in the Sampling Zones of Eastern China A modelling Study. *J. Scientia Agric. Sinica.* **1998**, *31*, 51–58.
51. Sultana, H.; Ali, N.; Iqbal, M.M.; Khan, A.M. Vulnerability and Adaptability of Wheat Production in Different Climatic Zones of Pakistan under Climate Change Scenarios. *Clim. Chang.* **2009**, *94*, 123–142. [CrossRef]
52. Lebel, S.; Fleskens, L.; Forster, P.M.; Jackson, L.S.; Lorenz, S. Evaluation of in Situ Rainwater Harvesting as an Adaptation Strategy to Climate Change for Maize Production in Rain-fed Africa. *Water Resour. Manag.* **2015**, *29*, 4803–4816. [CrossRef]
53. Rasiuba, T. Water Budget Water Use Efficiency in Agriculture in Olifants Catchment. Master's Thesis, University of the Witwatersrand, Johannesburg, South Africa, 2007.
54. Chimonyo, V.G.P.; Mabhaudhi, T.; Modi, A.T. Assessment of Sorghum-Cowpea Intercrop System under Water-Limited Conditions Using a Decision Support Tool. *Water SA* **2016**, *42*, 316–327. [CrossRef]
55. Shrestha, L.; Shrestha, N.K. Assessment of Climate Change Impact on Crop Yield and Irrigation Water Requirement of Two Major Cereal Crops (rice and Wheat) in Bhaktapur District, Nepal. *J. Water Clim. Chang.* **2017**, *8*, 320–335. [CrossRef]

climate

MDPI

Article

Impacts of Agroclimatic Variability on Maize Production in the Setsoto Municipality in the Free State Province, South Africa

Abubakar Hadisu Bello [1,*], Mary Scholes [1] and Solomon W. Newete [1,2]

[1] School of Animal, Plant and Environmental Sciences, University of the Witwatersrand, Private Bag X3, Johannesburg 2050, South Africa; Mary.Scholes@wits.ac.za (M.S.); NeweteS@arc.agric.za (S.W.N.)

[2] Agricultural Research Council—Institute for Soil, Climate and Water 600, Belvedere Street, Private Bag X79, Arcadia, Pretoria 0001, South Africa

* Correspondence: 1364152@students.wits.ac.za; Tel.: +27-733457473

Received: 20 October 2020; Accepted: 7 December 2020; Published: 14 December 2020

Abstract: The majority of people in South Africa eat maize, which is grown as a rain-fed crop in the summer rainfall areas of the country, as their staple food. The country is usually food secure except in drought years, which are expected to increase in severity and frequency. This study investigated the impacts of rainfall and minimum and maximum temperatures on maize yield in the Setsoto municipality of the Free State province of South Africa from 1985 to 2016. The variation of the agroclimatic variables, including the Palmer stress diversity index (PSDI), was investigated over the growing period (Oct–Apr) which varied across the four target stations (Clocolan, Senekal, Marquard and Ficksburg). The highest coefficients of variance (CV) recorded for the minimum and maximum temperatures and rainfall were 16.2%, 6.2% and 29% during the growing period. Non-parametric Mann Kendal and Sen's slope estimator were used for the trend analysis. The result showed significant positive trends in minimum temperature across the stations except for Clocolan where a negative trend of 0.2 to 0.12 °C year^{-1} was observed. The maximum temperature increased significantly across all the stations by 0.04–0.05 °C year^{-1} during the growing period. The temperature effects were most noticeable in the months of November and February when leaf initiation and kernel filling occur, respectively. The changes in rainfall were significant only in Ficksburg in the month of January with a value of 2.34 mm year^{-1}. Nevertheless, the rainfall showed a strong positive correlation with yield (r 0.46, p = < 0.05). The overall variation in maize production is explained by the contribution of the agroclimatic parameters; the minimum temperature (R^2 0.13–0.152), maximum temperature (R^2 0.214–0.432) and rainfall (R^2 0.17–0.473) for the growing period across the stations during the study period. The PSDI showed dry years and wet years but with most of the years recording close to normal rainfall. An increase in both the minimum and maximum temperatures over time will have a negative impact on crop yield.

Keywords: agroclimatic variability; minimum and maximum temperatures; maize yield; rainfall patterns; Setsoto municipality; climate change; Free State province

1. Introduction

There is a global consensus that climate change trends are real, and a rapidly advancing threat to millions of livelihoods, by affecting agricultural activities, food security, water resources, health, social systems and the appropriate functioning of ecosystems Barros, Field [1]. Some studies forecast that the necessary increase in food production needs to be between 70 and 210% by 2050 and 2100, to ensure global food security [2,3]. Temperature and rainfall are very important factors that affect crop production [4], mainly affecting the duration of the growing season [5]. The relationship between temperature and rainfall is very variable across the globe [6], this finding is also true for South Africa, but the model projections for the next 20–50 years show that the eastern portion of the country will receive approximately the same rainfall with the western parts becoming significantly drier [7]. The relationship between temperature and rainfall is in most cases an inverse relationship; thus, the higher the temperature the lower the rainfall [8,9]. The study by Dasgupta, Morton [10] indicated that the mean global temperature has increased by 0.5 °C per annum. This rising temperature trend suggests that there is an increase in warm indices (hot days, hottest days) and a decrease in extreme cold indices (cold days, cold nights) [9]. Studies across the world show that minimum temperatures are increasing at a faster rate than the maximum temperatures which may be as a result of global warming [11–13].

Global warming affects climate change and increases the occurrence of extreme weather events including flooding and droughts [14]. The surface air temperatures in some areas of Africa have shown a steady increase of 0.03 °C annually [15]. The South African average air temperature has increased by 1.2 °C since the 1960s and the warming rate has increased at twice the global average rate [16,17]. Thus, understanding the underlying factors that influence the climatic change of the region could improve forecasting and limit the negative impacts in the region (Richard et al., 2001).

Agricultural production is susceptible to climate change variability in the Sub-Saharan region. Higher temperatures can decrease crop yields and animal production [18]. According to Scholes et al. (2015) for each one-degree Celsius rise in temperature, there is a 5% decrease in crop yield. Temperatures raised above optimal levels create biochemical challenges for plant cells, more especially the enzymes associated with the photosynthetic pathway. The southern and northern parts of Africa are expected to be about 4 to 6 °C hotter by 2080 and the precipitation is projected to decrease by 10–20% by this period (Collier et al., 2008). Derived variables, e.g., Palmer Stress Diversity index (PDSI), are used across the globe for monitoring meteorological drought as well as agricultural drought [19,20]. The meteorological component deals with changes in rainfall, whilst the agricultural drought component indicates changes in soil moisture. In this research, the self-calibrating PDSI (Sc-PDSI) proposed by Wells [21] was used as an indicator of agricultural drought, since we are interested in the soil moisture and potential evaporation without focusing on the impact of agricultural practices, including fertilizer applications and improved seed and water conservation measures on the yield of maize [22,23].

Maize (*Zea mays* L.) is the most common staple crop grown in Sub-Saharan Africa (SSA) [24]. It is a dominant component in the diets of most households in the region. On average, a decreasing trend of 10–20% in maize yield has been projected by 2050 for the tropics as a result of climate change [25]. Maize grows better at low to medium (20–28 °C) temperatures, because that allows for maximum radiation interception and optimal growth [26].

South Africa is amongst the ten highest maize producing countries in the world [27]. It produces an average of 12 million tons per year; contributing approximately 2% of the world's maize production [28]. The Free State province alone produces over 35% of the maize in South Africa [29]. Overall, the environmental conditions and natural resources of the Free State are conducive for maize production, but there are concerns of looming agro-climatological hazards which may have a detrimental effect on production [30]. This is supported by Smale and Jayne [31] who found that the output of maize production varies yearly in South Africa mostly due to climate variability. Since only 1% of the cultivated area uses irrigation for maize production [32], there is a particularly high reliance on rainfall and thus vulnerability to changing rainfall patterns and amounts.

This study investigated the impacts of agroclimatic variability on maize production in the district of the Setsoto Municipality in the Free State province of South Africa from 1985 to 2016. Droughts and extreme events are becoming more frequent and the drought characteristics are not well understood, at this particular local scale. Temperature and rainfall patterns are usually presented over an annual cycle but this study focusses on this important region, at the time scale of the growing season, October to April. The spatial variability in the temperature and rainfall trends is high which could negatively impact the maize yields for this area which are relatively low when compared with other maize growing locations. This district may be very close to the threshold where maize can no longer be grown, and this will have a major impact on rural poverty and unemployment. Currently, all stations studied were suitable for maize production, but the interaction of increasing temperatures with evapotranspiration into the future will make some areas in the Free State province less suitable for maize production [33,34].

2. Materials and Methods

2.1. Study Area

The Setsoto Municipality is under the administrative district of Thabo Mofutsanyane in the Free State province (Figure 1). The seasonal rainfall usually starts in October and ends in April with more than 80% of the rainfall occurring from October to March [35,36]. The soil type is shallow, loamy soil with moderate water holding capacity [37]. Soil degradation and overgrazing are prominent environmental problems which have not received adequate research attention [38].

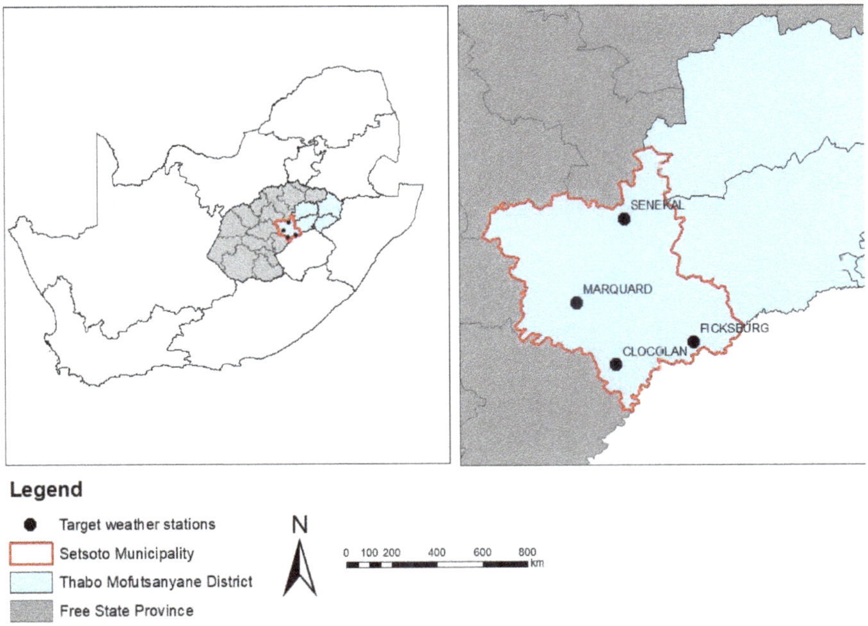

Figure 1. Map of the Free State province of South Africa showing the locations of the study area, the Setsoto Municipality and the target weather stations.

The availability of weather stations and completeness of data played important roles in the selection of the target stations within the municipality. This study targeted the weather stations in Clocolan, Marquard, Senekal and Ficksburg based on their spatial location and availability of data. There were five other stations nearby that were used for infilling missing data, these were selected based on the availability of data and proximity to the target stations (Table 1).

Table 1. List of weather stations used for this study with their longitudes, latitudes, elevation, duration of data availability and their data type. R denotes rainfall and T denotes both the minimum and maximum temperatures.

	Weather Stations	Latitude	Longitude	Elevation (m)	Data Type	Data Period (Years)
1	Senekal-AGR	−28.32200	27.6200	1433	R	40
2	Ficksburg	−28.82700	27.9040	1628	R&T	32
3	Marquard	−28.66500	27.4250	1497	R&T	40
4	Clocolan	−28.92108	27.5840	1602	R&T	36
5	Senekal-Driepan	−28.38900	27.5865	1587	R&T	31
6	Paul Roux	−28.29900	27.9480	1569	R	39
8	Lambertianin	−28.8200	27.5820	1646	R	32
7	Uintjieshoek	−28.5830	27.5200	1600	R	31

The average rainfall of Thabo Mofutsanyana is 600 mm per annum [39]. The province has the highest number of farming units in South Africa, with large areas of fertile and arable lands resulting in a significant proportion of the nation's agricultural production [40].

2.2. Data and Data Management

The daily maximum and minimum temperatures and the daily rainfall data of the study area for the period from 1985–2016 were obtained from the Agricultural Research Council (ARC) meteorological database and the South African Weather Service (SAWS). In this study, an agricultural year is defined from July to June of the following year. This allows the presentation of the growing period from October to April of the following year as a continuous record.

Meteorological data with the smallest number of missing data values (≤5%) were selected from stations within the municipality The UK method was used for the infilling of daily T_{max} and T_{min} values because of the technique's ability to accommodate the differences in altitude and its local effects. Missing rainfall data were estimated using the modified Inverse Distance Weighting method (IDWm), which allows for the influence of elevation on rainfall [41,42], missing rainfall, T_{min} and T_{max} values were less than 10% of the total data set, which satisfies the world meteorological organization (WMO) criteria for a robust climatic data analysis. Only stations with a complete data set having a duration of not less than 30 years were used for IDWm (Table 1).

Maize yield data (tons ha^{-1}) for the Setsoto Municipality for the period between 1985 and 2016 were obtained from the South African Department of Agriculture, Forestry and Fisheries [43] for the four areas except for Ficksberg where data were only available for 1985–2005. Most of the statistical analyses were computed using quantum XL 2016 and JASP 0.9.0.1 statistical software. Collection and availability of temperature, rainfall and yield data are very limited in South Africa due to the lack of infrastructure and compliance, this is a common problem especially in SSA. It would have been ideal if these data could have been used together with other variables e.g., measurements of evaporation and radiation but again these data are not collected by the South African Weather Service nor by the farmer's unions.

The self-calibrating PDSI (Sc-PDSI) was calculated using monthly temperature and precipitation. A detailed description of the fairly complex calculation of the Palmer index consisting of five steps is published in several journals [21,44–46]. The Sc-PDSI accounts for all the constants contained in the PDSI and includes a methodology in which the constants are calculated dynamically based upon the characteristics present at each station location. The self-calibrating nature of Sc-PDSI is developed for each station and changes based upon the climate regime of the location. It has wet and dry scales. The index was calculated for three decades as well as for the entire data set from 1985–2016. According to Palmer [44], the range of the monthly index time series is between −4 and +4. Negative (positive) PDSI values indicate dry (wet) periods, while those near-zero presume a state that is close to the average rainfall. The Palmer hydrological drought index (PHDI), is used to assess long-term moisture supply. The Sc-PDSI was calculated using a program developed by researchers in URL https://github.com/Sibada/scPDSI.

2.3. Climatic Trend Analysis

The non-parametric Mann Kendall (MK) test [47] was used to determine the significance of the climate trends, because the climatic data were not independent and normally distributed. The seasonal trends for T_{min}, T_{max} and Rainfall during the growing period with yield data were determined using a linear regression model. The free and open software package developed by the Finnish Meteorological Institute (MAKESENS) (https://en.ilmatieteenlaitos.fi/makesens) was used for the Mann Kendall (MK) test and Sen's slope estimator. The Sen's slope estimator allows for the significance of the trend to be analyzed. The MK test is robust, simple and frequently used in climate, environmental and hydrological studies [13,48–51]. The Sen's slope is a robust estimate of the underlying trend.

2.4. The Crop Yield Anomalies and Correlation with Climate Variables

The Pearson correlation coefficient which has proven to be an appropriate method for gaining insights into this type of study [52] was used to determine the relationship between maize yield and climatic variables. The data were detrended before performing linear regressions which prevents periodicity in the data. T_{min} and T_{max} anomalies and rainfall anomalies were correlated with detrended yield values to investigate the impacts of agroclimatic variables on maize production for the period of the study. Detrended yield values were used, for only the growing months (October–April), the coefficient of variance (CV) and standard deviation (SD) were calculated. The CV shows the variability of data around the mean of the population $CV = \mu/\sigma$ where: σ = standard deviation, μ = mean, the variability of the data is determined using CVs presented as a percentage. The standard diversion measures the dispersion of the dataset as relative to its mean. It is the square root of variance.

3. Results and Discussion

3.1. Variation in the Minimum and Maximum Temperatures during the Growing Period (October–April)

The average mean annual T_{min} of the area is presented in Table 2. The range of the average mean T_{min} was from 10.4 °C to 14.2 °C and for T_{max} was from 25.6 °C to 28.6 °C. The lowest T_{min} of 5.6 °C and the T_{max} of 10.0 °C were found in Ficksburg. The highest T_{min} and T_{max} recorded during the growing period in Clocolan were 16.4 °C and 31.2 °C, respectively. The CV of the T_{min} and T_{max} was between 5.8% to 16.0% and 3.8% to 8.3%, respectively (Table 2).

Table 2. The mean, minimum, maximum, SD and CV (%), for the minimum and maximum temperatures during the growing period (°C) in the Setsoto Municipality for the period between 1985 to 2016.

	T_{min} (°C)					T_{max} (°C)				
Stations	Mean	Min	Max	SD	CV	Mean	Min	Max	SD	CV
Marquard	11.6	10	13.4	0.7	6.2	27.1	24.7	29.5	1.1	4
Clocolan	14.2	9.9	16.4	2.1	14.9	28.6	24.1	31.2	2.4	8.3
Senekal	12.1	9.4	13.4	0.7	5.8	27.7	25.1	30.4	1.1	3.8
Ficksburg	10.4	5.6	13.9	1.7	16	11.6	10	13.4	0.7	6.2

3.2. The Growing Period Rainfall from 1985 to 2016

The average rainfall for the growing period in Setsoto ranged from 540.71 mm to 632.38 mm with CV ranging from 21 to 29% (Table 3). Ficksburg had the highest rainfall during the growing period (1154.10 mm) while Marquard had the lowest rainfall (204.1 mm). The patterns of rainfall variations of the growing period were similar between Senekal and Marquard and Clocolan and Ficksburg with only observed differences of about 3% between them. The rainfall of the growing period accounts for approximately 88% of the annual rainfall (Table 3).

Table 3. Rainfall (mm) during the growing period October–April (mean, minimum, maximum, standard deviation and coefficient of variation) in Setsoto Municipality (1985–2016).

Stations	Average Rainfall in Growing Period (mm)					Annual Rainfall (mm)				
	Mean	Min	Max	SD	CV	Mean	Min	Max	SD	CV
Marquard	540.7	204.1	969.5	158.5	29	613.4	259.1	1029.7	178.2	29
Clocolan	593.2	329.6	888.7	122.9	21	677.1	386.5	1074.9	149.4	22
Senekal	569.9	310	952	149.9	26	645	386.8	1019.2	167.4	26
Ficksburg	632.4	359.2	1154.1	151.4	24	718.1	397.6	1224.1	168.8	23

Mean annual rainfall over the Setsoto municipality ranged from 613 mm to 718 mm (Table 3). The summer months from October to April account for most of the annual rainfall in the municipality. The highest annual rainfall values observed were in Ficksburg with 1224 mm (Table 3). The lowest value ranged from 259 mm to 397 mm. Ficksburg had the highest mean annual rainfall (718.1 mm) followed by Clocolan (677.1 mm), while the lowest was recorded in Marquard (613.4 mm) followed by Senekal (645 mm) The CV of the annual rainfall was very high ranging from 34 to 45 (Table 3).

3.3. Maize Crop Production 1985—-2016

The average maize yield for Setsoto from 1985 to 2016 ranged from 1.96 tons ha^{-1} to 2.89 tons ha^{-1}. The highest maize yield achieved during this period was in 2016 with 6.18 tons ha^{-1} in Clocolan, while the lowest of 0.10 tons ha^{-1} was recorded in 1991 in Senekal (Figure 2). The maize yield CVs over this period was between 37.8% and 46.2% per annum, with a standard deviation of between 0.91 and 1.31 tons ha^{-1} across the municipality (Table 4).

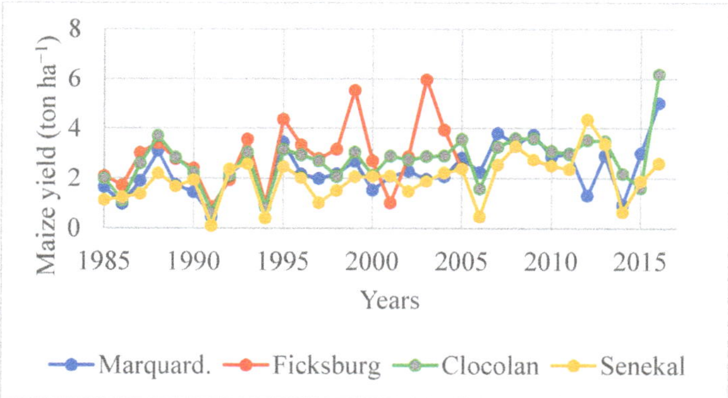

Figure 2. The annual maize yield (tons ha^{-1}) for the four stations in the Setsoto municipality from 1985 to 2016 used for this study.

Table 4. The average maize yield (tons ha^{-1}) for the four stations in Setsoto Municipality from 1985–2016 used for this study.

	Average Maize Yield (tons ha^{-1})				
Stations	Mean	Min	Max	SD	CV
Marquard	2.33	0.47	5	0.98	41.93
Clocolan	2.72	0.64	6.18	1.03	37.75
Senekal	1.96	0.1	4.34	0.91	46.19
Ficksburg	2.89	0.85	5.96	1.33	46.21

The dataset available for Ficksburg in this study was only for 20 years (1985–2005), as opposed to 32 years in the other three weather stations. Each station showed high inter-annual variation in yield. All seem to overlap at least in the first few years (1985 to 1995). The yield in Ficksburg showed the highest inter-annual variation between 1995 and 2005 (Figure 2).

3.4. *Climate Trend Analysis*

3.4.1. Minimum and Maximum Temperature Trends

The Clocolan monthly and the growing period minimum temperatures showed a negative trend at the 0.001 significance except in the months of November and April which showed a negative trend at a significance level of 0.05. The values of the Sen's slope were all less than zero (Table 5). In Senekal the T_{min} did not show any trend for the period of the study except for the month of January, where an increase of 0.02 °C year^{-1} was reported, compared to the increasing trend of 0.05 °C per annum shown in Ficksburg at a significance level of 0.05. In Marquard the T_{min} trend showed a positive trend for the months of October, November and December at the rates of 0.09, 0.09 and 0.06 °C increase year^{-1}, respectively during the growing period (0.01 significance level). The February, March, April and the growing period trends were negative with decreases of minimum temperatures of 0.1, 0.2, 0.25 and 0.05 °C year^{-1} (Table 5).

Table 5. Setsoto monthly growing period minimum temperature annual trends during the growing period from 1985–2016. Mann Kendall (MK) trend (Test Z) and Sen's slope estimate (Q).

Months	Marquard			Clocolan			Senekal			Ficksburg		
	Test z	Q	R^2	Test z	Q	R^2	Test z	Q	R^2	Test z	Q	R^2
OCT	2.72 **	0.09	0.25	−3.5 ***	−0.18	0.55	0.62	0.01	0	−0.05	0	0
NOV	2.64 **	0.09	0.25	−2.47 *	−0.15	0.35	−0.1	−0.01	0	−0.15	−0.01	0
DEC	3 **	0.06	0.36	−3.61 ***	−0.12	0.42	0.73	0.01	0.02	0.58	0.02	0
JAN	0.36	0.01	0	−3.51 ***	−0.12	0.43	1.64 *	0.02	0.1	2.3 *	0.05	0.1
FEB	−3.71 **	−0.1	0.34	−3.34 ***	−0.12	0.46	0.89	0.02	0.05	1.61	0.07	0.06
MAR	−3.91 **	−0.23	0.6	−3.57	−0.2	0.5	0.97	0.02	0.04	1.49	0.05	0.03
APR	−4.25 **	−0.25	0.61	−2.39	−0.15	0.35	−1.43	−0.04	0.06	0.84	0.03	0
GP	−3.52 **	−0.05	0.42	−3.7 ***	−0.14	0.42	1.39	0.01	0.01	1.51	0.03	0.03

NB: *** denotes significance when alpha = 0.001, ** denote significance when alpha = 0.01 and * denote significance when alpha = 0.05.

A commonly occurring pattern in climate change studies shows minimum temperatures to be increasing globally and more particularly in Sub-Saharan Africa [53]. The trends were very variable, all stations showing increases, except Clocolan which showed an overall decrease. The projected mid-altitude minimum temperature increases for subtropical Africa is 2.6 °C century^{-1} [54]. The data were very variable by the month and in Marquard, there is a significant increase in the trend of T_{min} in the months of October, November and December, likewise in January in Senekal and Ficksburg. These data are very difficult to explain. It is interesting to note that, T_{min} spatial–temporal variability is just outside the WMO 30 km radius used for justification of infilling of data. There are local factors such as vegetation cover, topography, slope and aspect of the area which affect the rainfall and temperature distribution. The IPCC (2014), states that provided the anthropogenic and greenhouse emissions remain at 2014 levels, these results fall within the projected century temperature increases of 3 °C, but only for extreme events [54].

In the months of October and November in Marquard, Senekal and Ficksburg the growing period T_{max} showed an increasing trend ranging from 0.04 to 0.10 °C year^{-1} at various levels of significance (Table 6). In Clocolan, T_{max} showed a decreasing trend in the months of March and April by 0.16 and 0.14 °C year^{-1} (0.05 significance level).

Table 6. Monthly Maximum Temperature (°C) annual trends during the growing period for the study period from 1985–2016. Mann Kendall MK Test Z denote Mann Kendall trend analysis test, and Q denotes 'the Sen's slope estimate' for the Setsoto municipality.

Months	Marquard			Clocolan			Senekal			Ficksburg		
	Test z	Q	R^2	Test z	Q	R^2	Test z	Q	R^2	Test z	Q	R^2
OCT	3.71 **	0.12	0.4	−0.68	−0.05	0.02	3.91 ***	0.12	0.38	3.02 **	0.11	0.26
NOV	1.9 +	0.08	0.11	−0.31	−0.03	0	2.38 *	0.08	0.15	2.09 *	0.1	0.14
DEC	0.76	0.04	0.04	0.13	0	0	0.44	0.02	0.03	1.52	0.06	0.12
JAN	0.26	0.01	0.01	−1.01	−0.04	0.02	−0.05	0	0.01	1.28	0.04	0.02
FEB	0.83	0.03	0.01	−1.1	−0.1	0.01	0.66	0.03	0.05	1.96 *	0.09	0.19
MAR	1.61	0.06	0.1	−2.01 *	−0.16	0.14	1.12	0.04	0.07	2.5 *	0.06	0.02
APR	1.1	0.05	0.03	−2.11 *	−0.14	0.06	0.7	0.04	0.01	−0.29	−0.01	0
GP	2.38 *	0.04	0.23	−1.23	−0.06	0.24	2.29 *	0.05	0.22	2.12 *	0.04	0.21

NB: + denote significance when alpha = 0.1, *** denote significance when alpha = 0.001, ** denote significance when alpha = 0.01 and * denote significance when alpha = 0.05.

The maximum temperatures over most of SSA are expected to increase above the global average [55]. The increasing trend of maximum temperature for Southern Africa is non-linear and its intensity is expected to increase drought and crop failure [14]. In this study, the maximum temperatures in the period between 1985 and 2016 showed an overall significant increase, during the maize growing period across the stations in the Setsoto municipality. The only months with significant decreases in T_{max} were March and April in Clocolan, while for the rest of the months either it remained unchanged or showed a significant increase (Table 6). The annual maximum temperatures increased by 0.08 °C year^{-1}, giving an increase of 2.56 °C for the entire study period of 32 years. These results also agree with the findings published by the IPCC (2014). The results also fall within the projected SSA temperature increases of 6.5 °C for the century [55–58].

3.4.2. Rainfall Trend Analysis

For all the stations used in this study only the month of January showed a positive trend of increasing rainfall in the Ficksburg station with 2.34 mm year^{-1} at a 0.05 significance level (Table 7). The rainfall trends for the study period of 32 years (from 1985 to 2016) in the Setsoto Municipality showed no significant changes. This statement applies to the seasonal distribution of the rainfall, the total amounts of rainfall and yearly distributions. The only significant data found were for the month of January in Ficksburg, where the rainfall significantly increased by 2.34 mm year^{-1} (Table 7). Rainfall in the Free State province shows high variability with the patterns, distribution, intensity and duration of rainfall varying spatially and temporally across different scales [59].

Table 7. Monthly rainfall (mm) and its annual trends during the growing period from 1985–2016 for the Setsoto municipality. MK Test Z denotes Mann Kendall trend analysis test, and Q denotes 'the Sen's slope estimate'.

Months	Marquard			Clocolan			Senekal			Ficksburg		
	Test z	Q	R^2	Test z	Q	R^2	Test z	Q	R^2	Test z	Q	R^2
OCT	−1.44	−1.09	−1.38	−1.38	−1.38	0.049	−1.36	−1.00	−1.31	−1.31	−0.88	0.014
NOV	0.63	0.69	−0.68	−0.68	−0.41	0.002	−0.05	−0.08	−0.26	−0.26	−0.39	0.039
DEC	0.00	0.00	0.97	0.97	1.12	0.031	0.19	0.33	0.65	0.65	0.52	0.039
JAN	−0.02	−0.02	1.12	1.12	1.62	0.042	1.62	2.11	2.06	2.06 *	2.34	0.179
FEB	0.73	0.60	0.44	0.44	0.43	0.011	−0.78	−0.59	−1.04	−1.04	−0.69	0.011
MAR	−0.94	−1.05	−0.99	−0.99	−0.64	0.037	−0.58	−0.56	−0.41	−0.41	−0.46	0.033
APR	−1.09	−0.74	−1.09	−1.09	−0.46	0.005	−0.10	−0.05	−0.44	−0.44	−0.45	0.134
GP	−1.36	−3.22	0.05	0.05	0.11	0.002	0.19	1.05	0.21	0.21	0.55	0.027

NB: * denote significance when alpha = 0.05.

3.5. Maize Yield Trends

Maize yield showed a positive trend in the three stations (Marquard, Clocolan and Senekal) increasing by different magnitudes. The maize yield in Marquard and Clocolan showed a positive trend increasing by 0.05- and 0.039-tons $ha^{-1}y^{-1}$, respectively. In Senekal, maize yield showed an increasing trend of 0.043 tons ha^{-1} (Table 8).

Table 8. Annual maize yield trends during the study period from 1985–2016. MK Test Z denotes the Mann Kendall trend analysis test, and Q denotes 'the Sen's slope estimate'.

	Test Z	Q	R^2
Marquard	2.76 **	0.050	0.218
Clocolan	2.45 **	0.039	0.196
Senekal	2.92 *	0.043	0.183
Ficksburg	1.27	0.054	0.119

NB: ** denote significance when alpha = 0.01 and * denote significance when alpha = 0.05.

Agroclimatic and maize yield variability in Sub-Saharan Africa (SSA) depends on the interactions between the combination of temperature, rainfall, and adaptive strategies [60]. The results from this study agree with other studies in SSA particularly with respect to temperatures and yield [61–66]. There were positive trends in all the stations for maize yield from 1985 to 2016 (Table 8). Marquard had the highest increasing trend of 0.05 tons ha^{-1} $year^{-1}$, followed by Senekal with 0.043 tons ha^{-1} $year^{-1}$ and Clocolan with 0.039 tons ha^{-1} $year^{-1}$. This general positive trend agrees with those found by [40] on a comparative analysis of maize yields for South Africa. The average maize yield for Setsoto during the period of this study was between 1.96 tons ha^{-1} to 2.89 tons ha^{-1} per year with an inter-annual variability between 38–46% (Table 4). Even though no agronomic data are available for these locations, it seems logical that some of these increases could have been accounted for by changed farming practices e.g., the addition of more inorganic fertilizers and changed maize varieties. The maize yield in the Setsoto municipality is below the free-state provincial average maize yield of 3.8 tons ha^{-1} [67] Maize production is said to be economically viable if 3.6 tons ha^{-1} is produced [40,67], the data from this study showed that maize yield is below this limit. The yield trends in this study were low and it is only marginally economical to produce maize in these areas. The contribution to GDP from farming in the Setsoto municipality is decreasing [68,69] and it has been suggested that some farms are no longer being planted with maize or alternate crops. Yield variability was high across the stations, with Senekal having the highest variability of 46.1% per year and it also recorded the lowest yield among the stations.

3.6. Maize Yield Correlation with Climatic Variables

3.6.1. De-trended Maize Yield Correlation with rainfall, T_{min} and T_{max} Anomalies

The Pearson correlation coefficient (r) and confidence interval levels of 0.05, 0.01 and 0.001 were used in this study to determine the relationship between yield and agroclimatic variables. Rainfall was positively correlated with yield during the growing period in Clocolan and with T_{max} in Senekal (r = 0.46 and 0.48 respectively) (p = 0.008 and 0.0005 respectively) (Table 9). In November, only the T_{min} in Marquard correlated with yield ($r = 0.39, p < 0.027$). During the month of January, the yield at this station is positively correlated with T_{min} (r = 0.37 and p = 0.038 at 0.05 confidence level).

Table 9. The correlation matrix for the monthly and growing period T_{min}, T_{max} and Rainfall variables with Maize yield in the three stations (the fourth station, Ficksburg, lacks sufficient data for analysis) of the Setsoto municipality from 1985–2016 (* $p < 0.05$, ** $p < 0.01$, *** $p < 0.001$). GP denotes growing period.

		Marquard			Clocolan			Senekal		
		Pearson's *r*	***p*-Value**	**VS-MPR⁺**	**Pearson's *r***	***p*-Value**	**VS-MPR⁺**	**Pearson's**	***p*-Value**	**VS-MPR⁺**
GP	T_{min}	0.03	0.87	1.00	0.12	0.52	1.00	0.02	0.90	1.00
	T_{max}	−0.18	0.33	1.01	0.12	0.51	1.00	**−0.40** *	0.02	4.43
	Rainfall	0.03	0.86	1.00	**0.46** **	0.01	8.97	0.23	0.20	1.14
OCT	T_{min}	0.07	0.69	1.00	−0.12	0.52	1.00	−0.03	0.89	1.00
	T_{max}	0.06	0.74	1.00	0.02	0.93	1.00	−0.02	0.93	1.00
	Rainfall	0.01	0.95	1.00	0.22	0.24	1.08	0.15	0.40	1.00
NOV	T_{min}	**0.42** *	0.02	5.69	0.04	0.84	1.00	0.07	0.69	1.00
	T_{max}	0.07	0.73	1.00	0.07	0.70	1.00	0.08	0.69	1.00
	Rainfall	−0.08	0.67	1.00	0.11	0.55	1.00	0.18	0.32	1.01
DEC	T_{min}	**0.42** *	0.02	5.69	0.05	0.80	1.00	−0.05	0.78	1.00
	T_{max}	0.09	0.61	1.00	0.23	0.20	1.14	−0.09	0.62	1.00
	Rainfall	−0.04	0.83	1.00	−0.18	0.33	1.01	−0.05	0.80	1.00
JAN	T_{min}	0.37	0.04	3.04	−0.22	0.23	1.10	0.25	0.16	1.24
	T_{max}	−0.35	0.05	2.38	−0.02	0.90	1.00	**−0.37** *	0.04	2.95
	Rainfall	0.05	0.77	1.00	0.22	0.23	1.09	0.25	0.17	1.22
FEB	T_{min}	−0.07	0.71	1.00	0.15	0.42	1.00	0.20	0.28	1.03
	T_{max}	**−0.51** **	0.00	19.83	0.13	0.49	1.00	**−0.42** *	0.02	5.43
	Rainfall	**0.45** *	0.01	7.64	**0.68** ***	<0.001	2118.11	0.17	0.34	1.00
MAR	T_{min}	−0.20	0.27	1.04	−0.28	0.13	1.40	0.02	0.93	1.00
	T_{max}	−0.03	0.87	1.00	−0.09	0.63	1.00	**−0.47** **	0.01	11.76
	Rainfall	−0.19	0.29	1.03	0.00	0.99	1.00	−0.12	0.51	1.00
APR	T_{min}	−0.15	0.41	1.00	−0.08	0.67	1.00	−0.20	0.27	1.04
	T_{max}	0.03	0.86	1.00	0.00	1.00	1.00	−0.17	0.36	1.00
	Rainfall	−0.16	0.38	1.00	−0.19	0.29	1.02	0.07	0.69	1.00

The minimum temperatures were correlated with maize yield only for the Marquard station in the months of November and February, this relationship was also found to be the case in studies conducted by Adisa, Botai [70]. Temperature drives the physiological and morphological development of the maize plant, with each process requiring a different minimum and maximum temperature. For instance, the study by Sanchez, Rasmussen [71] showed that leaf initiation needs a minimum of 7 °C, while shoot growth takes place above 14 °C and root growth above 13 °C. These minimum temperature conditions were not met for all cases except for the leaf initiation process in November (Table 5 above). However, in January the minimum temperature requirements for leaf initiation and shoot and root growth were met even for the late planting cultivars. Minimum temperatures, especially in November, seem to be critical for the early establishment and growth of the seedlings which ultimately influences the yield. The correlation and the regression analyses provided evidence for the significance of the minimum temperature on yield in Marquard, especially in the months of November and January. However, the November minimum temperature trend showed an increase of 0.09 °C per annum (see Table 5 above), which showed an increase of 1% in T_{min} in November increasing the yield by 0.274 tons ha^{-1} in Marquard. Climate change predictions for semi-arid regions of SSA have changed from earlier studies which gave values of 1.6 °C to recent projections of above 2.4 °C by 2050, depending on emissions and other anthropogenic activities [72]. Increasing trends in minimum temperatures are predicted for SSA, and extreme climate events, especially the frequency and severity could negatively impact yields [73].

The February T_{max} was negatively correlated with yield in Marquard and positively in Senekal ($r = -0.49$ and 0.657; $p = 0.005$ and <0.001 and 835.835, respectively) at 0.01 and 0.001 confidence levels, respectively. Similarly, the February rainfall in Marquard was positively correlated with yield ($r = 0.42$, $p = 0.018$) at 0.05 confidence level. There was also a strong correlation between them in Clocolan ($r = 0.69$, $p < 0.001$ and) in the month of February, while in March, the T_{max} in Senekal showed a positive correlation (r = 0.4512 $p = 0.003$) at 0.01 confidence level with yield (Table 9).

The results from this study showed that the maximum temperatures for the entire growing season were significantly correlated with maize yield only for Senekal. This was as a result of the significant correlation in the months of February and March. The stations of Clocolan and Ficksburg showed no correlation between the T_{max} and maize yield, while those in Marquard showed a significant negative correlation. The results in Marquard were also similar to other studies which showed that temperatures above 30 °C have a negative impact on maize production in southern Africa [74]. Senekal had the lowest maximum temperatures and a 1% increase of T_{max} in the months of February, March and the entire growing period (October–April) could increase the maize yield by 0.029, 0.408 and 0.536 tons ha^{-1} (Table 9). On the other hand, Marquard had the highest maximum temperatures and a 1% increase of T_{max} could decrease maize yield by 0.290 tons ha^{-1}. Lobell, Bänziger [74] showed that a 1% increase of maximum temperature above the optimal temperature for growth under drought stress could result in a maize yield decline of 1.7%. Clocolan had the highest mean T_{max} value and SD value of 28.6 °C and 2.4 °C respectively. There are several other studies that showed that high temperatures, together with soil and plant water stress lead to a decline in crop yield [75,76]. Maize yield in Marquard will be most vulnerable to water stress if the maximum temperatures continue to increase, especially at the anthesis stage, where the optimal temperature is 32 °C and the maximum tolerable T_{max} is 36 °C [58]. Muchow (1990) showed that temperatures outside the range of 13–32 °C decrease the yield by shortening the period of the kernel filling. These conditions also apply in Marquard with high February maximum temperatures which prevailed when kernel filling would have taken place if planting took place in November.

3.6.2. Maize Yield Relationship with Rainfall, Minimum and Maximum Temperature Anomalies

The monthly minimum, and maximum temperatures, as well as the rainfall that showed a significant correlation with maize yield (see Table 9 above) were subjected to regression analysis. The yield was the dependent variable while monthly T_{min}, T_{max} and rainfall were the independent variables used across the different stations of the Setsoto Municipality. The influence of the T_{min} on maize yield during the months of November and January in Marquard were significant ($p < 0.00027$ and $p < 0.038$, respectively) (Table 10). The T_{max} during the month of February showed a significant negative impact on maize yield when regression analysis was conducted ($p < 0.005$, $R^2 = 0.23$) whilst for the same month, rainfall showed a positive impact on the maize yield in Marquard. An increase of one unit of rainfall in (mm) can increase the yield by 0.0921 tons ha^{-1} (Table 10).

Table 10. A summary of regression results between detrended maize yield and the climatic (T_{min}, T_{max} and Rainfall) anomalies. Note: p = p-value at 0.05.

Months	Marquard			Clocolan			Senekal		
	Intercept	***p***	**R^2**	**Intercept**	***p***	**R^2**	**Intercept**	***p***	**R^2**
Nov T_{min}	0.274	0.0027	0.152	Nil	Nil	Nil	Nil	Nil	Nil
Jan T_{min}	0.572	0.038	0.135	Nil	Nil	Nil	Nil	Nil	Nil
Feb T_{max}	−0.290	0.005	0.238	Nil	Nil	Nil	0.0290	0.000	0.432
Mar T_{max}	Nil	Nil	Nil	Nil	Nil	Nil	0.408	0.003	0.262
GP T_{max}	Nil	Nil	Nil	Nil	Nil	Nil	0.005	0.008	0.214
GP Rainfall	Nil	Nil	Nil	0.005	0.008	0.214	Nil	Nil	Nil
Feb Rainfall	0.0094	0.018	0.174	0.015	0.000	0.472	Nil	Nil	Nil
GP	Nil	Nil	Nil	Nil	Nil	Nil	Nil	Nil	Nil

In Senekal, maximum temperatures in the months of February, March as well as the entire growing period (October–April) had a significantly positive impact on the maize yield ($p < 0.05$) (Table 10). In February, for every increase in degree Celsius of T_{max} above the base temperature led to an increase of the yield by 0.3459 tons ha^{-1} $year^{-1}$, while an increase in T_{max} in March and the whole season of the growing period (October–April) led to an increase of maize yield by 0.367 and 0.592 tons ha^{-1} respectively in Senekal (Table 10).

The effect of rainfall during the growing period and the month of February in Clocolan, showed a significant and positive relationship with the maize yield ($p < 0.05$) ($R^2 = 0.214$ and 0.472, respectively). An increase in rainfall by a unit (mm), increased the yield from 0.1028 to 0.1179 tons ha^{-1} $year^{-1}$ (Table 10).

3.6.3. Self-Calibrating Palmer Drought Stress Index

The average Self-calibrating Palmer Drought Severity index (ScPDSI) values for the growing period October–April are shown in Table 11. The first decade (1885–1994) had normal rainfall in Ficksburg, with a dry period in Clocolan and a wet period in Marquad and Senekal. The second decade showed three of the stations having a dry period and in the third decade, again three stations showed a dry period, with an extremely dry period being measured in Marquad. These decadal data support the maize yield data shown in Figure 2 with the first decade having the least variable maize yield.

Table 11. The average Self calibrating Palmer Drought Severity index (Sc_PDSI) values.

Period	Index	Stations			
		Marquard	**Ficksburg**	**Clocolan**	**Senekal**
1985–1994	Sc-PDSI	1.170776	0.097271	−1.52128	1.924278
1995–2004	Sc-PDSI	1.959666	−0.61539	−0.31812	−2.03299
2005–2016	Sc-PDSI	−3.02268	−1.11227	2.235037	−0.10628

Rainfall is a key driver of yield [77]. The amount of rainfall in the month of February was particularly strongly correlated (with $r = 0.69$) with yield in Clocolan and Marquard, adding further support to earlier evidence that the rainfall and temperatures in February have a strong influence on yield. The rainfall received in Clocolan had the lowest variability (CV 21%) when compared with the other stations (CVs up to 49%). Clocolan receives an average rainfall of 593 mm, which was similar to the 500 mm rainfall reported by for the eastern part of the Free State province. The CV associated with the total rainfall of 21–49% across the four stations was high and if either the total rainfall decreases, or variability increases then the risk of crop failure will increase. The results in this study support the findings of [78] who identified November as critical for the start of the growing season in Senekal. Maize planted later than November becomes susceptible to the frost from May onwards before the crops reach maturity [36] and expose the crop to increased rainfall variability. Maize planted in early November, will allow for maximum tasseling and grain-filling in February, which is the most sensitive period for water stress, even more, sensitive than the early establishment stages [79]. This study showed that a 1% increase in the rainfall amount in February and the overall growing period can increase the yield by 0.015- and 0.005-tons ha^{-1} respectively (Table 4). In most African countries agricultural production depends solely on rainfall pattern, distribution and duration [80,81]. This study confirmed the research by who indicated that high variability of rainfall threatens rain-fed agriculture in South Africa. These findings are similar to other previous work showing declining rainfall patterns in southern Africa.

4. Conclusions

The T_{min} and T_{max} trends showed variation across the weather stations used in this study. For instance, the T_{min} in Clocolan, showed a declining trend throughout the growing period between October and April, while in Marquard the minimum temperature increased between October and December. The maximum temperature was consistently increasing in all the stations except for Clocolan, where a decline was only reported for the month of March. The November and February trends are important for maize production that involves planting (leaf initiations, leaf and root growth) and development (tasseling and grain filling) of maize, respectively. The entire growing period (October–April) minimum and maximum temperatures for the period from 1985 to 2016, varied across the four different stations of the Setsoto municipality. The increasing minimum and maximum temperatures in all the stations of this study showed that: (1) where the minimum temperature is currently too low for optimal growth, an increase in these temperatures will increase yield and (2) the overall increase in both the minimum and maximum temperatures over time can negatively impact yield, but the magnitude of the effect is dependent on when exactly the increases are taking place during the growing season. November and February have been highlighted as specific times at which the crop is most at risk.

The changes in rainfall were significant only in Ficksburg in the month of January with a value of 2.34 mm year^{-1} Nevertheless, the rainfall showed a strong positive correlation with yield (r 0.46, $p \leq 0.05$). This study indicates that the rainfall variability is increasing in parts of the study area, which could be attributed to several global and regional rainfall phenomena. There were some periods where it did appear that the yield was below average, similarly, there were periods from 2006–2012, where the yield was above the average maize yield per hectare (2.42 tons ha^{-1}). There are some concerns, especially in the Senekal area, that it will be no longer economically viable for maize production. Yield is not just a product of climatic variables, but also a combination of other agronomic factors. The average rate of increase of yield in the Setsoto Municipality is 0.044 tons ha^{-1} per annum across the stations.

The strongest positive correlation (46–68%) with yield and rainfall was during the growing period in Clocolan. The changes in minimum temperature are having two different effects on the yield in the area where: if it is colder, the yield will be negatively impacted; if it is getting warmer, where the minimum temperature has previously limited yield, the yield will be positively impacted. Increasing maximum temperatures still shows no negative impacts on maize yield except for a single month of February in Marquard. Palmer drought stress indices should be explored further to help support more accurate forecasting. This study serves as an important baseline of the impacts of agroclimatic variables on maize yield at this local scale which is a key area of production. Farmers cannot make rapid decisions about farming practices, where to plant or whether to sell the land. This study contributes to raising awareness about the risk of ongoing maize production in this area.

Author Contributions: A.H.B., collected and analyzed the data and wrote the initial draft of the paper. M.S. and S.W.N. conceptualized the study and acted as supervisors of the postgraduate student, H.A. M.S. did a significant amount of reanalysis and interpretation of the data and writing of the manuscript. S.N. provided some data and did some checking of the manuscript. All authors have read and agreed to the published version of the manuscript.

Funding: This research was funded by Prof Mary Scholes, who holds a Research Chair in Systems Analysis which is funded by the Department of Science and Innovation and the National Research Foundation in South Africa. The grant number is 101057.

Conflicts of Interest: The authors declare that there is no conflict of interest.

References

1. Barros, V.; Field, C.; Dokke, D.; Mastrandrea, M.; Mach, K.; Bilir, T.E.; Ebi, K.L.; Estrada, Y.O.; Genova, R.C.; Girma, B.; et al. *Climate Change 2014: Impacts, Adaptation, and Vulnerability-Part B: Regional Aspects-Contribution of Working Group II to the Fifth Assessment Report of the Intergovernmental Panel on Climate Change*; Cambridge University Press: New York, NY, USA, 2014.
2. Clay, J. Freeze the footprint of food. *Nat. Cell Biol.* **2011**, *475*, 287–289. [CrossRef]
3. Gornall, J.; Betts, R.; Burke, E.; Clark, R.; Camp, J.; Willett, K.; Wiltshire, A. Implications of climate change for agricultural productivity in the early twenty-first century. *Philos. Trans. R. Soc. B Biol. Sci.* **2010**, *365*, 2973–2989. [CrossRef]
4. Hatfield, J.L.; Boote, K.J.; Kimball, B.A.; Ziska, L.H.; Izaurralde, R.C.; Ort, D.; Thomson, A.M.; Wolfe, D. Climate impacts on agriculture: Implications for crop production. *Agron. J.* **2011**, *103*, 351–370. [CrossRef]
5. Olesen, J.; Bindi, M. Consequences of climate change for European agricultural productivity, land use and policy. *Eur. J. Agron.* **2002**, *16*, 239–262. [CrossRef]
6. Asam, S.; Callegari, M.; Matiu, M.; Fiore, G.; De Gregorio, L.; Jacob, A.W.; Menzel, A.; Zebisch, M.; Notarnicola, C. Relationship between spatiotemporal variations of climate, snow cover and plant phenology over the Alps—An Earth observation-based analysis. *Remote. Sens.* **2018**, *10*, 1757. [CrossRef]
7. Engelbrecht, F.; Adegoke, J.; Bopape, M.-J.; Naidoo, M.; Garland, R.M.; Thatcher, M.; McGregor, J.; Katzfey, J.; Werner, M.; Ichoku, C.; et al. Projections of rapidly rising surface temperatures over Africa under low mitigation. *Environ. Res. Lett.* **2015**, *10*, 085004. [CrossRef]
8. Tyson, P.D.; Dyer, T.G.; Mametse, M. Secular changes in South African rainfall: 1880 to 1972. *Q. J. R. Meteorol. Soc.* **1975**, *101*, 817–833. [CrossRef]
9. Aguilar, E.; Barry, A.A.; Brunet, M.; Ekang, L.; Fernandes, A.; Massoukina, M.; Mbah, J.; Mhanda, A.; Nascimento, D.J.D.; Peterson, T.C.; et al. Changes in temperature and precipitation extremes in western central Africa, Guinea Conakry, and Zimbabwe, 1955–2006. *J. Geophys. Res. Space Phys.* **2009**, *114*. [CrossRef]
10. Dasgupta, P.; Morton, J.F.; Dodman, D.; Karapinar, B.; Meza, F.; Rivera-Ferre, M.G.; Toure Sarr, A.; Vincent, K.E. Rural areas. In *Climate Change 2014: Impacts, Adaptation, and Vulnerability. Part A: Global and Sectoral Aspects. Contribution of Working Group II to the Fifth Assessment Report of the Intergovernmental Panel on Climate Change*; Cambridge University Press: Cambridge, UK; New York, NY, USA, 2014; pp. 613–657.
11. Nicholson, S.E. Climate and climatic variability of rainfall over eastern Africa. *Rev. Geophys.* **2017**, *55*, 590–635. [CrossRef]
12. Kruger, A.C.; Sekele, S.S. Trends in extreme temperature indices in South Africa: 1962–2009. *Int. J. Clim.* **2013**, *33*, 661–676. [CrossRef]
13. Hulme, M. Rainfall changes in Africa: 1931–1960 to 1961–1990. *Int. J. Climatol.* **1992**, *12*, 685–699. [CrossRef]
14. Tolba, M.K.; Cutajar, M.Z.; Pyhälä, M.; Ehhalt, D.H.; Obasi, G.O.P.; Killmann, W.; Briceño, S.; Virji, H.; Odada, E.; Olago, D.; et al. Climate change and Africa. *Clim. Change Afr.* **2008**, *24*, 337–353. [CrossRef]
15. Conway, D.; Mould, C.; Bewket, W. Over one century of rainfall and temperature observations in Addis Ababa, Ethiopia. *Int. J. Clim.* **2004**, *24*, 77–91. [CrossRef]

16. Van Wilgen, N. Climate Change: Briefings from Southern Africa. *Trans. R. Soc. South Afr.* **2016**, *71*, 205–206. [CrossRef]
17. Moeletsi, M.E.; Walker, S.; Landman, W.A. ENSO and implications on rainfall characteristics with reference to maize production in the Free State Province of South Africa. *Phys. Chem. Earth Parts A/B/C* **2011**, *36*, 715–726. [CrossRef]
18. Dube, S.; Scholes, R.J.; Nelson, G.C.; Mason-D'Croz, D.; Palazzo, A. South African food security and climate change: Agriculture futures. *Econ. Open-Access Open-Assess. E-J.* **2013**, *7–9*. [CrossRef]
19. Wu, B.; Ma, Z.; Yan, N. Agricultural drought mitigating indices derived from the changes in drought characteristics. *Remote Sens. Environ.* **2020**, *244*, 111813. [CrossRef]
20. Wang, W.; Ertsen, M.W.; Svoboda, M.D.; Hafeez, M. Propagation of drought: From meteorological drought to agricultural and hydrological drought. *Adv. Meteorol.* **2016**, *2016*, 1–5. [CrossRef]
21. Wells, N.; Goddard, S.; Hayes, M.J. A self-calibrating Palmer drought severity index. *J. Clim.* **2004**, *17*, 2335–2351. [CrossRef]
22. Dai, A.; Trenberth, K.E.; Qian, T. A global dataset of Palmer drought severity index for 1870–2002: Relationship with soil moisture and effects of surface warming. *J. Hydrometeorol.* **2004**, *5*, 1117–1130. [CrossRef]
23. Dai, A. Drought under global warming: A review. *Wiley Interdiscip. Rev. Clim. Chang.* **2011**, *2*, 45–65. [CrossRef]
24. Hansen, J.W.; Mason, S.J.; Sun, L.; Tall, A. Review of seasonal climate forecasting for agriculture in Sub-Saharan Africa. *Exp. Agric.* **2011**, *47*, 205–240. [CrossRef]
25. Jones, P.G.; Thornton, P. Croppers to livestock keepers: Livelihood transitions to 2050 in Africa due to climate change. *Environ. Sci. Policy* **2009**, *12*, 427–437. [CrossRef]
26. Hammer, G.L.; Van Oosterom, E.; McLean, G.; Chapman, S.C.; Broad, I.; Harland, P.; Muchow, R.C. Adapting APSIM to model the physiology and genetics of complex adaptive traits in field crops. *J. Exp. Bot.* **2010**, *61*, 2185–2202. [CrossRef]
27. Landman, W.A.; Engelbrecht, F.; Hewitson, B.; Malherbe, J.; Van Der Merwe, J. Towards bridging the gap between climate change projections and maize producers in South Africa. *Theor. Appl. Clim.* **2017**, *132*, 1153–1163. [CrossRef]
28. FAO/FAOSTAT. *FAO Agriculture Department, Agricultural Production, Livestock Primary. Total World Meat 1970–2010, Food Balance Sheets, Japan, USA*; FAO: Rome, Italy, 2012.
29. Blignaut, J.; Ueckermann, L.; Aronson, J. Agriculture production's sensitivity to changes in climate in South Africa. *South. Afr. J. Sci.* **2009**, *105*, 61–68. [CrossRef]
30. De Jager, J.; Potgieter, A.; Berg, W.V.D. Framework for forecasting the extent and severity of drought in maize in the Free State Province of South Africa. *Agric. Syst.* **1998**, *57*, 351–365. [CrossRef]
31. Smale, M.; Jayne, T. Maize in Eastern and Southern Africa: Seeds: Of success in retrospect. EPTD (Environment and Production Technology Division) Discussion Paper no. 97. International Food Policy Research Institute. 2014. Available online: www.fao.org/docs/eims/upload/166420 (accessed on 10 December 2020).
32. Mukhala, E.; Groenewald, D. Experiences and perceptions of black small-scale irrigation farmers in the Free State. *S. Afr. J. Agric. Ext.* **1998**, *27*, 1–18.
33. Cammarano, D.; Valdivia, R.O.; Beletse, Y.G.; Durand, W.; Crespo, O.; Tesfuhuney, W.A.; Jones, M.R.; Walker, S.; Mpuisang, T.N.; Nhemachena, C.; et al. Integrated assessment of climate change impacts on crop productivity and income of commercial maize farms in northeast South Africa. *Food Secur.* **2020**, *12*, 659–678. [CrossRef]
34. Abubakar, H.B.; Newete, S.W.; Scholes, M.; Bello, A.H. Drought characterization and trend detection using the reconnaissance drought index for Setsoto Municipality of the Free State Province of South Africa and the impact on maize yield. *Water* **2020**, *12*, 2993. [CrossRef]
35. Moeletsi, M.; Moopisa, S.G.; Walker, S.; Tsubo, M. Development of an agroclimatological risk tool for dryland maize production in the Free State Province of South Africa. *Comput. Electron. Agric.* **2013**, *95*, 108–121. [CrossRef]
36. Moeletsi, M.; Walker, S. Rainy season characteristics of the Free State Province of South Africa with reference to rain-fed maize production. *Water SA* **2012**, *38*, 775–782. [CrossRef]
37. Hensley, M.; Le Roux, P.; Du Preez, C.; Van Huyssteen, C.; Kotze, E.; Van Rensburg, L. Soils: The Free State's agricultural base. *South Afr. Geogr. J.* **2006**, *88*, 11–21. [CrossRef]

38. Buso, N. Municipal Commonage Administration in the Free State Province: Can Municipalities in the Current Local Government Dispensation Promote Emerging Farming? October 2003. Available online: http://repository.hsrc.ac.za/handle/20.500.11910/8230 (accessed on 10 December 2020).
39. Beletse, Y.G.; Durand, W.; Nhemachena, C.; Williams, P.A.; Tesfuhuney, W.A.; Jones, M.R.; Teweldemedhin, M.Y.; Gamedze, S.M.; Bonolo, P.M.; Jonas, S.; et al. Projected impacts of climate change scenarios on the production of maize in Southern Africa: An integrated assessment case study of the Bethlehem District, Central Free State, South Africa. In *Handbook of Climate Change and Agroecosystems: The Agricultural Model Intercomparison and Improvement Project Integrated Crop and Economic Assessments, Part 2*; Imperial College Press: London, UK, 2015; Volume 4, pp. 125–157.
40. Adisa, O.M.; Botai, C.M.; Botai, J.O.; Hassen, A.; Darkey, D.; Tesfamariam, E.; Adisa, A.F.; Adeola, A.M.; Ncongwane, K.P. Analysis of agro-climatic parameters and their influence on maize production in South Africa. *Theor. Appl. Clim.* **2018**, *134*, 991–1004. [CrossRef]
41. Golkhatmi, N.; Sanaeinejad, S.; Ghahraman, B.; Pazhand, H. Extended modified inverse distance method for interpolation rainfall. *Int. J. Eng. Invent.* **2012**, *3*, 57–65.
42. Viale, M.; Garreaud, R.D. Orographic effects of the subtropical and extratropical Andes on upwind precipitating clouds. *J. Geophys. Res. Atmos.* **2015**, *120*, 4962–4974. [CrossRef]
43. DAFF. *South African Fertilizers Market Analysis Report 2015. DAFF Report*; DAFF: Pretoria, South Africa, 2015.
44. Palmer, W.C. *Meteorological Drought*; Research paper no. 45; US Weather Bureau: Washington, DC, USA, 1965.
45. Zhai, J.; Su, B.; Krysanova, V.; Vetter, T.; Gao, C.; Jiang, T. Spatial variation and trends in PDSI and SPI indices and their relation to streamflow in 10 large regions of China. *J. Clim.* **2010**, *23*, 649–663. [CrossRef]
46. Vicente-Serrano, S.M.; Beguería, S.; López-Moreno, J.I. Comment on "Characteristics and trends in various forms of the Palmer Drought Severity Index (PDSI) during 1900–2008" by Aiguo Dai. *J. Geophys. Res.* **2011**, *116*, D19112. [CrossRef]
47. Sneyers, R. On the use of statistical analysis for the objective determination of climate change. *Meteorol. Z.* **1992**, 247–256. [CrossRef]
48. Bandyopadhyay, N.; Bhuiyan, C.; Saha, A.K. Heat waves, temperature extremes and their impacts on monsoon rainfall and meteorological drought in Gujarat, India. *Nat. Hazards* **2016**, *82*, 367–388. [CrossRef]
49. Ewert, F.; Rötter, R.P.; Bindi, M.; Webber, H.; Trnka, M.; Kersebaum, K.C.; Olesen, J.E.; Van Ittersum, M.K.; Janssen, S.; Rivington, M.; et al. Crop modelling for integrated assessment of risk to food production from climate change. *Environ. Model. Softw.* **2015**, *72*, 287–303. [CrossRef]
50. Mahony, M.; Hulme, M. Modelling and the nation: Institutionalising climate prediction in the UK, 1988–1992. *Minerva* **2016**, *54*, 445–470. [CrossRef]
51. Pablos, M.; Martínez-Fernández, J.; Sanchez, N.; González-Zamora, Á. Temporal and spatial comparison of agricultural drought indices from moderate resolution satellite soil moisture data over Northwest Spain. *Remote. Sens.* **2017**, *9*, 1168. [CrossRef]
52. Milošević, D.; Savić, S.; Stojanović, V.; Popov-Raljić, J. Effects of precipitation and temperatures on crop yield variability in Vojvodina (Serbia). *Ital. J. Agrometeorol.-Riv. Ital. Agrometeorol.* **2015**, *20*, 35–46.
53. Russo, S.; Marchese, A.F.; Sillmann, J.; Immé, G. When will unusual heat waves become normal in a warming Africa? *Environ. Res. Lett.* **2016**, *11*, 054016. [CrossRef]
54. Haverkort, A.J.; Franke, A.C.; Engelbrecht, F.A.; Steyn, J.M. Climate change and potato production in contrasting South African agro-ecosystems 1. Effects on land and water use efficiencies. *Potato Res.* **2013**, *56*, 31–50. [CrossRef]
55. IPCC. *IPCC, 2014: Climate Change 2014: Synthesis Report. Contribution of Working Groups I. II and III to the Fifth Assessment Report of the Intergovernmental Panel on Climate Change*; IPCC: Geneva, Switzerland, 2014; p. 151.
56. Dezfuli, A.K.; Zaitchik, B.F.; Gnanadesikan, A. Regional atmospheric circulation and rainfall variability in South Equatorial Africa. *J. Clim.* **2015**, *28*, 809–818. [CrossRef]
57. Lana, M.A.; Vasconcelos, A.C.F.; Gornott, C.; Schaffert, A.; Bonatti, M.; Volk, J.; Graef, F.; Kersebaum, K.C.; Sieber, S. Is dry soil planting an adaptation strategy for maize cultivation in semi-arid Tanzania? *Food Secur.* **2018**, *10*, 897–910. [CrossRef]
58. Eggert, B.; Berg, P.; Haerter, J.; Jacob, D.; Moseley, C. Temporal and spatial scaling impacts on extreme precipitation. *Atmos. Chem Phys.* **2015**, *15*, 5957–5971. [CrossRef]

59. Thomas, A.C.; Pershing, A.J.; Friedland, K.D.; Nye, J.A.; Mills, K.E.; Alexander, M.A.; Record, N.R.; Weatherbee, R.; Henderson, M.E.; Drinkwater, K. Seasonal trends and phenology shifts in sea surface temperature on the North American northeastern continental shelf. *Elem. Sci. Anthr.* **2017**, *5*. [CrossRef]
60. Shaw, R.H. Climate requirement. Corn and corn improvement. *Agronomy* **1988**, *18*, 609–638. [CrossRef]
61. Bergamaschi, H.; Dalmago, G.A.; Bergonci, J.I.; Bianchi, C.A.M.; Müller, A.G.; Comiran, F.; Heckler, B.M.M. Water supply in the critical period of maize and the grain production. *Pesqui. Agropecuária Bras.* **2004**, *39*, 831–839. [CrossRef]
62. Maziya-Dixon, B. *Nigeria Food Consumption and Nutrition Survey 2001–2003: Summary*; IITA: Ibadan, Nigeria, 2004.
63. Parry, M.L.; Rosenzweig, C.; Iglesias, A.; Livermore, M.; Fischer, G. Effects of climate change on global food production under SRES emissions and socio-economic scenarios. *Glob. Environ. Change* **2004**, *14*, 53–67. [CrossRef]
64. Thornton, P.; Jones, P.G.; Alagarswamy, G.; Andresen, J. Spatial variation of crop yield response to climate change in East Africa. *Glob. Environ. Change* **2009**, *19*, 54–65. [CrossRef]
65. Adam, S.; Agatsiva, J.L.; Akwany, P.; Arunga, M.; Bagine, R. *Nature's Benefits in Kenya: An Atlas of Ecosystems and Human Well-being*; World Resources Institute: Washington DC, USA; Nairobi, Kenya, 2007.
66. Walker, N.; Schulze, R. Climate change impacts on agro-ecosystem sustainability across three climate regions in the maize belt of South Africa. *Agric. Ecosyst. Environ.* **2008**, *124*, 114–124. [CrossRef]
67. Marais, L.; Cloete, J. *Patterns of Territorial Development and Inequality from South Africa's Periphery: Evidence from the Free State Province*; Working Paper Series; RIMISP: Bloemfontein, South Africa, 2016.
68. Morakile, G. Survey on Preferred Supplier Base Mechanism for Smallholder Farmers/Cooperatives to Derive Better Access to Government Market. Master's Thesis, University of South Africa, Pretoria, South Africa, 2018.
69. Sánchez, B.; Rasmussen, A.; Porter, J.R. Temperatures and the growth and development of maize and rice: A review. *Glob. Change Biol.* **2014**, *20*, 408–417. [CrossRef]
70. Cairns, J.E.; Hellin, J.; Sonder, K.; Araus, J.L.; MacRobert, J.F.; Thierfelder, C.; Prasanna, B.M. Adapting maize production to climate change in sub-Saharan Africa. *Food Secur.* **2013**, *5*, 345–360. [CrossRef]
71. IPCC. *The Physical Science Basis. Contribution of Working Group I to the Fourth Assessment Report of the Intergovernmental Panel on Climate Change*; IPCC: Geneva, Switzerland, 2007; p. 996.
72. Lobell, D.B.; Bänziger, M.; Magorokosho, C.; Vivek, B. Nonlinear heat effects on African maize as evidenced by historical yield trials. *Nat. Clim. Change* **2011**, *1*, 42–45. [CrossRef]
73. Tesfaye, K.; Kruseman, G.; Cairns, J.E.; Zaman-Allah, M.; Wegary, D.; Zaidi, P.; Boote, K.J.; Rahut, D.; Erenstein, O. Potential benefits of drought and heat tolerance for adapting maize to climate change in tropical environments. *Clim. Risk Manag.* **2018**, *19*, 106–119. [CrossRef]
74. Steward, P.; Dougill, A.J.; Thierfelder, C.; Pittelkow, C.M.; Stringer, L.C.; Kudzala, M.; Shackelford, G. The adaptive capacity of maize-based conservation agriculture systems to climate stress in tropical and subtropical environments: A meta-regression of yields. *Agric. Ecosyst. Environ.* **2018**, *251*, 194–202. [CrossRef]
75. Scholes, M.; Scholes, R.B.; Lucas, M. *Climate Change: Briefings from Southern Africa*; NYU Press: New York, NY, USA, 2015.
76. Moeletsi, M.; Walker, S. A simple agroclimatic index to delineate suitable growing areas for rainfed maize production in the Free State Province of South Africa. *Agric. For. Meteorol.* **2012**, *162*, 63–70. [CrossRef]
77. Midgley, G.; Chapman, R.; Mukheibir, P.; Tadross, M.; Hewitson, B.; Wand. S.; Schulze, R.; Lumsden, T.; Horan, M.; Warburton, M. *Impacts, Vulnerability and Adaptation in Key South African Sectors. An input into the Long Term Mitigation Scenarios Process Cape Town: Energy Research Centre*; University of Cape Town: Cape Town, South Africa, 2007.
78. Buckle, C. *Weather and Climate in Africa*; Longman: Harlow, UK, 1996; p. 321.
79. Hunter, R.D.; Meentemeyer, R.K. Climatologically aided mapping of daily precipitation and temperature. *J. Appl. Meteorol.* **2005**, *44*, 1501–1510. [CrossRef]
80. Fauchereau, N.; Trzaska, S.; Rouault, M.; Richard, Y. Rainfall variability and changes in Southern Africa during the 20th century in the global warming context. *Nat. Hazards* **2003**, *29*, 139–154. [CrossRef]

81. Jury, M.R. Climate trends in southern Africa. *South Afr. J. Sci.* **2013**, *109*, 1–11. [CrossRef]

Publisher's Note: MDPI stays neutral with regard to jurisdictional claims in published maps and institutional affiliations.

climate

MDPI

Article

Trends of Climate Change and Variability in Three Agro-Ecological Settings in Central Ethiopia: Contrasts of Meteorological Data and Farmers' Perceptions

Dula Etana [1,2,*], Denyse J. R. M. Snelder [3], Cornelia F. A. van Wesenbeeck [4] and Tjard de Cock Buning [2]

1 College of Development Studies, Addis Ababa University, P.O. Box 1176 Addis Ababa, Ethiopia
2 Athena Institute of Research, Vrije Universiteit Amsterdam, De Boelelaan 1085, 1081 HV Amsterdam, The Netherlands; tjard.de.cockbuning@vu.nl
3 Centre for International Cooperation, Vrije Universiteit Amsterdam, De Boelelaan 1105, 1081 HV Amsterdam, The Netherlands; d.j.r.m.snelder@vu.nl
4 Centre for World Food Studies, Vrije Universiteit Amsterdam, De Boelelaan 1105, 1081 HV Amsterdam, The Netherlands; c.f.a.van.wesenbeeck@vu.nl
* Correspondence: etanad29@gmail.com or d.e.gebremariyam@vu.nl

Received: 31 August 2020; Accepted: 19 October 2020; Published: 22 October 2020

Abstract: Using gridded daily temperature and rainfall data covering 30 years (1988–2017), this study investigates trends in rainfall, temperature, and extreme events in three agro-ecological settings in central Ethiopia. The Mann Kendall test and Sen's slope estimator were used to examine the trends and slope of changes in climate indices. The profile of farmers whose perception converges with or diverges from meteorological data was characterized using polling. The average annual temperature has increased by 0.4 and 0.3 °C per decade in the lowland and midland areas, respectively. Average annual rainfall has increased only in the midland areas by 178 mm per decade. Farmers' perception of increasing temperature fairly aligns with meteorological data. However, there is a noticeable difference between farmers' perception of rainfall and meteorological data. The perception of farmers with poor economic status, access to media, and higher social capital aligns with measured trends. Conversely, the perception of economically better-off and uneducated farmers diverges from meteorological data. Accurate perception is constrained by the failure of the traditional forecast methods to describe complex weather variabilities and lack of access to down-scaled weather information. The findings highlight the importance of availing specific and agro-ecologically relevant weather forecasts to overcome perceptual problems and to support effective adaptation.

Keywords: temperature; rainfall; drought; weather; livelihood

1. Introduction

It is increasingly becoming apparent that climate change, spatio-temporal variability, and extreme events are issues of concern in Africa due to exceptionally high vulnerability [1,2]. Africa is warmer than it was 100 years ago and this warming not only continues but also will accelerate to a rate between 2 and 6 °C in 100 years [3]. Unlike average temperature that is projected to increase across the continent, there is diversity in the pattern of rainfall [4]. According to the projection of the Intergovernmental Panel on Climate Change [5], there will be a reduction in rainfall in Northern and Southern Africa, but an increase in Eastern Africa at the end of the 21st century. Model projections show varying results for Western Africa, yet most of them indicate a wetter core rainfall season with a small delay in the rainy season by the end of the century. There is also a projected increase in extreme temperatures and rainfall in Africa [5]. Due to current and future trends of climate change, it is expected that there will be a decline in the area suitable for agriculture, a shortening of the length of growing seasons, and diminishing crop yields [6].

Although Ethiopia is no exception to the problem of climate change, studies at different spatial scales show contrasting patterns of rainfall and temperature. Rainfall in Ethiopia is highly heterogeneous showing a wide range of patterns with no clear direction of change [1]. Another study [7] reported no significant change in annual rainfall at the national level but a significant decline in the *kiremt* rainfall (i.e., long rainy season between June and September) for south-western and central Ethiopia. A study in south-western Ethiopia [8] reported decreasing trends of rainfall. Another national-level study showed varied rainfall patterns in different parts of the country [9]. It confirmed significantly decreasing trends of *kiremt* and annual rainfall in northern, north-western, and western parts of the country. However, an increasing trend of annual rainfall was observed at a limited number of locations in eastern Ethiopia. For central Ethiopia, an increase in annual and *kiremt* rainfall but a decrease in *belg* rainfall [short rainy season between March and May] [10]. In the upper Blue Nile basin of Ethiopia, an insignificant increasing trend of annual rainfall was reported [11].

Regional- and local-level analyses show different trends of temperature. Based on climate model projections, one study [1] reported warming in all seasons across the country with relatively modest differences between regions. Another study [8] found upward trends in temperature for south-western Ethiopia. Both increasing and decreasing trends of temperature records were reported in different parts of the upper Blue Nile basin [11]. Both mean annual maximum and mean annual minimum temperature were increasing in the northern, central, and southern parts of the basin but decreasing trend was observed in the western part. A study in central Ethiopia [12] reported significantly increasing trends in annual maximum and minimum temperatures for midland and lowland areas. In north-central Ethiopia, significantly increasing trends of mean and minimum average temperature were reported [13]. There is also evidence that extreme events are becoming common in the country [9,14,15] and considerably vary by eco-environments [14,16].

The identification of the micro patterns of changes in climate variables is not sufficient to address climate problems. Farmers' perceptions of changes in climate variables is also important for climate risk management and agricultural adaptation [17–19]. The way farmers respond to climate change and variability (CCV) depends on how they perceive the problems. Perception motivates action, which suggests that failure to recognize CCV as a livelihood threat might reduce concern and hinder action. Farmers make adaptation decisions based on their perceptions of changes in the climate variables [19]. The use of autonomous adaptation strategies specifically depends on farmers' perceptions of local weather conditions. The convergence of perception with and divergence from observed trends also determine the type and time of taking actions. Farming decisions to be made and adaptation actions to be taken are more likely to be effective if there is a convergence between objective measurements and subjective assessments. However, it is not easy for farmers to have an accurate perception of changes in climate variables. The difficulty emanates from the fact that climate change is a long-term process, whereas farmers' perception refers to short-term experience relying on memories [17].

Given these challenges, studies show contrasting results on whether farmers can accurately perceive actual changes in local climate variables. In general, farmers' perception of an increase in temperature aligns with meteorological records [17,20,21]. However, for rainfall, studies show divergence between perception and records [18,19,21,22]. In Ethiopia, despite heavy reliance on rain-based economic activities and the absence of relevant information, studies linking perception to meteorological data are quite limited. One study [23] found that increased temperature and declining rainfall were the most widely held perceptions among farmers in the central highlands of Ethiopia. However, the result was not compared with meteorological data to validate the accuracy of farmers' subjective assessments. Another study [24] investigated farmers' perception in northern Ethiopia and found a divergence between perception of declining rainfall and rainfall measurements. Most of the previous studies compare farmers' perception with observed results from the nearest weather station due to lack of temperature and rainfall data at the household level However, this comparison is less precise as all farmers around the stations will have the same measurement values of the climate variables. Besides, previous studies comparing perception and actual measurements did not specifically show the features of farmers whose perceptions converge with or diverge from observed meteorological data. Our study addresses these gaps by using a novel approach to better integrate measured changes in climate variables with a farmers' perception survey. The integration of perception and the measurements adds new insights into the current literature beyond simply presenting the percentage of farmers who are wrong or right. Whereas climate refers to average weather conditions over a long period of time, weather shows short-term atmospheric conditions. The meteorological data were employed to investigate both long-term changes and short term annual/seasonal variabilities. Although farmers commonly observe short-term atmospheric conditions in their farming operations, they can also perceive long-term changes to make adjustments to their livelihood practices. Hence, both concepts, climate and weather, are used in this study.

Understanding how climate variables are changing at the local level is important for planning appropriate adaptation strategies and boosting agricultural productivity [25]. However, no discernible and consistent patterns of change in and congruence between climate variables and perceptions can be established from these studies. Ethiopia is known for its highly diverse topography with altitudinal differences ranging from 125 m below sea level to 4620 m above sea level [7]. Given the high spatial variation in topography, analysis at the national or regional level masks local variations in temperature and rainfall and hence are of limited use to farmers seeking local solutions to manage the effects of climate change and variability. Hence, downscaling the level of analysis to meaningful geographic units makes the measurements more informative [7] and the information more relevant for farmers to plan for proactive adaptation responses. The ways local climate changes are understood by farmers are equally important in motivating adaptation. Therefore, this study has dual objectives. The first objective is the investigation of agro-ecological differences and temporal changes in rainfall and temperature as well as associated extreme events in topographically diverse areas in central Ethiopia. The second is comparison of the results of meteorological data with farmers' perceptions to discern convergences and divergences. The perceptions of farmers are assessed against the statistical results as the congruence/incongruence between the two has implications for risk management and adaptation decision making. The study contributes to the scant literature on the agro-ecological comparison of climate change and variability. Besides, by integrating meteorological data with a farmers' perception survey, our study further provides insights on farmers' understanding of the local weather conditions and its alignment with observed trends of climate variables.

2. Data and Methods

2.1. Description of the Study Areas

Oromia National Regional State is one of the regions in Ethiopia most vulnerable to climate change and variability. This study covers three districts in Oromia region in central Ethiopia (Kembibit, Kuyu, and Boset) dominantly representing, respectively, highland (H), midland (M), and lowland (L) areas (Figure 1). This agro-ecological classification is mainly based on altitudinal variations that have a strong impact on temperature and rainfall and consequently on agricultural land uses, mainly crop production. Highland, midland, and lowland cover altitudinal ranges of 2300–3200, 1500–2300, and 500–1500 m above sea level, respectively. Kembibit district covers the total area of about 928 km^2. It lies between 9°12′–9°32′ N latitude and 39°04′–39°33′ E longitude. The agro-climatic zone of the district is mainly highland (temperate) with pocket areas found in mid-altitude (sub-tropical) areas. Kuyu district is located between 9°35′–9°49′ N latitude and 38°03′–38°31′ E longitude. Its total area is 994.7 km^2. Owing to its altitudinal range, the district constitutes three agro-climatic zones (temperate, subtropical, and tropical areas) with a dominant sub-tropical climate. Boset district, which covers a total area of 1514.1 km^2, is found between 8°25′–8°50′ N latitude and 39°16′–39°50′ E longitude. Most parts of the area lie between altitudinal ranges of 1000–1500 m above sea level, with a dominantly tropical climate.

The three areas are characterized by a bimodal rainfall distribution with a short *belg* rainy season, and a long *kiremt* rainy season. About 85% of the population in these areas live in rural areas, with livelihoods being mainly dependent on crop and livestock production. Owing to differences in temperature and rainfall distribution, these three agro-ecological settings are also characterized by fairly distinct crop production patterns. Sorghum and teff are the dominant types of crops produced in the lowland areas. Maize (*Zea Mays*), sorghum (*Sorghum bicolor*), *teff* (*Tef eragrostis*), wheat (*Triticum*), and oil seeds are dominantly produced in the midland areas. In the highland areas, barley and pulses are extensively produced. Farmers follow subsistence means of living that most of the products are used for home consumption. There is a high risk of yield reduction or crop failure during years of adverse weather conditions, threatening their food security. Consequently, the problem of food insecurity is widespread, and a sizeable proportion of the population is supported by the Productive Safety Net Program and emergency food aid. The program is implemented by the government of Ethiopia with the support of development partners in areas prone to chronic food insecurity to help the poor build assets and improve their livelihoods and, eventually, become food self-sufficient and resilient to shocks. The problem of food insecurity in the three study areas is related to declining agricultural productivity induced by adverse weather conditions and other socio-economic problems, such as shortage of farmland, land degradation, and limited use of improved agricultural technologies. The vulnerability of these areas is further compounded by deforestation, population pressure, lack of alternative livelihood options, and poor rural infrastructure.

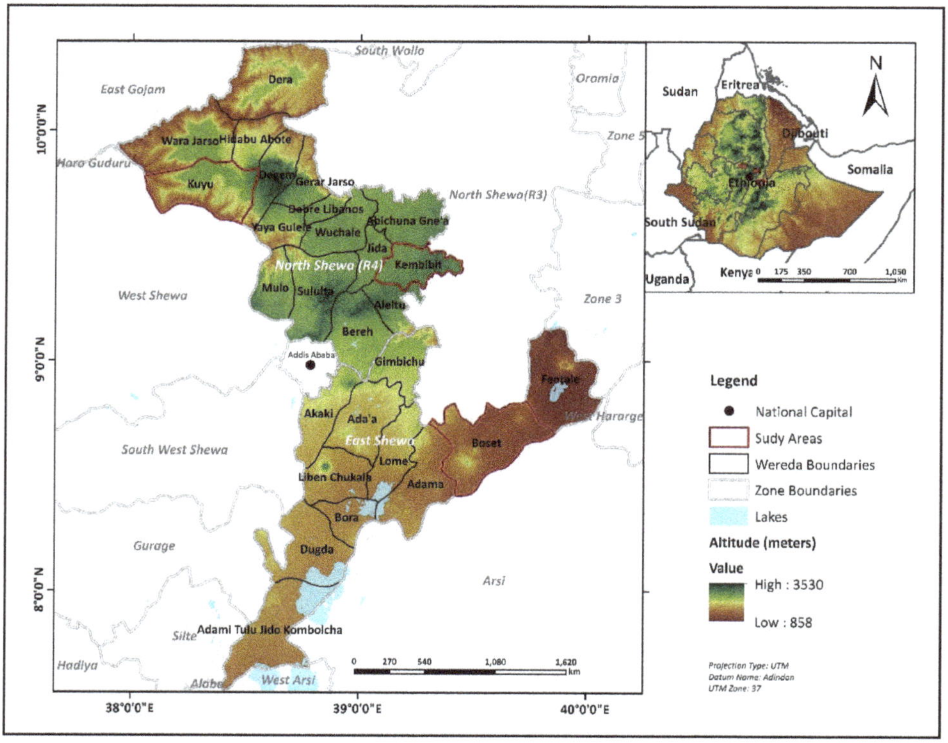

Figure 1. Geographical location and topography of the study areas.

2.2. Sample Size and Sampling Techniques

The sample size of the study was determined using a sample size calculation for a finite population [26]. The computation was made with the assumptions of 95% confidence interval; 5% level of significance; and 60% of households perceiving climate change and using adaptation strategies. Taking the population size of one of the districts, the sample size was calculated to be 270 households. Considering each district as an independent unit, the total sample size was 810 households. A multi-stage sampling technique was used to identify sample households. The three districts and nine kebeles (lowest administrative unit in Ethiopia) were selected through purposive sampling at the first and second stages, respectively. The selection was made based on the consideration of similarity of livelihood systems and prevalence of climate-related risk factors. At the last stage, sample households were selected using a simple random sampling technique from the list of households living in each kebele. Purposive sampling techniques were employed to identify focus group discussants and key informants.

2.3. Sources of Data and Methods of Data Collection

The data used in this study were obtained mainly from the National Meteorology Agency of Ethiopia and smallholder farmers. The study used gridded daily data of rainfall, maximum temperature (T_{max}), and minimum temperature (T_{min}) of one grid point in each agro-ecological setting covering the period of 30 years, 1988 to 2017. The dataset has a spatial resolution of 4km that combines station observations and satellite data from the European Organization for the Exploitation of Meteorological Satellites (EUMETSAT) and the US National Aeronautics and Space Administration (NASA). The use of a gridded dataset was necessitated by the limited availability of weather stations [12] and the problem of missing records of rainfall and temperature values in observation data [13], which reduces the validity of time-series trend analysis derived from incomplete data.

Primary data were collected between February and August 2018 from the heads (male or female heads primary responsible for making decisions and generating means of living primarily from agricultural activities) of the sampled smallholder farming households using survey questionnaires and focus group discussions. A paper-based survey questionnaire consisting of close-ended questions was used to collect data on farmers' perceptions of changes in climate variables in their localities. The questionnaire was pilot tested to assure completeness and clarity. Enumerators who have prior data collection experience as well as accustomed to the study areas were recruited and trained on the content of the questionnaire as well as on techniques of interviewing to collect the survey data. On-spot checking of the questionnaires was made to ensure completeness. In addition, skip rules and ranges were introduced to the data entry software to generate automated error reports during data entry. Furthermore, the accuracy of the entered data was assessed by running frequencies and cross-tabulations, and wrong entries were corrected. Qualitative data were collected using focus group discussions (FGDs). It was used to capture farmers' understanding of climate change and variability and the use of weather information to make farming decisions. Each group constituting seven to twelve members, four focus group discussions were conducted in each study district. The discussion, which took an hour on average and conducted in a local language, was moderated by the corresponding author and guided by open-ended questions. The audio-recorded discussion was first transcribed and then translated to English before textual analysis.

2.4. Definition and Measurement of Variables

Farmers' perceptions of changes in temperature and rainfall was measured by asking a close-ended question on whether temperature or rainfall is increasing, decreasing, or observed no change during the past 15 years. Data on the perceptions of extreme events were generated by asking the respondents a yes/no question on whether they had observed the occurrence of a range of climate events (drought, flood, snowfall, frost, delayed onset of rainfall, early termination of rainfall, and waterlogging) during the last 15 years. Given that perception is a function of the demographic and socio-economic characteristics of households, they were considered as explanatory variables to discern the convergence or divergence of farmers' perceptions from the observed meteorological trends. These characteristics were: age of household head (young—20–39; adult—40–59; old—60+), sex of household head (male, female), educational level of household head (no education, primary or above), size of land owned (small—<1 hectare; medium—1–2 hectare; large—≥2 hectare) economic status (low, medium, high, which was classified based on possession of farming tools and household equipment), access to media (no access at all, had access at least once a week which was determined based on farmer's access to a radio or television or newspaper), and social capital (low, medium, high). Social capital was measured on a four-point scale using twelve questions emphasizing household heads' participation in community-based organizations, trust and reciprocity, and contact with locally based formal institutions. The questions were internally consistent to measure social capital (alpha = 0.78). Economic status and social capital were grouped into three classes using the cumulative square root of the frequency method.

2.5. Methods of Data Analyses

2.5.1. Data Quality Assessment

Preliminary assessment of the dataset was performed to ensure that temperature and rainfall data were of acceptable quality. Quality control functions of the ClimPACT2 software [27] were used for automated detection of erroneous data through generation of statistical summary and visual inspection of plots. The results showed that duplicate dates were not found; repeated maximum and minimum temperature values were not observed; negative precipitation values were not present in the dataset; too large values of precipitation (>200 mm) and temperature (>50 °C) were not observed; no large jumps in maximum and minimum temperature values (i.e., temperature difference with the previous day is ≥20 °C) were found; there was no record in which the maximum temperature was lower than the minimum temperature; and no missing value was found for each variable. Quality assessment was followed by the homogeneity test for each meteorological station to identify multiple step change points that could exist in a time series data. The RHtests_dlyPrcp package in R was used for the testing and homogenization of daily precipitation data [28]. Likewise, the RHtestsV4 software package was used to detect and adjust for multiple change points in temperature data that may have first-order autoregressive errors [29]. The monthly series was tested first and the result was used to test the daily series. We used a base period of 1990–2015 and the homogeneity tests were made without using reference series [30]. In the homogeneity tests, we found statistically significant discontinuty in maximum temperature in the lowland area and in minimum temperature in the lowland and midland areas. Adjustments to these daily data were applied using the quantile-matching algorithm [31], and adjusted data were used as homogenized data for trend analysis and the calculation of indices.

2.5.2. Measurement of Variability

The Standardized Anomaly Index (*SAI*) was calculated to discern variation in average temperature across the years. It was calculated using the following formula:

$$SAI = \frac{T_A - T_M}{\delta}$$

where T_A refers to average temperature of a year; T_M shows long-term (1988–2017) mean average temperature; and δ is standard deviation of the long-term average temperature. The annual Rainfall Anomaly Index (*RAI*) was used to identify years and seasons of positive and negative anomalies [32]. It was computed as follows for positive and negative anomalies, respectively:

$$RAI = +3\left(\frac{RF - M_{RF}}{M_{H10} - M_{RF}}\right) \textit{ and } RAI = -3\left(\frac{RF - M_{RF}}{M_{L10} - M_{RF}}\right)$$

where *RF* is the amount of rainfall during a particular year; M_{RF} is the mean rainfall of the observation period (1988–2017); M_{H10} is mean rainfall of the 10 highest values during the observation period; and M_{L10} is the mean of the lowest 10 values of the period of record.

2.5.3. Measurement of Extreme Events

Extreme climate indices were computed using the ClimPACT2 software package in R [27]. The temperature-related extreme indices used in this study were FD (number of frost days with daily Tmin < 0 °C), CSDI (cold spell duration indicator), WSDI (warm spell duration indicator), DTR (diurnal temperature range), TXx (hottest day—maximum value of daily Tmax), TNx (hottest night—maximum value of daily Tmin), TXn (coolest day—minimum value of daily Tmax), TNn (coolest night—minimum value of daily Tmin), TN10p (percentage of cold nights during which daily Tmin is less than 10th percentile), TN90p (percentage of warm nights during which daily Tmin is greater than 90th percentile), TX10p (percentage of cold days during which daily Tmax is less than 10th percentile), and TX90p (percentage of warm days during which daily Tmax is greater than 90th percentile). Rainfall-related extreme indices considered in this study were PRCPTOT (annual total wet day precipitation), CDD (Consecutive Dry Days with precipitation of less than 1 mm), CWD (Consecutive Wet Days with precipitation of at least 1 mm), R10mm (number of heavy precipitation days with at least 10 mm), R20mm (number of very heavy precipitation days with at least 20 mm), R95p (very wet day precipitation where the annual sum of daily precipitation is greater than 95th percentile), Rx1day (maximum 1-day precipitation), and Rx5day (maximum 5-day precipitation). All indices were calculated on an annual basis. For the definition and computation of each index, see Alexander and Herold [26].

Among extreme events, drought was measured using the Standardized Precipitation Evapotranspiration Index (SPEI). It was used to assess yearly patterns of drought in the study areas. Unlike other precipitation-based indices, SPEI is multi-scalar since it integrates the effects of temperature and precipitation [33]. SPEI is computed from climatic water balance which is the difference between precipitation and potential evapotranspiration (PET). Depending on the availability of data, PET was computed using Hargreaves equation that make use of precipitation, maximum temperature, and minimum temperature [34]. SPEI package in R was used for computation of annual indices.

2.5.4. Trend Analysis

The Mann Kendall (MK) test, which is a non-parametric test used to analyze monotonic trends of changes in hydro-meteorological data, was used to examine trends in seasonal and annual temperature and rainfall as well as temperature and rainfall extremes [35,36]. Positive and negative values of MK test results indicate increasing or decreasing monotonic trends, respectively. The magnitude of changes in the trends of rainfall and temperature data was determined using Theil-Sen's slope estimator. The MK test statistic, S, was calculated as:

$$s = \sum_{i=1}^{n-1} \sum_{j=i+1}^{n} sgn\left(X_j - X_i\right)$$

where X_i and X_j refer to the annual values of the climate variables in years i and j, respectively. For time-series data with significant autocorrelation, the modified MK test was used. In this procedure, bias-corrected prewhitening, involving transformation of an autocorrelated sequence into an uncorrelated one before trend testing, was used [37]. This technique enhances the effectiveness of prewhitening in trend analysis by eliminating under- or overestimation of the autocorrelation parameter within the limits of sampling variations [37].

2.5.5. Onset and Cessation of the Rainy Season

The date of onset of *kiremt* rain was determined using a minimum threshold daily precipitation of 1 mm and a total of at least 20 mm of rainfall accumulated in three consecutive days after June 1 [38].

For the *belg* season, due to highly erratic rainfall, onset was defined as a total of 10 mm of rain in three consecutive days after 1 March. To avoid the mislabeling of a false start, the additional criterion was used that the three-days cumulative total was not followed by a dry spell of at least ten consecutive days within 30 days. Daily rainfall of less than 1 mm was considered as a dry spell. The end of the rainy season was identified as the first day after 1 September of each year when the water balance, estimated using R-INSTAT, falls to 0 which causes water stress to crops. Assuming that it is the level which determines the occurrence of severe water stress to crops [39], the maximum soil water holding capacity was set to be 100 mm and evaporation was set at 5 mm per day. The length-of-growth period (LGP) is the duration of time between the onset and cessation dates. The probability of exceedance was calculated using RAINBOW software to determine early, normal, and late onset and cessation dates of rainfall [40].

2.5.6. Analysis of Convergence and Divergence Using Polling

The values of the climate variables were computed for each household. This was performed using GAMS software in three stages. First, using the latitudinal and longitudinal locations of three meteorological stations and the households, the Manhattan distance between each household and the three stations was calculated. Second, based on these metrics, the nearest station, the second nearest station, and the third nearest station were identified. Following the Inverse Distance Weighted interpolation process, we took the inverse of the distance values and normalized these values to sum to 1 to calculate the weighting factors corresponding to the three stations. Lastly, the values for each household were calculated by multiplying the observed meteorological data of the three stations by the respective weighting scores of the households.

Then, the polling method was applied to discern the profiles of households whose perception converges with or diverges from the meteorological data. Polling is a multivariate analysis technique involving a joint analysis of a large number of integer-valued explanatory variables using the maximum likelihood prediction method [41]. It is used to jointly evaluate the roles of different variables in predicting the likelihood of convergence or divergence between meteorological data and perceptions. The joint empirical frequency distribution is defined from observed values of the explanatory variables. Then, conditional frequency distributions are derived from this joint distribution by partitioning the answers by, e.g., S respondents indexed s into a vector y of a dependent variable and a vector x of explanatory variables, taking the frequencies of y conditional on x [41–43].

$$Conditional\ frequency = \frac{m_{yx}}{\sum_{y \epsilon Gx} m_{yx}} \qquad Coverage = \frac{m_{yx}}{\sum_{x} m_{yx}}$$

where m is the mass of the observations, Ys and Xs show integer coded values of the dependent and explanatory variables, respectively. The conditional frequencies show probability estimates of y given profile x. Hence, the set of most probable characteristics associated with each x value (the "winner") has the highest probability of having the desired y outcomes (convergence or divergence). The coverage of a profile x is the mass of a class within profile x divided by the total mass of the relevant group. The edge of the winning profile over the runner up (i.e., the second best guess) is the ratio of their maximum likelihood probabilities (i.e., the share of the population covered by the most likely profile relative to the share covered by the runner-up). Selection of the best profile from the set of explanatory variables was based on the coverage and edge of each combination. In addition to the observed and perceived climate variables, all possible combinations of four explanatory variables were used to identify the profiles of households whose perceptions converge with or diverge from the meteorological results.

3. Results

3.1. Long-Term Trends of Temperature and Rainfall for Different Agro-Ecological Settings

3.1.1. Trends of Temperature

The average annual temperature of the lowland, midland, and highland areas during the observation period was 22.1, 15.5, and 14.6 °C, respectively. As shown in Table 1, the average annual temperature significantly increased by 0.4 °C per decade in the lowland and by 0.3 °C per decade in the midland areas. The result suggests that the increase in average annual temperature in the lowland areas was related to significant increases in average maximum temperatures, whereas, in the midland areas, there was a significant increase in both average maximum and average minimum temperatures. The seasonal pattern shows increasing trends in average annual temperature during both *belg* and *kiremt* seasons in all areas. Increases in average maximum temperatures seem to have caused significant increases in the average temperature of the *belg* season in the lowland areas and average temperatures of the *kiremt* season in the midland areas. In other cases, this significant increasing trend was partly related to an increase in the average minimum temperature.

Table 1. Agro-ecological differences in seasonal and annual trends of temperature and rainfall (1988–2017).

Place	Variable	*Belg*		*Kiremt*		Annual	
		MK	Slope	MK	Slope	MK	Slope
Lowland	Tmin	0.016	0.003	0.269 *	0.039	0.154	0.023
	Tmax	0.615 ***	0.090	0.384	0.060	0.616 ***	0.068
	Tavr	0.333 **	0.032	0.366 **	0.044	0.438 ***	0.042
Midland	Tmin	0.306 *	0.023	0.223	0.015	0.407 *	0.027
	Tmax	0.145	0.022	0.315 *	0.027	0.320 *	0.025
	Tavr	0.319 *	0.021	0.255 *	0.019	0.434 **	0.030
Highland	Tmin	0.497 ***	0.129	0.453 **	0.078	0.409 *	0.065
	Tmax	0.044	0.006	−0.159	−0.028	−0.009	−0.001
	Tavr	0.269 *	0.057	0.347 *	0.044	0.241	0.033
Lowland	Rainfall	0.103	0.661	−0.002	−0.028	−0.039	−0.344
Midland		0.379 *	4.538	0.591 **	15.443	0.621 ***	24.784
Highland		−0.136	−0.707	−0.021	−0.699	−0.062	−2.022

* $p < 0.05$, ** $p < 0.01$, *** $p < 0.001$.

3.1.2. Trends of Rainfall

The MK test result shows contrasting patterns of the trends of annual rainfall (Table 1). In the lowland and highland areas, the decline in rainfall was not statistically significant, indicating no trend in rainfall time series. Conversely, there was a significant increasing trend in annual rainfall in the midland area. In this area, *belg* and *kiremt* rainfall significantly increased at respective rates of 45 and 154 mm per decade.

3.2. Annual and Seasonal Variability of Temperature and Rainfall

3.2.1. Temperature Variability

There was noticeable inter-annual and seasonal variability in the average temperature of the study areas (Figure 2). In the lowland and midland areas, the annual average temperature was less than the overall average of the observation period during the first two decades. However, the anomaly indices were positive in recent years, suggesting that the areas are warming. The pattern in the midland areas shows warmer years during the first one and half decade and colder years recently. During most of the years since 2001, the average annual temperature was consecutively lower than the 30 years average. The annual variability of average temperature was very high during the *belg* season compared to the *kiremt* season in all areas.

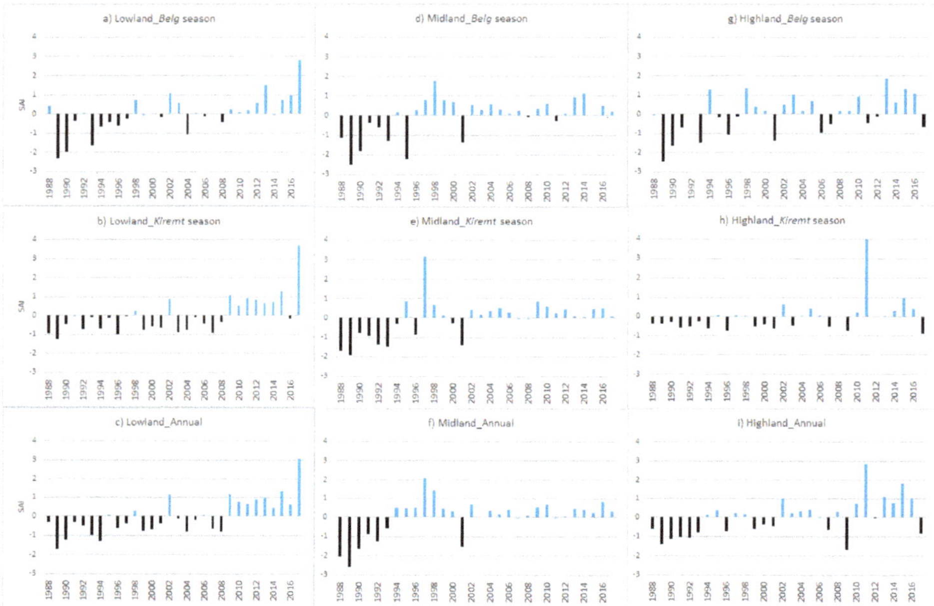

Figure 2. Trends of standardized anomaly index of average annual temperature by agro-ecological settings and season (1988–2017).

3.2.2. Rainfall Variability

The long-term average annual rainfall of the midland areas was 1185 mm followed by the highland and the lowland areas with respective amounts of 785 and 509 mm. There was noticeable inter-annual and inter-seasonal variability of rainfall (Figure 3). Negative rainfall anomalies were commonly observed during the *belg* season in all agro-ecological settings, being more apparent in the highland areas in recent years. There were also several years of below-average rainfall during the main farming season in the three areas. The *kiremt* season in all areas was characterized by yearly differences in rainfall anomalies. Consecutive years of below-average *kiremt* rainfall was observed mainly in the highland areas. Annual rainfall was below 30 years average in about 47% of the observations in the lowland areas during both *belg* and *kiremt* seasons. In the midland and highland areas, below-average *kiremt* rainfall was observed during 47% and 60% of the observations, respectively.

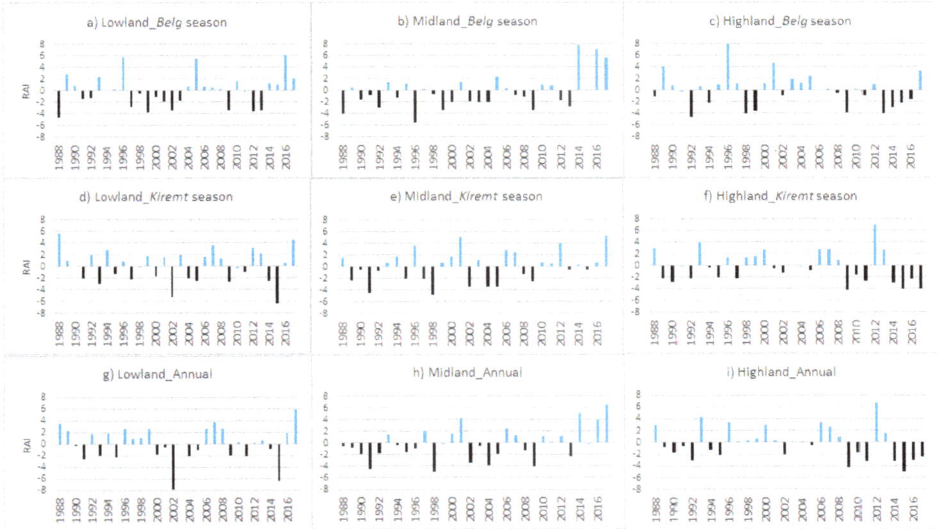

Figure 3. Trends of annual rainfall anomaly index by agro-ecological settings and season (1988–2017).

3.3. *Analysis of Extreme Events*

3.3.1. Trends of Extreme Precipitation Indices

Annual total wet-day precipitation (PRCPTOT) has significantly increased by about 25 mm per year in the midland areas (Table 2). Moreover, the number of consecutive wet days significantly increased in the midland areas by about three days every decade. The annual number of days with precipitation of at least 10 and 20 mm has increased in these areas respectively by about 9 and 4 days per decade. Similarly, there was a significant increase in maximum 1-day (RX1day) and maximum 5-day (RX5day) precipitation in the midland areas by about 9 and 19 mm per decade, respectively. These figures denote the higher chances of occurrence and increasing trend of flooding in these areas. On the other hand, there was a significant increase in the number of consecutive dry days by 23 days per decade in the highland areas. The amount of maximum 1-day precipitation has significantly decreased by 3.3 mm a decade in the highland areas. There were no significant trends in any of the precipitation indices in the lowland areas.

Table 2. Trends of precipitation indices by agro-ecological settings (1988–2017).

Variables	Lowland		Midland		Highland	
	MK	Slope	MK	Slope	MK	Slope
PRCPTOT	−0.037	−0.369	0.621 ***	25.037	−0.076	−2.180
CDD	0.090	0.750	−0.097	−0.667	0.364 *	2.300
CWD	−0.120	−0.053	0.291 *	0.333	−0.114	−0.087
R10mm	−0.044	−0.001	0.616 ***	0.945	0.021	0.001
R20mm	0.134	0.001	0.389 **	0.394	−0.069	0.001
R95p	0.116	0.625	0.461 ***	10.529	−0.092	−0.667
RX1day	0.126	0.162	0.397 *	0.880	−0.299 *	−0.333
RX5day	0.225	0.400	0.500 ***	1.909	−0.007	−0.022

* $p < 0.05$, ** $p < 0.01$, *** $p < 0.001$.

3.3.2. Trends of Extreme Temperature Indices

The results in Table 3 show an increasing trend of extreme temperature indices. Annual maximum daily maximum temperature (TXx) has significantly increased by approximately 0.4 °C per decade in both the lowland and highland areas. Likewise, minimum daily maximum temperature (TXn) has significantly increased by 0.8 °C per decade in the lowland areas. Annual maximum daily minimum temperature (TNx) has significantly increased in the midland and highland areas. In the lowland areas, a significant decrease in the percentage of cold days (TX10p) by about 6% was noted. Conversely, the annual percentage of warm days (TX90p) has increased significantly by 7.7% in the lowland and by 5.2% per decade in the highland areas. The percentage of cold days (TN10p) has declined in all areas, but the decline was statistically significant in the lowland (by 2.5%) and midland (by 9.4%) areas. The percentage of warm nights (TN90p) has significantly increased in the midland and highland areas at 4.9% per decade. Consequent to changes in daily maximum and minimum temperature, there was a change in DTR. It significantly increased in the lowland areas by about 0.4 °C per decade but significantly decreased in the midland areas by about 0.7 °C per decade. The decline in cold spell (CDSI) was significant only in the midland areas. Contrariwise, the number of warm spell days (WSDI) has significantly increased in both the lowland and highland areas.

Table 3. Trends of extreme temperature indices by agro-ecological settings (1988–2017).

	Lowland		Midland		Highland	
	MK	**Slope**	**MK**	**Slope**	**MK**	**Slope**
FD	-	-	−0.032	−0.065	0.248	0.001
CSDI	−0.025	0.001	−0.417 **	−0.636	−0.153	0.001
WSDI	0.490 **	0.375	−0.017	0.001	0.470 ***	0.001
DTR	0.366 **	0.036	−0.419 **	−0.074	0.076	0.007
TXx	0.419 **	0.044	0.140	0.026	0.258 *	0.036
TXn	0.490 ***	0.081	−0.247	−0.057	0.162	0.035
TNx	0.205	0.020	0.400 **	0.122	0.397 **	0.06
TNn	0.110	0.012	0.116	0.059	0.149	0.019
TX10p	−0.571 ***	−0.613	0.071	0.133	−0.221	−0.194
TX90p	0.596 ***	0.768	−0.020	−0.012	0.473 ***	0.519
TN10p	−0.292 *	−0.249	−0.582 ***	−0.937	−0.193	−0.205
TN90p	0.131	0.160	0.436 ***	0.485	0.384 **	0.488

* $p < 0.05$, ** $p < 0.01$, *** $p < 0.001$.

3.3.3. Trends of Drought

Figure 4 shows the yearly patterns of drought measured using SPEI. The number of drier years was increasing in the lowland and highland areas recently, whereas the reverse was observed in the midland areas. In the lowland areas, the years before 2000 were mainly wet. Since 2000, most of the years were drier and it was observed consecutively between 2001 and 2005 as well as between 2009 and 2015. In the highland areas, although it was not as frequent as the lowland areas, there were intermittent dry years. However, it has occurred frequently since 2009. The pattern in the midland areas is somewhat different. Although dry years were occurring frequently and consecutively during the first two decades of observation, increase in the number of wet years was observed during the last decade. There were only two drier years since 2009 in the midland areas, whereas it was observed six times in both lowland and highland areas. Drier years constituted 53% of the years of observation in the lowland and midland areas, and 47% in the highland areas.

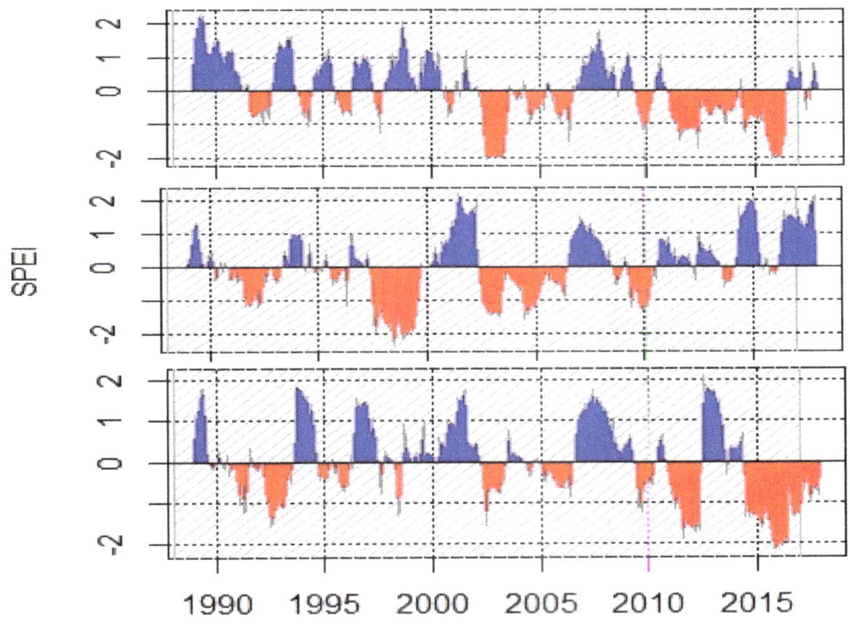

Figure 4. Trends of the Standardized Precipitation Evapotranspiration Index (SPEI) in the lowland (**top**), midland (**middle**), and highland (**bottom**) areas.

3.4. Onset and Cessation of Rainfall

There was a failure of *belg* rain (no consecutive three days rainfall total of 10 mm) for one year in lowland and midland areas and for five years in the highland areas (Table 4). Most of the observation years in the highland and lowland areas were characterized by false onset in which the beginning of rainfall was followed by more than ten days of a dry spell during the subsequent 30 days. It was only in one-third of the years of observation that there was a proper onset of *belg* rain in the lowland and highland areas. This season is also characterized by a small number of rain days and a prolonged period of dry spells, some being longer than 50 days in all areas.

Table 4. Characteristics of *belg* season in three agro-ecological settings.

Indicators		Lowland	Midland	Highland
Onset (number of years)	Failure of *belg*	1	1	5
	False onset	19	5	15
	Proper onset	10	24	10
Months of onset (number of years)	March	2	10	2
	April	4	12	4
	May	4	2	4
	Cessation	1st week of May	1st week of May	1st week of May
Length of dry spell days	Mean	30	25	34
	Longer	50	54	58
	Shorter	17	11	17
Number of rain days	Mean	11	21	11
	Minimum	3	3	2
	Maximum	25	39	22

There were noticeable inter-annual and agro-ecological differences in the onset and cessation of *kiremt* rain (Figure 5). The date of onset was generally early in the midland areas and late in the lowland areas. In the midland areas, it was as early as the 153rd day of the year (1 June) in 2001 and as late as the 183rd day of the year (1 July) in 1989. In the lowland areas, early onset was observed in 1996 on the 158th day of the year (6 July). Very late onset was observed on the 231st day (18 August) in 2015. The average date of onset in the lowland areas was 12 July. In the highland areas, the early date of onset was observed on 5 June 1996. Rainfall mainly stopped in September in the lowland areas (245th–274th day). Although the end of the *kiremt* season was observed to be in September in a few years and exceptionally late (the first week of November) in 1999 and 2000, October was the main ending time in the midland areas. In the highland areas, the season variably ended either in September or October. Owing to these differences in the dates of onset and cessation, the length-of-growth period (LGP) varied between years and agro-ecological settings. Since there was late onset and early termination, LGP was shorter in the lowland areas. On the other hand, early onset and late cessation elongated LGP in the midland areas. Although the LGP noticeably varies across years, most of the major cereal crops produced in the areas such as sorghum, maize, and teff require longer period.

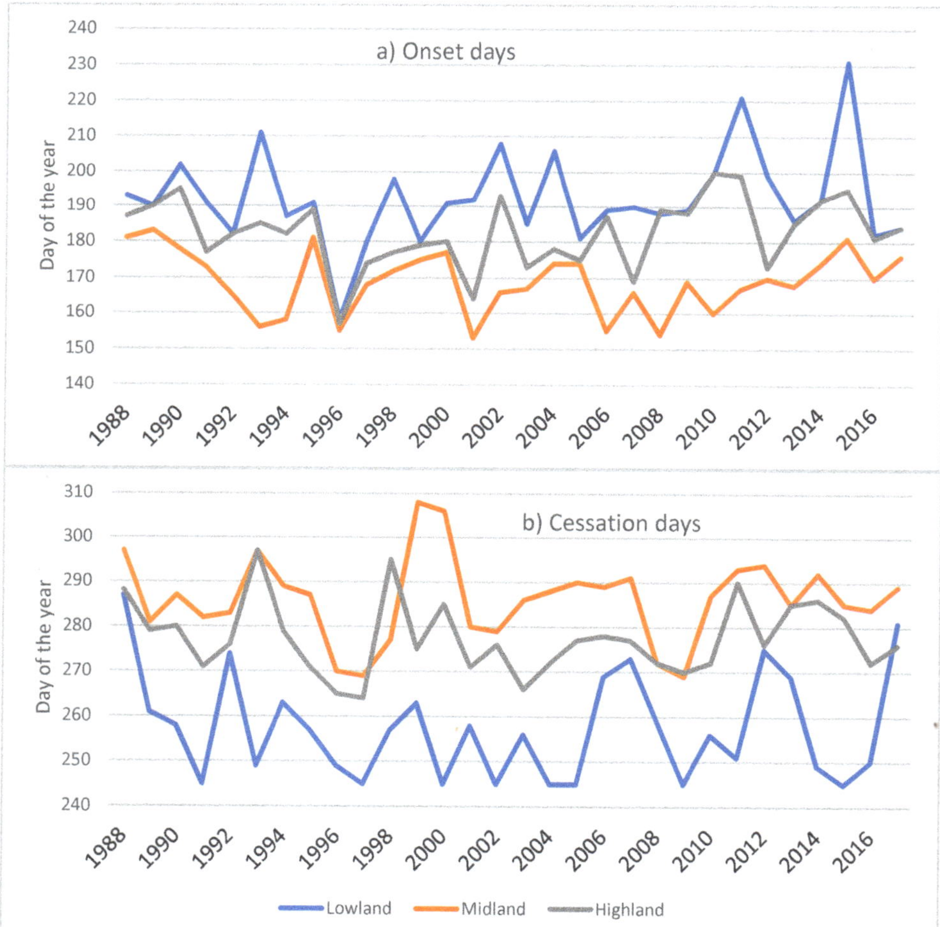

Figure 5. Trends of onset and cessation days of rainfall by agro-ecological settings.

Following previous works [44], the probability of exceedance of 75%, 50%, and 25% were used to categorize the time of onset and cessation of *kiremt* rainfall to be early, normal, or late. The time of onset of rainfall was normal in 18, 15, and 14 years of observation in the lowland, midland, and highland areas, respectively. During the remaining years, it was either early or late. The time of cessation was early for 12 years in the lowland areas, but it ceased at a normal time for 16 years in the midland and for 12 years in the highland areas. As shown in the joint consideration of time of onset and cessation in Figure 6, it was only in about one-fourth of the observation period that the time of onset and cessation of rainfall was normal in the lowland and midland areas. In the highland areas, early onset was followed by either early or normal cessation during most of the years. In about one-tenth of the years of observation, the rain started late and stopped at a normal time. The time of onset and the time of cessation that are assumed to be favorable for agricultural activities (early onset and late cessation, normal onset and normal cessation, normal onset and late cessation) were observed during a few years only; this holds for all areas.

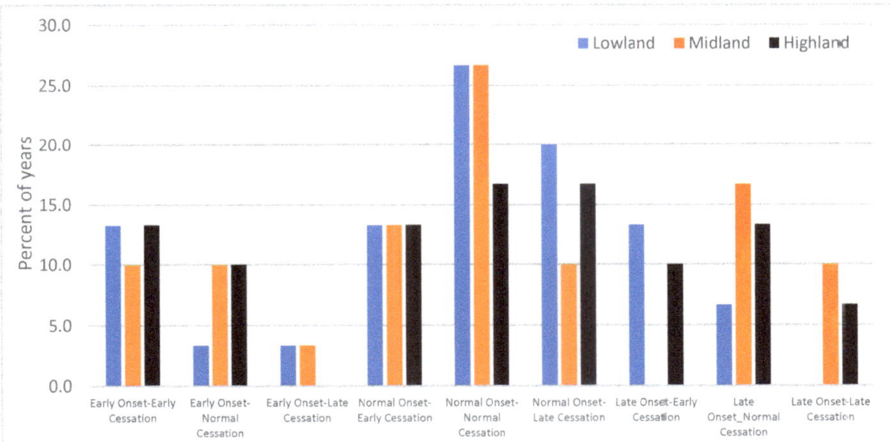

Figure 6. Distribution of time of onset and cessation of *kiremt* rainfall by agro-ecological settings.

3.5. Farmers' Perceptions of Climate Change, Variability, and Extreme Events

In the household survey, the major climate-related events reported by respondents were climate variability (delayed onset and early termination of rainfall) and the occurrence of extreme events (Figure 7). The values in the Figure show the percentage of farmers responding "yes" to the question on the perceived occurrence of the extreme events. More than 90% of the household heads in each area reported delayed onset and early termination of rainfall. The percentage of household heads who reported drought was the highest in the lowland areas and it declined consistently as altitude increases. Frost and waterlogging were mainly the problem of farmers in the highland areas. Compared to the other two areas, the percentage of household heads who reported flood and snowfall was higher in the midland areas.

The problem that was commonly raised during the FGDs in all areas was lack of rainfall. Farmers in the lowland areas stated the problem as follows: *"it is lack of rainfall that makes us inferior to other people. Our neighbors in the other kebeles play with water. But, in this kebele, it is lack of rainfall that makes us and our children jobless; that changes our skin color; that changes our hair color to grey before we get old"* [FGD-L-9]. Farmers in another village in the lowland area further explained that they are not able to benefit from their fertile land due to lack of rainfall saying that: *"if there is rain, the hair even grows on the bare head of a person, let alone on this land. There is a lack of rainfall"* [FGD-L-10]. In particular, lack of rainfall is most pronounced during the *belg* season in the three areas, due to which farmers indicated that they are forced to produce only once a year during the *kiremt* season, abandoning the production of *belg* crops. Farming in *kiremt* is also affected by delayed onset and early termination of rainfall. As pointed-out by farmers, *"the rain falls late after sowing time passes . . . and due to early termination of rainfall, the farmlands get dry and crops do not grow very well"* [FGD-M-7].

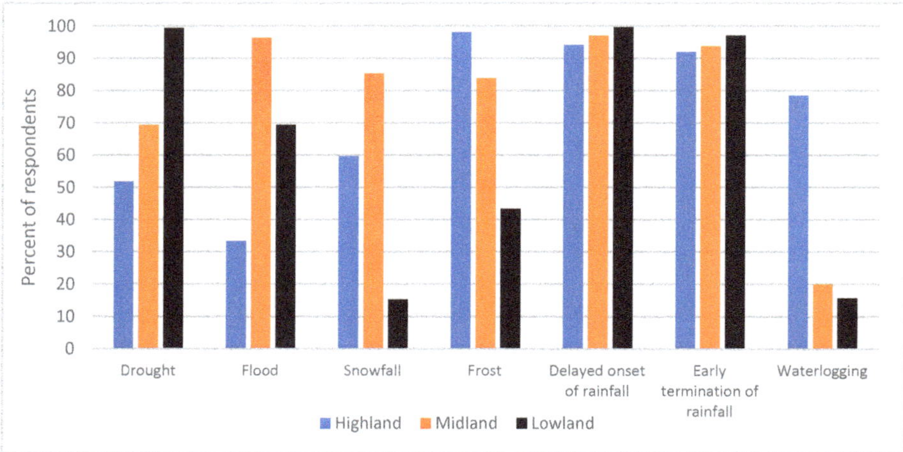

Figure 7. Distribution of households' perceptions of climate variability and extreme events by agro-ecological settings.

Similar to the results of the household survey, evidence from the FGDs indicated that extreme events occur in the study areas with varying magnitude. Drought was boldly stated by residents of the lowland areas as follows: *"It is this problem [drought] that made us lag behind; that wasted our age; that depleted our resources. We have wide farmland; we are healthy. Our major bottleneck is drought . . . . drought made as beggars"* [FGD-L-9]. Frost is mainly raised as a problem in the highland areas. Farmers explained that *"it [frost] comes when the crop matures. When it comes, our effort of one year is damaged in one day"* [FGD-H-3]. According to farmers' observation, though the cold period begins in November, it has become colder than in the past and the cold period starts as early as September. The other problem identified by farmers in the highland areas was waterlogging. The frequently mentioned extreme event in the midland areas was heavy rainfall in summer, which causes erosion and exacerbates the problem of a landslide. Farmers explained that *"in the past, during the rainy season, there were foggy days with drizzle rainfall for the whole day that was conducive for agriculture. Now, rain falls heavily and erodes our soil, which also becomes a cause for a landslide"* [FGD-M-5].

Farmers have developed traditional methods of forecasting weather conditions and making farming decisions. Owing to the absence of established means of knowledge transmission, these methods are not generally known in some villages. Farmers revealed that the local-knowledge-based traditional forecast system involving the observation of various signals is known only by few elderlies and that there is a generation gap in valuing the roles of these traditional forecasts in the usual farming activities. The information obtained from the traditional forecast is not considered to be dependable for farming decision making, as farmers said, as the observed weather condition deviates from the predictions based on traditional knowledge and expectations. Consequently, observing traditional signals is not an assurance that rain will come or will come at the expected time. The limited role of traditional methods of the forecast increases the demand for modern weather information for farming activities and climate risk management. However, the farmers noted that they do not have access to weather information. When it is available through media broadcasts, it is often reported at a higher spatial scale which, according to farmers, does not show the local weather condition and hence is not relevant for farming decisions. In addition, information on the expected time of onset and cessation of rainfall is generally missing in the weather forecasts of higher spatio-temporal resolution.

3.6. Convergence and Divergence Between Meteorological Data and Farmers' Perceptions

With no noticeable difference between the three agro-ecological settings, most of the farming households were headed by young persons and adults (Table 5). Most of the households were headed by males. About two-thirds of the households had no formal education. Nearly one-third of the farmers had less than one hectare of land, whereas the percent of households owing greater than two hectares was relatively higher in the highland areas. Half of the farming households had medium economic status. In the highland areas, close to half of the farmers were economically better-off, whereas, in the midland areas, only about one-in-ten households had higher economic status. While most of the farmers had moderate social capital, higher percentage of farmers in the highland areas had high social capital. Slightly more than half of the farmers had no access to media.

Table 5. Percentage distribution of households' demographic and socio-economic characteristics by agro-ecological settings.

Variables	Categories	N	All Households	Highland	Midland	Lowland
Age of household head	20–39	298	36.8	32.9	29.5	37.6
	40–59	313	38.6	31.3	32.6	36.1
	≥60	199	24.6	37.2	40.2	22.6
Sex of household head	Male	702	86.7	34.3	32.8	32.9
	Female	108	13.3	26.9	37.0	36.1
Education of household head	No education	547	67.5	36.0	30.2	33.8
	Primary and above	263	32.5	27.8	39.9	32.3
Size of land owned (in hectare)	<1	284	35.1	33.8	34.9	31.3
	1–2	222	27.4	19.8	41.4	38.7
	≥2	304	37.5	42.8	26.0	31.2
Economic status of a household	Low	180	22.2	21.7	51.7	26.7
	Medium	403	49.8	29.8	36.2	34.0
	High	227	28.0	48.9	13.7	37.4
Social capital of a household	Low	244	30.1	29.9	27.0	43.0
	Medium	360	44.4	30.8	38.9	30.3
	High	206	25.4	41.7	31.1	27.2
Head's access to media	No access	437	54.0	30.9	44.6	24.5
	At least once a week	373	46.0	36.2	20.1	43.7
Total		810	100	33.3	33.3	33.3

In general, the change in temperature was correctly perceived as about three-fourth of the farmers correctly perceived that temperature was increasing (Table 6). However, for rainfall, perception and actual results aligned for only 5% of the farmers. While the measurement showed no significant change in the amount of rainfall during 30 years of observation, most of the farmers (62.2%) perceived that it was either decreasing (55.9%) or increasing (6.3%). In areas where rainfall was increasing, the perception of 32.7% of the farmers was not consistent with the actual result as they perceived that it was decreasing (32.3%) or that there was no change (0.4%). The occurrence of drought was correctly perceived by half of the farmers. The perception of about one-third of the farmers converged with the meteorological result concerning the occurrence of flood. Variability in the time of onset and cessation of rainfall was correctly perceived by most of the farmers. Close to two-thirds of the farmers correctly perceived late onset of rainfall. Likewise, the perception of more than half of the farmers (55.2%) converged with the meteorological data that there was an early cessation of rainfall during the *kiremt* season. However, although the meteorological data showed otherwise, slightly more than one-third of the farmers wrongly perceived that there were late onset and early cessation of *kiremt* rainfall.

Table 6. Percentage distribution of variation of farmers' perceptions of climate change and variability by observed meteorological results (n = 810).

Measured Variables	Categories	Farmers' Perceived Changes	
		Yes	No
Temperature change	Yes	76	17.1
	No	6.8	0
Rainfall change	Yes	0.7	32.7
	No	62.2	4.3
Drought occurrence	Yes	50.4	16.3
	No	23.1	10.2
Flood occurrence	Yes	32.2	1.2
	No	34.1	32.5
Late onset of rain	Yes	63.7	3.0
	No	33.2	0.1
Early cessation of rain	Yes	55.2	4.0
	No	39.0	1.9

Table 7 shows the profiles of farmers whose perceptions converges with and diverges from observed rainfall and temperature records. In the midland areas, the perception of farmers converged with the statistical result showing a significant increase in rainfall over the 30 years. Likewise, male household heads and those who had access to media had more accurate perception of rainfall trends. The share of farmers with correct perception of increasing rainfall in this winning profile was 34%. In the highland areas, farmers' perception converged with the meteorological results that there was no change in the amount of rainfall. Old-age farmers and those who had access to media correctly perceived that there was no change in the trend of rainfall. On the other hand, with the highest share of farmers with diverging perception included in this winning profile (86%), there was a divergence between meteorological results and perception among farmers with no education, no access to media, a large size of land (≥2 ha), and medium economic status. Male household heads, those who had medium social capital, and those with access to media had a correct perception of an increase in temperature that converges with meteorological results. Farmers residing in the lowland areas also correctly perceived an increasing trend of temperature. A wrong perception of temperature was observed among farmers with no education, no access to media, a medium or large size of land, and medium economic status. Temperature perception that diverges from the meteorological result was also noticed among farmers residing in the midland areas.

Table 7. Winning profiles of convergence and divergence between measurement and farmers' perceptions of rainfall and temperature changes.

	Variables and Winning Profiles				Coverage	Edge
Rainfall						
Yes-Yes	Mrain—Increasing; Prain—Increasing					
a	PWATLOG	SEX	MEDIA	AGRO	34%	1.0
b	No	Male	Yes	Midland		
No-No	Mrain—No change; Prain—No change					
a	AGE	SEX	MEDIA	AGRO	14%	1.27
b	Old	Male	Yes	Highland		
No-Yes	Mrain—No change; Prain—Increasing					
a	EDUC	LAND	ECON	MEDIA	86%	1.22
b	No	Large	Medium	No		
Yes-No	Mrain—Increasing; Prain—Decreasing					
a	EDUC	LAND	ECON	MEDIA	17%	1.22
b	No	Large	Medium	No		

Table 7. *Cont.*

	Variables and Winning Profiles				Coverage	Edge
Temperature						
Yes-Yes	Mtemp—Increasing; Ptemp—Increasing					
a	SEX	SOCAP	MEDIA	AGRO	10%	1.15
b	Male	Medium	Yes	Lowland		
No-Yes	Mtemp—No change; Ptemp—Increasing					
a	EDUC	LAND	MEDIA	AGRO	21%	1.43
b	No	Medium	No	Midland		
Yes-No	Mtemp—Increasing; Ptemp—Decreasing					
a	EDUC	LAND	ECON	MEDIA	19%	2.30
b	No	Large	Medium	No		

a—Variables; b—Winning profiles; Mrain—Measured rainfall; Prain—Perceived rainfall; Mtemp—Measured temperature; Ptemp—Perceived temperature; PWATLOG—Perceived waterlogging; AGE—Age of household head; SEX—Sex of household head; EDUC—Educational level; MEDIA—Access to media; ECON—Economic status of households; LAND—Size of landholding; SOCAP—Social capital; AGRO—Agro-ecological setting.

The profiles of households whose perceptions of drought and flood was consistent with observed trends of drought and flood are shown in Table 8. Drought was computed using the Standardized Precipitation Evapotranspiration Index, whereas flood was measured using a proxy indicator of change in the number of heavy precipitation days (R10mm). Accurate perception of drought varies by agro-ecological settings. While farmers in the lowland areas perceived the occurrence of drought, which was convergent with the meteorological result, the absence of drought was correctly noticed by farmers in the midland areas. Adults, males, as well as farmers with access to media at least once a week and higher social capital correctly perceived the occurrence of drought. However, the perception of farmers in the highland areas diverged from the meteorological result. Although the observation showed drought occurrence, it was not perceived by farmers. The drought perception of farmers who had no education, no access to media, owned a large size of land, and medium economic status diverged from the meteorological data. The coverage this profile has of the relevant farmers was 27%. The likelihood of having the highest probability to have this diverging perception was also relatively higher (1.95). Convergence and divergence in flood perception also vary by agro-ecological settings. In the midland areas, farmers' perception of a flood as a key problem was confirmed by the meteorological data. In addition, the perception of farmers with at least primary-level education, owned a small size of land, and medium social capital on the occurrence of flood aligned with the observed data. The perception of farmers in the highland areas also converged with the meteorological data that there was no flood. In the lowland areas, although farmers perceived the occurrence of flood, it was not supported by meteorological data. Lack of education, lack of access to media, being a young household head, and medium economic status further characterizes households with an inaccurate perception of flood occurrence. The share of farmers failing to recognize the actual occurrence of flood in the winning profile was 40%. It is also worth noting that there was heterogeneity among farmers from the same agro-ecological settings as there were farmers from the midland areas who had an inaccurate perception of drought and flood occurrence.

The profile of households with convergent and divergent perceptions of the time of onset and cessation of *kiremt* rainfall is shown in Table 9. Accurate perception of the occurrence of late onset of rainfall was observed among male household heads, owners of the small size of land, households with medium social capital, and residents of the midland areas. Eleven percent of the farmers with the right prediction are characterized by this winning profile. The perception of a late onset of rainfall that deviates from the meteorological result was observed among farmers in the lowland and highland areas. In the lowland areas, farmers perceived a late onset of rainfall, which was not consistent with the actual measurement. In the highland areas, although the meteorological data showed a late onset of rainfall, farmers' perception diverged from this. In addition, lack of education, low social capital, lack of access to media, young household heads, and ownership of medium size of land characterized households whose perception diverged from the meteorological data. Early cessation of *kiremt* rainfall was accurately perceived by adults, males, owners of the small size of land, medium social capital, and poor farmers. Although the share of farmers included in the profile was smaller, lack of education, ownership of a large size of land, lack of access to media, and medium economic status characterized farmers whose perception of the time of cessation of rainfall diverged from the observed meteorological data. However, as indicated by the edge value, it is with higher certainty that the combination of these variables characterizes the winning profile.

Table 8. Winning profiles of convergence and divergence between measurement and farmers' perceptions of the occurrences of drought and flood.

	Variables and Winning Profiles				Coverage	Edge
Drought						
Yes-Yes	Mdrought—Yes; Pdrought—Yes					
a	SEX	SOCAP	MEDIA	AGRO	16%	1.37
b	Male	Medium	Yes	Lowland		
No-No	Mdrought—No; Pdrought—No					
a	AGE	ECON	SOCAP	AGRO	12%	1.43
b	Adult	Medium	High	Midland		
No-Yes	Mdrought—No; Pdrought—Yes					
a	EDUC	ECON	MEDIA	AGRO	22%	1.05
b	No	Medium	No	Midland		
Yes-No	Mdrought—Yes; Pdrought—No					
a	EDUC	LAND	MEDIA	AGRO	27%	1.95
b	No	Large	No	Highland		
Flood						
Yes-Yes	Mflood—Yes; Pflood—Yes					
a	EDUC	LAND	SOCAP	AGRO	13%	1.10
b	Primary+	Small	Medium	Midland		
No-No	Mflood-No; Pflood-No					
a	EDUC	SOCAP	MEDIA	AGRO	14%	1.41
b	No	Medium	Yes	Highland		
No-Yes	Mflood—No; Pflood—Yes					
a	EDUC	ECON	MEDIA	AGRO	14%	1.19
b	No	Medium	No	Lowland		
Yes-No	Mflood—Yes; Pflood—No					
a	AGE	SEX	ECON	AGRO	40%	1.32
b	Young	Male	Medium	Midland		

a—Variables; b—Winning profiles; Mdrought—Measured occurrence of drought; Pdrought—Perceived occurrence of drought; Mflood—Measured occurrence of flood; Pflood—Perceived occurrence of flood.

Table 9. Winning profiles of convergence and divergence between measurement and farmers' perceptions of the time of onset and cessation of rainfall.

	Variables and Winning Profiles				Coverage	Edge
Late onset						
Yes-Yes	MLateOnset—Yes; PLateOnset—Yes					
a	SEX	LAND	SOCAP	AGRO	11%	1.28
b	Male	Small	Medium	Midland		
No-Yes	MLateOnset—No; PLateOnset—Yes					
a	EDUC	LAND	SOCAP	AGRO	12%	1.1
b	No	Medium	Low	Lowland		
Yes-No	MLateOnset—Yes; PLateOnset—No					
a	AGE	EDUC	MEDIA	AGRO	17%	1.32
b	Young	No	No	Highland		
Early cessation						
Yes-Yes	MEarlyCessation—Yes; PEarlyCessation—Yes					
a	SEX	LAND	SOCAP	MEDIA	10%	1.23
b	Male	Small	Medium	Yes		
No-No	MEarlyCessation—No; PEarlyCessation—No					
a	AGE	SEX	ECON	MEDIA	2%	1.0
b	Adult	Male	Poor	No		
No-Yes	MEarlyCessation—No; PEarlyCessation—Yes					
a	EDUC	LAND	ECON	MEDIA	8%	1.09
b	No	Large	Medium	No		
Yes-No	MEarlyCessation—Yes; PEarlyCessation—No					
a	EDUC	LAND	ECON	MEDIA	3%	2.02
b	No	Large	Medium	No		

a—Variables; b—Winning profiles; MLateOnset—Measured late onset of rainfall; PLateOnset—Perceived late onset of rainfall; MEarlyCessation—Measured early cessation of rainfall; PEarlyCessation—Perceived early cessation of rainfall.

4. Discussion

In Ethiopia, long-term changes in climate conditions, the inter-annual and seasonal variability of temperature and rainfall, and the frequency of occurrence of extreme events are detrimental for agricultural activities and food security. The results of the analyses of temperature and rainfall time series data reveal a variety of changes in climate conditions of the study areas and notable differences between the agro-ecological settings. The findings generally show increasing warming, annual and seasonal rainfall variability, increasing extreme events, variation in rainfall onset and cessation dates, and convergence and divergence between measured variables and perceptions.

The average temperature of the study areas is increasing which reflects the rising global mean temperature. Such increasing trends of temperature in Ethiopia are also reported in other studies [12,16]. Concerning rainfall, we found a significantly increasing trend in the midland areas but no trend in the highland and lowland areas. Like other parts of Ethiopia [1,12], the study areas are characterized by inter-annual and intra-seasonal rainfall variability. A shift in rainfall anomalies each year indicate the repeated occurrence of rainfall deficits during the farming seasons. In addition, *belg* season is characterized by either total failure of rainfall or false start, both referring to a lack of rainfall to undertake farming activities. Although easterly winds from the Indian Ocean and shifts in the Inter-Tropical Convergence Zone are the main underlying factors for rainfall variation in Ethiopia [10], the diverse topography of the country plays a crucial role in the variability of temperature and rainfall distribution across agro-ecological areas.

The effect of high variability in the amount and distribution of *belg* rain on the livelihood of smallholder farmers in Ethiopia is noticeable for various reasons. First, since the season comes after a long dry season, *belg* rain is crucial for water availability, the production of *belg* crops, and the growth of pasture for livestock. Second, rainfall variability during the *belg* season constrains farmers' options to produce *belg* crops [10]. Although *belg* crops are important for farmers to bridge the time until the harvest of summer crops without significant food shortage, the risk of planting these crops is very high due to prolonged dry spells and short growth periods. As noted by farmers, these result in crop failure, lower crop productivity, and the abandoning of production of *belg* crops, ultimately increasing vulnerability to food insecurity. Third, the poor performance of *belg* rain affects crop production activities during the subsequent main rainy season by influencing the soil moisture and thereby the time of planting long-duration crop varieties such as maize and sorghum [15]. Variability in the amount of rainfall and the time of onset and cessation are also challenging for farmers as they cannot follow conventional farming calendars. Variability or failure of rainfall further exacerbates under- and/or unemployment due to loss of farming days.

There are also challenges associated with reliance on *kiremt* rainfall for crop production. Due to yearly variation in the time of onset and cessation, there is high uncertainty in farmers' decisions of types of crops to be produced and time of planting. In the lowland areas, for instance, owing to the normal or late onset and early cessation, LGP is shorter and the rain stops before the ripening of crops. Consequently, farmers harvest substantially lower yields or there is a complete failure of crops. In the midland areas, too, early termination of rainfall at the beginning of September makes crops infertile. Farmers in the highland areas would benefit from early rainfall and early planting as crops are harvested earlier. However, late onset results in late planting, which makes crops with longer-duration growth periods vulnerable to very cold weather that often starts in September/October and lasts until December, leading to an immense loss of yields.

Consistent with previous findings [2,14,16], the results suggest increasing warm days and nights and decreasing cold days and nights. An increase in extreme events causes changes to human systems much more than changes in average climate conditions [25]. Warming leads to higher rates of evaporation [1] and puts additional stresses on water resources [3], which, through a reduction of crop and livestock production, escalates livelihood vulnerability. There is also a risk of an increase in pests, weeds, and disease which affect both crop and livestock production [5]. The significant values of heavy (R10mm) and very heavy (R20mm) precipitation as well as maximum 1-day (RX1day) and maximum

5-day (RX5day) precipitation denote a high intensity rainfall in the midland areas. The occurrence of flooding, which was mentioned by farmers as one of their problems, is partly explained by the significant increase in heavy precipitation in the area. The effect of flooding is aggravated by the sloping topography of the area and lack of vegetation cover. The occurrence of landslides in the midland areas is also partly related to heavy rainfall. In the lowland and midland areas, frequent occurrences of CDDs and drier years have a deleterious effect on farming activities and farmers' livelihoods.

Both convergence and divergence are observed between farmers' perceptions and the results of meteorological data. Despite heterogeneity among farmers, the perception, of more than half of them, of temperature, the occurrence of drought, and the late onset and early cessation of rainfall was in unison with the meteorological data. There was a clear overlap between the perception that temperature is increasing and the statistically significant increasing trends of temperature data. This finding is congruent with many previous studies that showed consistency between perception and measurement of temperature [45]. However, there was variation, especially regarding rainfall trends. Farmers' perception of decreasing rainfall was not supported by statistical data. We found an increasing trend of rainfall in the midland areas but no significant change in the highland and midland areas. This finding is consistent with previous studies showing that farmers' perception of declining rainfall deviates from rainfall records [19,24,46]. Farmers' perception of trends of rainfall may not corroborate observed meteorological trends for various reasons. As noted in a previous study [20], farmers' perception of decreasing rainfall while it is not happening might show failure in the expected utility and availability heuristic. In line with the utilitarian perspective, farmers' perception of declining rainfall more reflects its livelihood impacts in terms of a decline in agricultural production and food security [18,22,24], which are also caused by factors other than climate change such as a decline in soil fertility and limited use of farm technologies [18,19]. Farmers' perception of declining rainfall might also arise from changes in the seasonality of rainfall and frequency of occurrence of extreme events instead of a change in the total amount of rainfall [46]. For farmers, change in rainfall is perceived as a process, not in terms of quantity [47]. They tend to base their perceptions of recent weather conditions and extreme events as well as on the wrong timing of heavy rainfall instead of long-term changes in average conditions [18,48]. When judging changes in rainfall, the time reference of farmers could be the period when rain is expected for planting, whereas the scientific analysis refers to long term or annual/seasonal changes [49]. Farmers also refer to the amount and distribution of rainfall during the cropping season to form perceptions.

Extreme events such as drought and rainfall variability are more accurately perceived by farmers. Drought takes a central position in the memory of people as it directly affects water and food availability [24], which contributed to a perception aligned with actual measurements. Farmers have good memories of extreme events that perceptions of their occurrence are more likely to be in tandem with observed meteorological data [18]. Although there are farmers whose perceptions deviate from the actual observation, the occurrence of late onset and early cessation of rainfall was correctly perceived by more than half of the farmers. Since the time of onset and cessation of rainfall is strongly related to farming activities, including the preparation of land for planting, farmers are highly likely to correctly recognize these changes. The convergence and divergence between perception and meteorological observation are strongly influenced by the agro-ecological contexts in which farmers undertake their farming activities. This shows that the consistency of perception with observed scientific trends depends on environmental differences in farmers' exposure to different climate variables. Farmers contextually define and characterize the weather conditions of a particular time and place based mainly on what they feel about the cropping season, entailing the important role of perceptual factors in framing their understanding of changes in climate variables.

Household characteristics account for both convergence and divergence between farmers' perceptions and meteorological data. We found that the perceptions of males, older farmers, and those with relatively higher social capital, access to media, and holding a small size of land converge with meteorological data. Male farmers' perception is aligned with meteorological data, which might be related to their better position to access information and primary responsibility to engage in farming activities. Proper recognition of changes in climate variables is based partly on the number of years of farming experience, meaning that older-age farmers have a more accurate perception than younger farmers [20,50]. Given the complexity of properly observing trends in weather conditions on the one hand and less reliance on traditional weather forecasts in the study areas on the other hand, higher social capital and exposure to mass media facilitate farmers' access to credible information that helps them form a correct perception of changes in local weather conditions [51]. Since their livelihood is most pronouncedly affected by adverse climate conditions, poor farmers are relatively well cognizant of changes in local weather conditions [50]. Conversely, misperceptions were noticed among economically better-off farmers. This is evident from the divergence of perceptions among farmers owning a large size of land and with medium economic status. Economically better-off farmers are more likely to generate their livelihoods from multiple sources that they are less dependent on weather-sensitive livelihood activities. Hence, they are likely to misperceive 'real' changes in climate variables. The results of our study also suggest that a lack of education contributes to the misperception of changes in weather conditions. Lack of education undermines access to varying sources of information and the cognitive ability to process information and make use of it to form an evidence-based perceptions.

Farming decisions and climate risk management plans partly depend on the availability of and access to reliable and relevant weather information. The use of traditional knowledge to forecast weather information is constrained by the high variability of the microclimate that made the forecasts less reliable. In the past, climate change occurred gradually and extreme events occur once in many years so that farmers can develop knowledge systems to adapt to. However, nowadays, the weather condition is highly variable not only between years and seasons but also within a day so that it has become difficult to describe the complex situation using the traditional systems that had been in use in the past. Although this is partly addressed through access to media which help farmers to have an accurate perception of changing weather conditions, there are also limitations associated with access to modern weather information. Weather stations are limited in number and unevenly distributed [12], the result of which fails to clearly show spatial differences of the micro-climate. Since the forecast is also made at a higher spatial scale and on a seasonal basis [52], it is less useful for farming decisions at the local level due to highly diverse topography. In addition, there is a lack of information on the time of onset and cessation of rainfall, which is important for decisions on planting time. Farming and adaptation decisions in an uncertain environment and without access to specific and reliable weather information are challenges for risk management. Besides, the lack of specific meteorological information contributes to farmers' incorrect perceptions of local weather changes [24].

5. Conclusions

Climate change and variability as well as the accuracy of farmers' perceptions of these changes are decisive for agricultural activities and the effectiveness of the livelihood strategies pursued by farmers. Geographical location as well as seasons have a great impact on the trends of changes in climate variables, occurrence of extreme events, and the accuracy of farmers' perceptions. All the three agro-ecological settings in this study are challenged by climatic factors that are either the same across all or vary between them. The increasing average maximum and average annual temperature, increasing warm extremes, and decreasing cold extremes denote that the study areas are warming. An increase in warm extremes and recurrent occurrence of drier years are the problems in the lowland and highland areas whereas heavy precipitation is observed in the midland areas. The effect of climate-related events of diverse nature are expected to be severe in the study areas. While midland areas face severe consequences of heavy rainfall, lowland and highland areas are highly challenged by a relatively small amount of rainfall, higher inter-annual variability, a shorter crop growth period, and a longer duration of dry spells. Rainfall variability, particularly during the short rainy season, is the major constraint in these areas, resulting in reliance on crop production once a year during the long rainy season which is also characterized by yearly variation in the time of onset and cessation. In spite of the accurate perception of increasing temperature, most farmers inaccurately perceived declining rainfall. Lower economic status, access to media, and higher social capital are associated with accurate perception. Perception diverges from actual trends among economically better-off farmers and households whose heads have no education. Although agro-ecological settings account for noticeable variation in the accuracy of perceptions of changes in climate variables, there is high heterogeneity among farmers in each agro-ecological setting. The divergence of farmers' perception from observed rainfall situation being highly likely to induce inaction, a lack of access to reliable weather information further undermines informed adaptation decision making. These are formidable challenges for smallholder farmers struggling to sustain their livelihoods as cropping calendar, the type of crops produced, and crop productivity are adversely affected by variable and uncertain climate conditions.

The observed changes in climatic variables have several implications for planning. First, reducing the impact of climate change requires the identification and implementation of adaptation strategies that are specifically suitable for the climate feature of each agro-ecological setting. For instance, variability in the distribution of rainfall brings to the fore the importance of water management as well as availing seeds that can be harvested in a short time or withstand water stress for effective adaptation. Second, the recurrence of climate variability and extreme events necessitates the expansion of alternative climate-resilient livelihood opportunities as a means to sustain food security. Third, increasing the availability of weather stations at the local level and enhancing the capacity to collect and analyze weather information increase the opportunity to anticipate the likely occurrences of weather-related risks and manage them through proactive measures. The deviation of farmers' perceptions from the observed changes might result in under- or over-estimating the impacts of changes and hampering their efforts for adaptation. In this vein, the dissemination of agro-ecologically specific, spatially interpolated, and locally relevant weather information is important to reach farmers and help them have accurate perceptions of the local weather conditions and make informed farming decisions and other livelihood choices. Specifically, it would be helpful for farmers to make proper farming adjustment and adaptation decisions if they have access to timely information not only on the amount and distribution of rainfall but also on the expected time of onset and cessation of rainfall during the cropping seasons. In addition to mass media (e.g., radio), the use of a cell phone and locally based formal (e.g., agricultural development agents, health extension workers) and informal (e.g., community-based organizations) structures would be useful to enhance farmers' access to reliable weather information. Furthermore, in spite of correctly perceiving changes in climate variables, since poor farmers lack the capacity to adapt, availing farm inputs that are tolerant to water stress and shorter crop growth period as well as improved production technologies increase their resilience to CCV.

Several issues remain unanswered. Farmers' perception of CCV is a necessary but not a sufficient condition to take adaptation actions [19]. Equally important is how they perceive the adverse effects of these changes on their livelihoods and the welfare of the community. The narrow focus on farmers' perceptions of changes in temperature and rainfall does not properly capture their comprehensive understanding of causes and consequences of climate change as well as possible responses, which are decisive to take action to minimize impacts. Farmers' understanding of local weather conditions is also rooted in socio-cultural factors. Hence, understanding farmers' holistic perspective on changing climate conditions as well as the underlying factors of variation in their perceptions requires further investigation. In addition, given the temporal changes in climate variables, adaptation decisions and the selection of adaptation strategies change across time. The dynamic interplay between climatic variables, households' vulnerability, and farmers' adaptation decision making is the subject of future inquiry.

Author Contributions: Conceptualization, D.E., D.J.R.M.S., C.F.A.v.W. and T.d.C.B.; methodology, D.E. and C.F.A.v.W.; software, D.E. and C.F.A.v.W.; validation, D.E., D.J.R.M.S., C.F.A.v.W. and T.d.C.B.; formal analysis, D.E. and C.F.A.v.W.; investigation, D.E.; data curation, D.E.; writing—original draft preparation, D.E.; writing—review and editing, D.E., D.J.R.M.S., C.F.A.v.W. and T.d.C.B.; visualization, D.E.; supervision, D.J.R.M.S., C.F.A.v.W. and T.d.C.B. All authors have read and agreed to the published version of the manuscript.

Funding: This research was funded by EP-NUFFIC, grant number R/002597.01. The APC was also funded by EP-NUFFIC.

Acknowledgments: The authors are very much grateful to the National Meteorology Agency of Ethiopia for the permission to use meteorological data. We would also like to thank the four anonymous reviewers for their valuable comments and suggestions which helped a lot in improving the paper.

Conflicts of Interest: The authors declare no conflict of interest.

References

1. Conway, D.; Schipper, E.L.F. Adaptation to climate change in Africa: Challenges and opportunities identified from Ethiopia. *Global Environ. Chang.* **2011**, *21*, 227–237. [CrossRef]
2. Omondi, P.A.O.; Awange, J.L.; Forootan, E.; Ogallo, L.A.; Barakiza, R.; Girmaw, G.B.; Fesseha, I.; Kululetera, V.; Kilembe, C.; Mbati, M.M. Changes in temperature and precipitation extremes over the Greater Horn of Africa region from 1961 to 2010. *Int. J. Climatol.* **2014**, *34*, 1262–1277. [CrossRef]
3. Hulme, M.; Doherty, R.; Ngara, T.; New, M.; Lister, D. African climate change: 1900–2100. *Clim. Res.* **2001**, *17*, 145–168. [CrossRef]
4. Kotir, J.H. Climate change and variability in Sub-Saharan Africa: A review of current and future trends and impacts on agriculture and food security. *Environ. Dev. Sustain.* **2011**, *13*, 587–605. [CrossRef]
5. Niang, I.; Ruppel, O.C.; Abdrabo, M.A.; Essel, A.; Lennard, C.; Padgham, J.; Urquhart, P. Africa. In *Climate Change 2014: Impacts, Adaptation, and Vulnerability. Part B: Regional Aspects. Contribution of Working Group II to the Fifth Assessment Report of the Intergovernmental Panel on Climate Change*; Barros, V.R., Field, C.B., Dokken, D.J., Mastrandrea, M.D., Mach, K.J., Bilir, T.E., Chatterjee, M., Ebi, K.L., Estrada, Y.O., Genova, R.C., et al., Eds.; Cambridge University Press: Cambridge, UK; New York, NY, USA, 2014; pp. 1199–1265.
6. Adhikari, U.; Nejadhashemi, A.P.; Woznicki, S.A. Climate change and eastern Africa: A review of impact on major crops. *Food Energy Secur.* **2015**, *4*, 110–132. [CrossRef]
7. Cheung, W.H.; Senay, G.B.; Singh, A. Trends and spatial distribution of annual and seasonal rainfall in Ethiopia. *Int. J. Climatol.* **2008**, *28*, 1723–1734. [CrossRef]
8. Jury, M.R.; Funk, C. Climatic trends over Ethiopia: Regional signals and drivers. *Int. J. Climatol.* **2013**, *33*, 1924–1935. [CrossRef]
9. Wagesho, N.; Goel, N.; Jain, M. Temporal and spatial variability of annual and seasonal rainfall over Ethiopia. *Hydrol. Sci. J.* **2013**, *58*, 354–373. [CrossRef]
10. Rosell, S. Regional perspective on rainfall change and variability in the central highlands of Ethiopia, 1978–2007. *Appl. Geogr.* **2011**, *31*, 329–338. [CrossRef]
11. Mengistu, D.; Bewket, W.; Lal, R. Recent spatiotemporal temperature and rainfall variability and trends over the Upper Blue Nile River Basin, Ethiopia. *Int. J. Climatol.* **2014**, *34*, 2278–2292. [CrossRef]
12. Alemayehu, A.; Bewket, W. Local spatiotemporal variability and trends in rainfall and temperature in the central highlands of Ethiopia. *Geogr. Ann. Ser. A Phys. Geogr.* **2017**, *99*, 85–101. [CrossRef]
13. Asfaw, A.; Simane, B.; Hassen, A.; Bantider, A. Variability and time series trend analysis of rainfall and temperature in northcentral Ethiopia: A case study in Woleka sub-basin. *Weather Clim. Extrem.* **2018**, *19*, 29–41. [CrossRef]
14. Esayas, B.; Simane, B.; Teferi, E.; Ongoma, V.; Tefera, N. Trends in extreme climate events over three agroecological zones of southern Ethiopia. *Adv. Meteorol.* **2018**, *2018*, 7354157. [CrossRef]
15. Gummadi, S.; Rao, K.P.C.; Seid, J.; Legesse, G.; Kadiyala, M.D.M.; Takele, R.; Amede, T.; Whitbread, A. Spatio-temporal variability and trends of precipitation and extreme rainfall events in Ethiopia in 1980–2010. *Theor. Appl. Climatol.* **2018**, *134*, 1315–1328. [CrossRef]
16. Mekasha, A.; Tesfaye, K.; Duncan, A.J. Trends in daily observed temperature and precipitation extremes over three Ethiopian eco-environments. *Int. J. Climatol.* **2014**, *34*, 1990–1999. [CrossRef]
17. Hasan, M.K.; Kumar, L. Comparison between meteorological data and farmer perceptions of climate change and vulnerability in relation to adaptation. *J. Environ. Manag.* **2019**, *237*, 54–62. [CrossRef]
18. Osbahr, H.; Dorward, P.; Stern, R.; Cooper, S. Supporting agricultural innovation in Uganda to respond to climate risk: Linking climate change and variability with farmer perceptions. *Exp. Agric.* **2011**, *47*, 293–316. [CrossRef]

19. Simelton, E.; Quinn, C.H.; Batisani, N.; Dougill, A.J.; Dyer, J.C.; Fraser, E.D.; Mkwambisi, D.; Sallu, S.; Stringer, L.C. Is rainfall really changing? Farmers' perceptions, meteorological data, and policy implications. *Clim. Dev.* **2013**, *5*, 123–138. [CrossRef]
20. Foguesatto, C.R.; Artuzo, F.D.; Talamini, E.; Machado, J.A.D. Understanding the divergences between farmer's perception and meteorological records regarding climate change: A review. *Environ. Dev. Sustain.* **2020**, *22*, 1–16. [CrossRef]
21. Imran, M.; Shrestha, R.P.; Datta, A. Comparing farmers' perceptions of climate change with meteorological data in three irrigated cropping zones of Punjab, Pakistan. *Environ. Dev. Sustain.* **2020**, *22*, 2121–2140. [CrossRef]
22. Rao, K.; Ndegwa, W.G.; Kizito, K.; Oyoo, A. Climate variability and change: Farmer perceptions and understanding of intra-seasonal variability in rainfall and associated risk in semi-arid Kenya. *Exp. Agric.* **2011**, *47*, 267–291. [CrossRef]
23. Bewket, W. Climate change perceptions and adaptive responses of smallholder farmers in central highlands of Ethiopia. *Int. J. Environ. Stud.* **2012**, *69*, 507–523. [CrossRef]
24. Meze-Hausken, E. Contrasting climate variability and meteorological drought with perceived drought and climate change in northern Ethiopia. *Clim. Res.* **2004**, *27*, 19–31. [CrossRef]
25. Aguilar, E.; Aziz Barry, A.; Brunet, M.; Ekang, L.; Fernandes, A.; Massoukina, M.; Mbah, J.; Mhanda, A.; Do Nascimento, D.; Peterson, T. Changes in temperature and precipitation extremes in western central Africa, Guinea Conakry, and Zimbabwe, 1955–2006. *J. Geophys. Res. Atmos.* **2009**, *114*. [CrossRef]
26. Kothari, C.R. *Research Methodology: Methods and Techniques*; New Age International: New Delhi, India, 2004.
27. Alexander, L.; Herold, N. *ClimPACT2: Indices and Software. A Document Prepared on Behalf of the Commission for Climatology (CCI) Expert Team on Sector-Specific Climate Indices (ET-SCI)*; World Meteorological Organization: Geneva, Switzerland, 2016.
28. Wang, X.L.; Feng, Y. *RHtests_dlyPrcp User Manual*; Climate Research Division, Atmospheric Science and Technology Directorate, Science and Technology Branch, Environment Canada: Toronto, ON, Canada, February 2013; Volume 25, p. 2014.
29. Wang, X.L.; Feng, Y. *RHtestsV4 User Manual*; Climate Research Division Atmospheric Science and Technology Directorate Science and Technology Branch, Environment Canada: Toronto, ON, Canada, 2013. Available online: http://etccdi.pacificclimate.org/software.shtml (accessed on 25 June 2019).
30. Wang, X.L. Accounting for autocorrelation in detecting mean shifts in climate data series using the penalized maximal t or F test. *J. Appl. Meteorol. Climatol.* **2008**, *47*, 2423–2444. [CrossRef]
31. Wang, X.L.; Chen, H.; Wu, Y.; Feng, Y.; Pu, Q. New techniques for the detection and adjustment of shifts in daily precipitation data series. *J. Appl. Meteorol. Climatol.* **2010**, *49*, 2416–2436. [CrossRef]
32. Van Rooy, M. A Rainfall Anomally Index Independent of Time and Space. *Notos* **1965**, *14*, 43–48.
33. Vicente-Serrano, S.M.; Beguería, S.; López-Moreno, J.I. A multiscalar drought index sensitive to global warming: The standardized precipitation evapotranspiration index. *J. Clim.* **2010**, *23*, 1696–1718. [CrossRef]
34. Hargreaves, G.H.; Allen, R.G. History and evaluation of Hargreaves evapotranspiration equation. *J. Irrig. Drain. Eng.* **2003**, *129*, 53–63. [CrossRef]
35. Kendall, M. *Rank Correlation Methods*; Charles Griffin: London, UK, 1975.
36. Mann, H.B. Nonparametric tests against trend. *Econom. J. Econom. Soc.* **1945**, *13*, 245–259. [CrossRef]
37. Hamed, K. Enhancing the effectiveness of prewhitening in trend analysis of hydrologic data. *J. Hydrol.* **2009**, *368*, 143–155. [CrossRef]
38. Tesfaye, K.; Walker, S. Matching of crop and environment for optimal water use: The case of Ethiopia. *Phys. Chem. Earth Parts A/B/C* **2004**, *29*, 1061–1067. [CrossRef]
39. Stern, R.; Rijks, D.; Dale, I.; Knock, J. *INSTAT Climatic Guide*; Statistical Service Centre, University of Reading: Reading, UK, 2006.
40. Raes, D.; Willems, P.; Baguidi, F. RAINBOW—A software package for analyzing data and testing the homogeneity of historical data sets. In Proceedings of the 4th International Workshop on 'Sustainable management of marginal drylands', Islamabad, Pakistan, 27–31 January 2006.
41. Keyzer, M.; Pande, S. Classification by crossing and polling for integrated processing of maps and surveys An addendum to GRCP-software. In *Staff Working Paper, Centre for World Food Studies*; Vrije Universiteit: Amsterdam, The Netherlands, 2007.

42. Etana, D.; van Wesenbeeck, C.F.; de Cock Buning, T. Socio-cultural aspects of farmers' perception of the risk of climate change and variability in Central Ethiopia. *Clim. Dev.* **2020**, 1–13. [CrossRef]
43. van Wesenbeeck, C.F.; Sonneveld, B.G.; Voortman, R.L. Localization and characterization of populations vulnerable to climate change: Two case studies in Sub-Saharan Africa. *Appl. Geogr.* **2016**, *66*, 81–91. [CrossRef]
44. Araya, A.; Stroosnijder, L.; Habtu, S.; Keesstra, S.D.; Berhe, M.; Hadgu, K. Risk assessment by sowing date for barley (Hordeum vulgare) in northern Ethiopia. *Agric. For. Meteorol.* **2012**, *154–155*, 30–37. [CrossRef]
45. Mulenga, B.P.; Wineman, A.; Sitko, N.J. Climate trends and farmers' perceptions of climate change in Zambia. *Environ. Manag.* **2017**, *59*, 291–306. [CrossRef] [PubMed]
46. Kosmowski, F.; Leblois, A.; Sultan, B. Perceptions of recent rainfall changes in Niger: A comparison between climate-sensitive and non-climate sensitive households. *Clim. Chang.* **2016**, *135*, 227–241. [CrossRef]
47. Roncoli, C. Ethnographic and participatory approaches to research on farmers' responses to climate predictions. *Clim. Res.* **2006**, *33*, 81–99. [CrossRef]
48. Marx, S.M.; Weber, E.U.; Orlove, B.S.; Leiserowitz, A.; Krantz, D.H.; Roncoli, C.; Phillips, J. Communication and mental processes: Experiential and analytic processing of uncertain climate information. *Global Environ. Chang.* **2007**, *17*, 47–58. [CrossRef]
49. Ovuka, M.; Lindqvist, S. Rainfall variability in murang'a district, kenya: Meteorological data and farmers' perception. *Geogr.Ann. Ser. A Phys. Geogr.* **2000**, *82*, 107–119. [CrossRef]
50. Ayal, D.Y.; Leal Filho, W. Farmers' perceptions of climate variability and its adverse impacts on crop and livestock production in Ethiopia. *J. Arid Environ.* **2017**, *140*, 20–28. [CrossRef]
51. Clayton, S.; Devine-Wright, P.; Stern, P.C.; Whitmarsh, L.; Carrico, A.; Steg, L.; Swim, J.; Bonnes, M. Psychological research and global climate change. *Nat. Clim. Chang.* **2015**, *5*, 640. [CrossRef]
52. Korecha, D.; Sorteberg, A. Validation of operational seasonal rainfall forecast in Ethiopia. *Water Resour. Res.* **2013**, *49*, 7681–7697. [CrossRef]

MDPI
St. Alban-Anlage 66
4052 Basel
Switzerland
Tel. +41 61 683 77 34
Fax +41 61 302 89 18
www.mdpi.com

Climate Editorial Office
E-mail: climate@mdpi.com
www.mdpi.com/journal/climate

www.ingramcontent.com/pod-product-compliance
Lightning Source LLC
Chambersburg PA
CBHW040836250926
51850CB00007BA/1327
* 9 7 8 3 0 3 6 5 7 4 4 1 7 *